terra australis 30

Terra Australis reports the results of archaeological and related research within the south and east of Asia, though mainly Australia, New Guinea and island Melanesia — lands that remained *terra australis incognita* to generations of prehistorians. Its subject is the settlement of the diverse environments in this isolated quarter of the globe by peoples who have maintained their discrete and traditional ways of life into the recent recorded or remembered past and at times into the observable present.

Since the beginning of the series, the basic colour on the spine and cover has distinguished the regional distribution of topics as follows: ochre for Australia, green for New Guinea, red for South-East Asia and blue for the Pacific Islands. From 2001, issues with a gold spine will include conference proceedings, edited papers and monographs which in topic or desired format do not fit easily within the original arrangements. All volumes are numbered within the same series.

List of volumes in *Terra Australis*

terra australis 30

ARCHAEOLOGICAL SCIENCE UNDER A MICROSCOPE

Studies in Residue and ancient DNA Analysis in Honour of Thomas H. Loy

Edited by Michael Haslam, Gail Robertson, Alison Crowther, Sue Nugent and Luke Kirkwood

ANU
THE AUSTRALIAN NATIONAL UNIVERSITY

E PRESS

ANU

E PRESS

© 2009 ANU E Press

Published by ANU E Press
The Australian National University
Canberra ACT 0200 Australia
Email: anuepress@anu.edu.au
Web: http://epress.anu.edu.au

National Library of Australia Cataloguing-in-Publication entry

Title: Archaeological science under a microscope [electronic resource] :
               studies in residue and ancient DNA analysis in honour of Thomas
               H. Loy / editors, Michael Haslam ... [et al.].

ISBN: 9781921536847 (pbk.)   9781921536854 (pdf.)

Series: Terra Australis ; 30.

Subjects: Loy, Thomas H., 1942-2005
               Archaeology--Methodology.
               Festschriften--Australia.

Other Authors/Contributors:
               Haslam, Michael.

Dewey Number: 930.10285

Series Editor: Sue O'Connor

Typesetting and design: Silvano Jung

Back cover map: *Hollandia Nova*. Thevenot 1663 by courtesy of the National Library of Australia.
Reprinted with permission of the National Library of Australia.

Terra Australis Editorial Board: Sue O'Connor, Jack Golson, Simon Haberle, Sally Brockwell, Geoffrey Clark

TABLE OF CONTENTS

Dr Thomas H. Loy (1942-2005)

1

PREFACE

Michael Haslam[1] and Alison Crowther[2,3]

1. Leverhulme Centre for Human Evolutionary Studies, University of Cambridge, Cambridge CB2 1QH United Kingdom
Email: mah66@cam.ac.uk

2. School of Social Science, University of Queensland, Brisbane QLD 4072 Australia
3. Department of Archaeology, University of Sheffield, Northgate House, West Street, Sheffield S1 4ET United Kingdom

Dr Thomas H. Loy (1942-2005) was a master storyteller, an innovative archaeologist and an inspiring teacher. He was equally at home walking a survey line in the red dust of northern Australia as he was enthusiastically lecturing on starch identification to an undergraduate audience, or sitting back with a cold beer spinning tales of escapes from helicopter crashes and bear attacks in frozen Canada. With an Apple Mac and a microscope always somewhere close by, Tom dedicated his working life to understanding the world around him by systematically examining the details that others may have overlooked, and inventing new methods of doing so if none existed. His sudden passing in October 2005 deprived us of the opportunity to present these papers to Tom personally, but both the diversity of research represented and its global coverage stand as testament to an enduring legacy, appropriate for an archaeological pioneer who demonstrated the value of examining the smallest traces for answers to the biggest questions.

The genesis of this collection was a symposium held to honour Tom's memory on 19 August 2006 at the University of Queensland, Brisbane, where he was Senior Lecturer in the School of Social Science. Papers delivered at that event have been augmented with invited contributions from colleagues not able to be present on the day. The symposium was memorable for the breadth of research presented (from residues on the tools of Homo floresiensis to the DNA of Henry VIII's warship) as well as for a stone knapping demonstration and discussion by Colin Saltmere of the Dugalunji Aboriginal Corporation, with whom Tom had established a strong friendship through his final field project near Camooweal in northwest Queensland. Following Tom's lead, many of the presentations recognised the responsibilities we as archaeologists bear towards both the past and present people that we deal with, a responsibility that requires vigilance in getting our stories straight. That recognition continues in the pages that follow. The central theme of this volume lies in using the detailed information recovered from microscopic and molecular archaeology to tell the most accurate stories we can about the human past, and doing so in a manner that encourages never-ending inquiry about the further avenues we may follow.

Beginning with a reproduction of the keynote address given by Richard Fullagar at the symposium, the volume is divided into two main sections. The first is titled 'Principles: synthesis, classification and experiment' and includes overviews and experimental or collection-based studies that aim to strengthen the fields of microscopic residue and ancient DNA analysis by examining the underlying principles on which these disciplines operate. In soliciting papers, the editors aimed to present an integrated if broad snapshot of the microscopic residue analysis field as it stands in the first decade of the twenty-first century. The synthesis of recent South African work provided by Lombard and Wadley provides a clear indication of the promise for residue studies to contribute to the important issues of human evolution, including the advent of hunting and the definition of modern human behaviour. From an historical perspective, Haslam's review of microscopic residue study sample sizes likewise stresses the need for thoughtful application of residue results if the field is to reach its full potential.

Specific residue types such as starches, raphides and blood proteins all played prominent roles in Tom's career as a residue analyst. This work is continued, first in the investigations of Jones and Barton into residue taphonomy over timescales from weeks to millennia, and second by Lentfer and Crowther in establishing archaeobotanical databases that will bring new rigour to discussions of past plant-use practices in Indonesia and the Pacific Islands. Watson *et al.* report on a decade-long search by Tom Loy and his students for an ancient DNA analogue to use in maintaining inter-laboratory standards (a topic continued by Hlinka *et al.*), emphasising again both respect for ancient remains and the need for procedural scrutiny to ensure reliable results.

The second section, titled 'Practice: case studies in residue and ancient DNA analysis', presents a series of studies with coverage from Europe to the Americas and Australasia. The full gamut of microscopic residue work is on display, from rockshelters (Hardy and Svoboda; Robertson) to open sites (Cooper and Nugent; Fullagar *et al.*) and private collections (Field *et al.*), and a number of stone tool types and materials receive close attention. An integrated battery of tests employed on Mesoamerican ceramic artefacts by Matheson *et al.* mirrors those used by Tom in his own early blood protein work, and reveals possible ceremonial use of the examined vessels. All these studies demonstrate the importance of detailed specialised analyses for adding social value to objects either newly or previously recovered, a theme continued in the DNA studies of museum pieces large and small by Spiers *et al.* and Hartnup *et al.*. There are few places in the modern world where you can still hunt the mighty moa, and the ancient DNA laboratory is one of them.

Reflecting on the work in this collection it is evident that one of the central strengths of residue and ancient DNA analyses lies in bringing rigorous yet innovative science to bear on otherwise intractable problems. That said, amongst the researchers reporting here there appears little room for science done for its own sake, and the anthropological and humanist underpinnings of archaeology are very much apparent. For example, the DNA case studies echo Tom's knack of identifying a specific topic of interest to modern audiences and then using that starting point to address wider issues. Microscopic residue analysis has yet to experience the same global explosion of output as ancient DNA research, but even in this selection important commonalities are emerging. One of these is the influence of rapid desiccation of residues as a significant aid to their long-term preservation, a point reiterated in various contexts by Jones and Matheson *et al.* regarding blood proteins, Barton for starches and Cooper and Nugent for a variety of residues. Identifying these common outcomes allows for future targeted research agenda, and the approaches demonstrated in these pages take important steps towards such coordinated effort.

Much was said about Tom Loy during his lifetime, and we will conclude by saying just one thing more: Tom was not always right, and he knew it. He also knew that the best way forward was to use his generosity and enthusiasm for revealing past human lives to inspire others to find the answers he did not have the time to find himself. This volume is testament to the success of that vision.

ACKNOWLEDGEMENTS

Credit is always due to many when an edited volume actually comes to fruition. We would first like to extend our sincere gratitude to the authors, as without their perseverance, patience and prompt responses this collection would not exist. Second, the reviewers deserve great thanks for their insights and attention to detail. For their support of the publishing process we offer our appreciation to the staff at ANU E press and Sue O'Connor of the Australian National University. For financial support we thank the School of Social Science and the Aboriginal and Torres Strait Islander Studies Unit at the University of Queensland, as well as the University of Queensland Archaeological Services Unit. To everyone who attended the memorial symposium in August 2006, thank you: your demonstration of support for Tom and his work form a lasting tribute to a unique individual. Finally, we recognise that a researcher is a product not only of his own ideas, but also of the influences of each colleague and student that collaborates on, questions and builds upon the work that forms a career. For those who worked with and learned from Tom, and those who will never meet him, we hope that there remains something of his inquiring spirit within these pages.

2

STONES, STORIES AND SCIENCE

Richard Fullagar

Scarp Archaeology, 25 Balfour Road, Austinmer NSW 2515 Australia
Email: richard.fullagar@scarp.com.au

The following was presented as the Keynote Address at 'Archaeological Science Under a Microscope: A symposium in honour of Tom Loy', held in Emmanuel College, The University of Queensland, on 19 August 2006.

Tom Loy died suddenly in October 2005. He left behind unfinished books and ongoing research projects mostly related to prehistoric residue analyses in collaboration with students working at the University of Queensland. A year or so down the track, several of these projects have come to fruition (as theses and numerous publications), and new directions have emerged. It is therefore appropriate and timely that the organising committee (Gail Robertson, Alison Crowther, Luke Kirkwood, Michael Haslam and Sue Nugent) pulled together this symposium, primarily to honour Tom, but also to reflect on the discipline he left behind, to ask about its latest developments and to examine where it's headed. That is the task of this symposium. My purpose here is not to put Tom's life under a microscope, but to briefly reflect on three strands of knowledge he pioneered: stone tool function, the stories and reconstructions based on them, and archaeological science.

I first met Tom in Victoria, BC, Canada, in 1983. He was seated at his large, old Reichert microscope, which reminded me of a modern telescope, like at Mt Stromlo. At the time, I was beginning a PhD thesis, at La Trobe University, to work on integrating use-wear and residue research. He was showing me the worn edges of stone tools with blood and hair residues, as clear as you can imagine, and all of which had just been published in the journal *Science*. Tom spent early years in the desert among the Navajo, and was trained in geology and consequently knew about lithology, as well as stone artefact technology. Although the artefacts he was showing me came from an arctic environment, he still enthused excitement because the organic tissues had survived so long – over thousands of years; because the details were so good – down to a splash of blood; and because hard evidence (from geology, biochemistry and biology) enabled precise conclusions – bison hairs, red blood corpuscles and a radiocarbon age. He was wildly enthusiastic about the potential of using plant and animal traces to work out ever more precisely how stone tools were used.

He spent most of his academic life developing residue analysis not just to find out about stone tool function but to find out what people did; and he did this in forensic detail. I think his primary concern was with people; at least what tools can tell us about people in the past. And stone artefacts were a major focus, although, as we all know, he studied residues wherever he could find them; on pottery, glass, bone, shell, skin, textiles, on ancient and modern materials. He promoted a kind of Stone Artefact Bank – stone artefacts collectively as a reserve of new information about resource use, blood lines, disease history, botanical landscapes and evolution.

In contrast with the mundane practice of tool-use, Tom was also deeply interested in Buddhist philosophy. Annie Dillard, who has written novels about human connections with nature, wrote a short story about a certain Larry, who was 'Teaching a Stone to Talk' – which is the title of the story. (Dillard said some profound things. One of my favorites is: Eskimo/Inuit: 'If I did not know about God, and I sin, would I go to hell?' Priest: 'No, not if you did not know.' Eskimo: 'Then why did you tell me?' She also quips that 'Nature's silence is her one remark and every flake of the world is a chip off that old mute and immutable block'.) Anyway, 'teaching a stone to talk' is a kind of ceremony or ritual for Larry, who is into meditation. So was Tom. But I think Tom was less interested in making a stone speak (although he would have liked the idea) than he was in the logic of science. He certainly liked making things, as well as talking – and he was a great raconteur. But he was not merely into 'squeezing blood from stones' (prophetically the title of an important paper by Glyn Isaac in 1977); seeing just how much we can get from a rock or a residue; seeking knowledge for its own sake. I think his search for knowledge entailed much more. As Isaac (1977:11) said in his paper, '(w)e need to concentrate our efforts on situations where the stones are only a part of a diverse record of mutually related traces of human behaviour and adaptation'.

Certainly a primary concern of Tom, as an archaeologist working with stones and bones, was what people made; what people actually did, on the ground, in the ground; hence his experiments with artifacts to test ideas about how people collected and gathered food. His experiments with bone artefacts to replicate Australopithecine extraction of *Hypoxis* African potato roots illustrate this endeavor. And somewhat in common with the late Rhys Jones (a former colleague of his at the Australian National University), I think Tom tried to get into the mind of prehistoric people. If he could get details of tool use right, he just might be able to test hypotheses about technological change, subsistence, exchange, ceremonies and perhaps even perceptions of landscape and society; how people saw the world.

As Jay Hall (then Head of Archaeology at the University of Queensland) has suggested previously, few studies by Tom fall short by a story. Most of his studies provide a detailed account of what might have happened; what people probably did; some account of human action and thinking. One of the best examples is the story of the man known as the Glacier Mummy, Tyrolean Iceman, or most commonly now, Ötzi. Tom introduced me to the archaeology of Ötzi, who was found in September 19, 1991 in the Ötztal Alps (just on the Italian side of the Austrian–Italian border as it finally turned out). Tom was among the first scientists contacted – in part because of his expertise and the fabulous preservation of organic tissue. Everything about Ötzi was intact, including his genitals. His woven grass cloak, shoes and bearskin hat indicated he might have been a shepherd, caught out in bad weather while moving his flock. However, artefacts found near his body – a bow, a quiver of arrows, a copper axe, a fire-making kit, a backpack and a flint dagger – suggested he may have been a hunter or even a warrior. Tom packed a microscope (a personal one belonging to Rowan Webb, now at the University of New England) and was flown to Austria. During my visits to the ANU, he had discussed how to record the usewear on the artefacts, and we decided to record use-wear and residues on the tool edges by taking acetate peels – which he brought back for me to examine, confirming a diagnostic polish from cutting highly siliceous plants. As his students know, Tom was a bit of a loner, but was remarkably generous in sharing his knowledge and involving others in his high profile research.

Ötzi is one of the greatest archaeological discoveries of the 20th century in part because his preservation is about as good as it gets. Archaeological, forensic, genetic and other molecular techniques are being pushed to the limit. But well-preserved mummies are found in the Peruvian Andes, the Egyptian pyramids, the bogs in the UK and Europe.

What makes Ötzi special? His antiquity is part of it; he was buried for longer than the others. And the fact that his belongings were not arranged artificially (as in a ritualised burial) provides us with a unique glimpse into everyday Neolithic life. Or does it? The mystery surrounding his death adds an extra dimension.

New evidence was previewed on video in 2003, and was fresh out of Tom's lab where he had analysed DNA preserved in blood films found on Ötzi's leather tunic, knife blade and one of the two arrows. Tom isolated the DNA fingerprints of four human individuals, and one of the individuals indicated was probably responsible for Ötzi's death. Another of the DNA profiles probably belonged to Ötzi. An unhealed stab wound in his right hand suggested he may have put up a fight, and study of blood pooling indicates he was moved before he died. Was he attacked by a gang? He was shot in the back with an arrow, but the shaft had been removed, so someone might have helped him. Who was with him? Why were valuable items of equipment like a copper axe not taken but left with him? Tom systematically traced the possibilities, like a crime scene investigator with new insights into motives and the likely sequence of events leading to Ötzi's death.

We need stories like these, not just because they appeal to the public but also because they help set up new hypotheses. The stories feed back to hard science; the scientific hypotheses that lead us to more detail; filling in the gaps about the life and times of Ötzi. This is important. We should get the facts right. And many of the papers at this conference show us the expanding array of current and new scientific approaches. I should also mention in this context that Tom Loy undertook the initial scans of the Kuk Swamp stone artefacts that revealed the first early evidence of starchy plant exploitation in the Papua New Guinea Highlands. I know this because I examined them with him at the ANU. We found remarkable preservation of starch, and it is these initial findings that led to further work at the Australian Museum and the University of Sydney confirming evidence for processing *Colocasia* taro and *Disocorea* yams in the Highlands of New Guinea 10,000 years ago.

Finally, apart to some extent from the stones, stories and the hard science, I want to mention some aspects of theory, and the role of Tom's lab. I would like to reinforce a new direction of stone artefact studies in Australia. Part of this new direction is drawing together a relatively new range of specialist studies like refitting or conjoining, reduction sequences, microwear and technological indicators of risk, a range of research in which the University of Queensland continues to play a key role, building on Tom's foundations.

Of course stone artefacts don't speak for themselves, but we are learning snippets of the conversation in large part because of the context. Stone artefacts as agents no doubt have an impact on human behaviour. Beautiful Kimberley points were extensively traded among Aboriginal groups in northwestern Australia and were emblems of social identity, craftsmanship and prestige. I am not sure that you would call the huge blocks that make up Stonehenge 'stone artefacts' in the normal sense of the term, but certainly the massive stone quarries and huge stone lithic scatters that mark the Australian landscape, are made of stone artefacts, and they must have signalled information of various kinds to Aboriginal people, including highly visible indicators of potential stone sources and locations of settlement. But these are like the beginning and end points in long lives of stone tools – long before the archaeologist picks them up. In between, are complex life histories that we are only beginning to understand. Several studies now, notably by Peter Hiscock (formerly UQ, now ANU) and Chris Clarkson (UQ) in Australia and Robin Torrence (Australian Museum) in Papua New Guinea, show how subtle changes in technological behavior, how and where stone artefacts are made, may be linked with other aspects of subsistence, resource use, settlement history and responses to risk. At different times, in some places, unretouched flakes are used for tasks in preference to standardised backed microliths that are used in other contexts. The need for tools at certain times and places means that particular materials and tool forms were preferred. At least this is an argument built on theoretical models of behaviour and detailed studies of flakes and cores.

After so many advances in our understanding of lithic technology, use-wear, residues and molecular biology, we are in a position to move well beyond compiling what George Odell called the laundry list of stone tool functions, by which he meant that usewear analysts 'analyse' the artefacts essentially by providing a list of what they were used for. We know that artefact forms were used for many tasks; for example, so-called 'points' are not always used as spear tips, they might be wood scrapers; so we have to look at what each artefact was used for in each assemblage to get an idea of what tasks were undertaken. Even so, there are some problematic assumptions in this approach. For example, it is usually implicit in this approach that all the artefacts were used at about the same time, yet we know this is not always true. The sequence of use is also important, as cores and flakes are frequently further sharpened and reduced. If we take away, or decouple, finished artefact form and function, can they be re-coupled in terms of reduction sequences? How do we link manufacturing and reduction stages with function? Can we ever only find out about tool function at the point of discard?

I suggest we need to rethink how artefact assemblages are sampled for functional studies. It would be extremely useful to sample assemblages with indicators of technological change to test whether they correlate with shifts in resource use or task composition. Does a technological shift in response to risk (say more backed microliths) correspond to different maintenance tasks? Of special significance will be usewear and residues on small flakes that have been broken from tool edges or the tiny retouch fragments from edge sharpening, rather than the discarded implement itself. It is on the platforms and dorsal surfaces of these sharpening flakes, only several millimetres in size, that we might expect to see records of tool use during earlier stages of reduction. Will sequences of use in tool life histories be the same at different times and places? What would we predict for different hominin species like Neanderthals and hobbits (*Homo floresiensis*)? This of course moves into theories of what constitutes modern human behaviour and warrants theories about particular kinds of activities, task performance and diagnostic archaeological indicators of past behaviours.

This is only one small aspect of archaeological science and theory to be explored in these papers, and I would like to finish by acknowledging how well Tom's lab, his students and colleagues are positioned to make advances in understanding stones, relating the stories, and further developing the science and the theory.

ACKNOWLEDGEMENTS

I thank the organizers and editors for the opportunity to present (and their help in revising) this Keynote Address. Parts of my reflections on Ötzi come from a publication following conversations with Tom Loy in 2004 (Fullagar, R. 2004. Ötzi spills his guts. Nature Australia 28(1):74-75).

REFERENCE

Isaac, G. 1977. Squeezing blood from stones. In R.V.S. Wright (ed.) *Stone Tools as Cultural Markers: Change, evolution and complexity*, pp.5-12. Canberra: AIATSIS / New Jersey: Humanities Press Inc.

3

THOMAS H. LOY PUBLICATIONS: 1978-2006

Loy, T. H. 1978. An archaeological application of seismic refraction profiling techniques. *Canadian Journal of Archaeology* 2: 155-164.

Loy, T. H. 1983. Prehistoric blood residues: detection on tool surfaces and identification of species of origin. *Science* 220: 1269-1271.

Loy, T. H. 1985. Preliminary residue analysis: AMNH specimen 20.4/509. In D. H. Thomas (ed.) *The Archaeology of Hidden Cave*, pp. 224-225. New York: Museum of Natural HIstory Press.

Loy, T. H. & D. E. Nelson 1986. Potential applications of the organic residues on ancient tools. In J. S. Olin & M. J. Blackman (eds.) *Proceedings of the 24th International Archaeometry Symposium*, pp. 179-185. Washington, D.C.: Smithsonian Institution Press.

Nelson, D. E., T. H. Loy, J. Vogel & J. Southon 1986. Radiocarbon dating blood residues on prehistoric stone tools. *Radiocarbon* 28(1): 170-174.

Loy, T. H. 1987. Recent advances in blood residue analysis. In W. R. Ambrose & J. M. J. Mummery (eds.) *Archaeometry: Further Australasian Studies*, pp. 57-65. Canberra: Australian National University.

Jones, R., R. Cosgrove, H. Allen, S. Cane, K. Kiernan, S. Webb, T. H. Loy, D. West & E. Stadler 1988. An archaeological reconaissance of Karst Caves within the southern forests region of Tasmania. *Australian Archaeology* 26: 1-23.

Loy, T. H. & A. R. Wood 1989. Blood residue analysis at Çayönü Tepesi, Turkey. *Journal of Field Archaeology* 16: 451-460.

Loy, T. H. 1990. When is a stone a tool? *Australian Natural History* 23(7): 584.

Loy, T. H. 1990. Getting blood from a stone. *Australian Natural History* 23(6): 470-479.

Loy, T. H., R. Jones, D. E. Nelson, B. Meehan, J. Vogel, J. Southon & R. Cosgrove 1990. Accelerator radiocarbon dating of human blood proteins in pigments from Late Pleistocene art sites in Australia. *Antiquity* 64: 110-116.

Loy, T. H. 1991. Prehistoric organic residues: recent advances in identification, dating and their antiquity. In W. Wagner & A. Pernicka (eds.) *Archaeometry '90: Proceedings of the 27th International Symposium on Archaeometry*, pp. 645-656. Boston: Birkhauser Verlag.

Loy, T. H. 1992. Destructive sampling in the analysis of rock art. *Rock Art Quarterly* 3(3-4): 1-8.

Loy, T. H. & B. L. Hardy 1992. Blood residue analysis of 90,000-year-old stone tools from Tabun Cave, Israel. *Antiquity* 66: 24-35.

Loy, T. H., M. Spriggs & S. Wickler 1992. Direct evidence for human use of plants 28,000 years ago: starch residues on stone artefacts from the northern Solomon Islands. *Antiquity* 66: 898-912.

Loy, T. H. 1993. The artefact as site: an example of the biomolecular analysis of organic residues on prehistoric tools. *World Archaeology* 25(1): 44-63.

Loy, T. H. 1993. Prehistoric organic residue analysis: the future meets the past. In M. Spriggs, D. E. Yen, W. Ambrose, R. Jones, A. Thorne & A. Andrews (eds.) *A Community of Culture: The people and prehistory of the Pacific*, pp. 56-72. Canberra: Australian National University.

Loy, T. H. 1993. On the dating of prehistoric organic residues. *The Artefact* 16: 46-49.

Loy, T. H. 1994. Methods in the analysis of starch residues on prehistoric stone tools. In J. G. Hather (ed.) *Tropical Archaeobotany: Applications and new developments*, pp. 86-114. London: Routledge.

Loy, T. H. 1994. Direct dating of rock art at Laurie Creek (NT), Australia: a reply to Nelson. *Antiquity* 68(258): 147-148.

Loy, T. H. 1994. Identifying species of origin from prehistoric blood residues: response. *Science* 266(5183): 299-300.

Loy, T. H. 1994. Residue analysis of artifacts and burned rock from the Mustang Branch and Barton sites (41HY209 and 41HY202). In R. A. Ricklis & M. B. Collins (eds.) *Archaic and Late Prehistoric Human Ecology in the Middle Onion Creek Valley, Hays County, Texas*, pp. 607-627. Austin, Texas: Texas Archaeological Research Laboratory, The University of Texas at Austin.

Loy, T. H. 1994. Direct dating of rock art at Laurie Creek. *Antiquity* 68(258): 147-148.

Loy, T. H. & K. I. Matthaei 1994. Species of origin determination from prehistoric blood residues using ancient genomic DNA. *Australasian Biotechnology* 4(3): 161-162.

Loy, T. H. 1997. Ultrapure water, is it pure enough? *Ancient Biomolecules* 1: 155-159.

Fullagar, R., T. H. Loy & S. Cox 1998. Starch grains, sediments and stone tool function: evidence from Bitokara, Papua New Guinea. In R. Fullagar (ed.) *A Closer Look: Recent Australian studies of stone tools*, pp. 49-60. Sydney: Archaeological Computing Laboratory, University of Sydney.

Loy, T. H. 1998. Blood on the axe. *New Scientist* 159(2151): 40-43.

Loy, T. H. 1998. Ice Man. *Nature Australia* 26: 60-63.

Loy, T. H. & E. J. Dixon 1998. Blood residues on fluted points from eastern Beringia. *American Antiquity* 63(1): 21-46.

Wolski, N. & T. H. Loy 1999. On the invisibility of contact: residue analyses on Aboriginal glass artefacts from Western Victoria. *The Artefact* 22: 65-73.

Matheson, C. D. & T. H. Loy 2001. Genetic sex identification of 9400-year-old human skull samples from Çayönü Tepesi, Turkey. *Journal of Archaeological Science* 28: 569-575.

Brown, T. & T. H. Loy 2002. Preliminary detection of haemoglobin from extinct mammals using capillary electrophoresis. *Tempus* 7: 205-212.

Hlinka, V., S. Ulm, T. H. Loy & J. Hall 2002. The genetic speciation of archaeological fish bone: a feasibility study from southeast Queensland. *Queensland Archaeological Research* 13: 71-78.

Haslam, M., J. Prangnell, L. Kirkwood, A. McKeough, A. Murphy & T. H. Loy 2003. A Lang Park mystery: analysis of remains from a nineteenth century burial in Brisbane. *Australian Archaeology* 56(1): 1-7.

Bordes, N., B. Pailthorpe, J. Hall, T. H. Loy, M. Williams, S. Ulm, X. Zhou & R. Fletcher 2004. Computational archaeology. *WACE4: Workshop on Advanced Collaborative Environments: Proceedings: September 23, 2004.*

Walshe, K. & T. H. Loy 2004. An adze manufactured from a telegraph insulator, Harvey's Return, Kangaroo Island. *Australian Archaeology* 58: 38-40.

Lamb, J. & T. H. Loy 2005. Seeing red: the use of Congo Red dye to identify cooked and damaged starch grains in archaeological residues. *Journal of Archaeological Science* 32: 1433-1440.

Loy, T. H. 2006. Optical properties of potential look-alikes. In R. Torrence & H. Barton (eds.) *Ancient Starch Research*, pp. 123-124. Walnut Creek, CA: Left Coast Press.

Loy, T. H. 2006. Starch on the axe. In R. Torrence & H. Barton (eds.) *Ancient Starch Research*, pp. 178-180. Walnut Creek, California: Left Coast Press.

Loy, T. H. 2006. Iodine-potassium-iodide test for starch. In R. Torrence & H. Barton (eds.) *Ancient Starch Research*, pp. 121-122. Walnut Creek, California: Left Coast Press.

Loy, T. H. 2006. Raphides. In R. Torrence & H. Barton (eds.) *Ancient Starch Research*, pp. 136. Walnut Creek, California: Left Coast Press.

Loy, T. H. & R. L. K. Fullagar 2006. Residue extractions. In R. Torrence & H. Barton (eds.) *Ancient Starch Research*, pp. 197-198. Walnut Creek, California: Left Coast Press.

4

THE IMPACT OF MICRO-RESIDUE STUDIES ON SOUTH AFRICAN MIDDLE STONE AGE RESEARCH

Marlize Lombard[1] and Lyn Wadley[2]

1. Institute for Human Evolution
University of the Witwatersrand
WITS 2050 South Africa
Email: Marlize.Lombard@wits.ac.za

2. Archaeology Department
School of Geography, Archaeology and Environmental Studies
University of the Witwatersrand
WITS 2050 South Africa

ABSTRACT

The Middle Stone Age of South Africa currently plays a central role in studies of the origins of symbolic behaviour. Micro-residue analyses on stone tools from sites with long Middle Stone Age sequences and good organic preservation are producing direct contextual evidence and detailed information about past technologies and associated behaviours. In this chapter we provide a brief and selected overview of some of our published contributions and demonstrate how micro-residue studies can now be used to assess hypotheses regarding hunting efficiency and hafting technologies. Compelling evidence is being produced that is contrary to the once-held notion that the Middle Stone Age shows little meaningful change through time. The cumulative results provide clear evidence for variability and change associated with anatomically modern humans. While our published work demonstrates our commitment to a multi-analytical approach to use-trace analysis, including micro-residue, use-wear and macrofracture analyses, we focus here on residues as it was also the focus of Tom Loy's research in South Africa.

KEYWORDS

Middle Stone Age, Howiesons Poort, Still Bay, Sibudu Cave, cognitive evolution, hunting technology, hafting technology

INTRODUCTION

There has been a marked increase in research interest in the African Middle Stone Age that spans the period of roughly 250 ka to 25 ka ago. The heightened awareness is due to new, multi-disciplinary data stimulating debate on the origins of anatomically and behaviourally modern humans, and while the archaeological record is nowhere near complete, progress is being made. Exciting but hotly contested interpretations of African origins now rival earlier interpretations of a Eurasian origin for modern humans (e.g. Marean and Assefa 2005; but also see d'Errico 2003; d'Errico *et al.* 2003; Shea 2003). Genetic and fossil evidence suggest that, from the anatomical perspective, humans in Africa were nearly modern by about 160 ka ago.

Fundamental questions are whether anatomical and cognitive modernity developed in tandem, and what criteria, if any, archaeologists should use to identify modern human behaviour

(Henshilwood and d'Errico 2005; Henshilwood and Marean 2003, 2006; Kuhn and Hovers 2006; McBrearty and Brooks 2000; Shea 2003; Wadley 2001, 2006a). New results seem to suggest that symbolic behaviour in Africa extends far earlier than the 50 ka 'Rubicon' (Henshilwood and Marean 2003; Marean and Assefa 2005; McBrearty and Brooks 2000; Mellars 1973, 1995, 2005, 2006; Minichillo 2005; Wadley 2006a, 2006b), and the Middle Stone Age of South Africa has claimed a central role in the quest for the origins of symbolic behaviour. It is against this background that the research of LW moved ever-deeper in time, first at Rose Cottage Cave in the Free State and more recently at Sibudu Cave in KwaZulu-Natal (Figure 1), and ML started exploring questions about Middle Stone Age hunting and hafting behaviour. These explorations and associated research questions and projects resulted in our approach to micro-residue analysis gaining its distinctive direction and momentum.

The first South African PhD in stone tool residue studies was completed in 2000 (Williamson 2000a) under the direction of Lyn Wadley and the late Tom Loy. This study introduced the basic principles and potential of residue analysis to Stone Age research on the sub-continent, where we have an approximately 2.5 million-year-old tool making tradition. From the start, our micro-residue work was deeply embedded in archaeological questions that have arisen during the course of excavations at Rose Cottage Cave and Sibudu Cave (see Gibson *et al.* 2004; Lombard 2004, 2005; Wadley, Williamson and Lombard 2004; Williamson 1996, 1997, 2000b, 2004, 2005). Some of these questions centred on the use and hafting of Howiesons Poort backed tools, the role of ochre during the Middle Stone Age and changes in subsistence strategies. Building on this initial work, experimentation, modern replication and blind testing became integral parts of our research design, without losing sight of the main goal – to improve our understanding of human behaviour during the Middle Stone Age (Hodgskiss 2006; Lombard *et al.* 2004; Lombard and Wadley 2007a, b; Pargeter 2007; Rots and Williamson 2004; Wadley 2005a, 2005b, 2005c, 2006a; Wadley and Lombard 2007; Wadley, Lombard and Williamson 2004).

Both Rose Cottage Cave and Sibudu Cave have yielded hundreds of thousands of stone tools occurring in varying contexts over time, and we have systematically excavated and curated tools with residue studies in mind. At both sites, but particularly at Sibudu Cave, we have extraordinary evidence for environmental and cultural change (Wadley 2006b). It is important to us to take a holistic view and to position micro-residue analysis within the broader framework of data derived from fauna, botanical remains and site sediments (for recent multi-disciplinary contributions from Sibudu Cave see Wadley 2006b and references therein). The micro-residue work on stone tools is closely linked to our multi-disciplinary approach to research on the other archaeological material, and is intended to crosscheck and strengthen behavioural interpretations. It is against this background that we are also consistently attempting to improve our methodology and interpretative skills.

Rose Cottage Cave appears to have been occupied, perhaps intermittently, over a period of about 90,000 years (Pienaar 2006; Soriano *et al.* 2007; Valladas *et al.* 2005). Sibudu Cave has a long series of Middle Stone Age occupations with stone tool assemblages that can be attributed to a pre-Still Bay phase at the base of the sequence, a Still Bay Industry, a Howiesons Poort Industry, a post-Howiesons Poort phase, and late and final Middle Stone Age phases. A preliminary OSL age for the Still Bay is calculated at ~ 73 ka, no ages for the Howiesons Poort are available yet, but a large suite of layers with post-Howiesons Poort assemblages has a weighted mean OSL age of 60.1 ± 1.5 ka, the late Middle Stone Age phase has a suite of ages with a weighted mean average OSL age of 49.7 ± 1.2 ka, and the final Middle Stone Age phase has an OSL age of about 37 ka (Jacobs *et al.* submitted; Wadley and Jacobs 2006) (Figure 2). The Howiesons Poort Industry, which contains many backed tools such as segments, appears to shift seamlessly into the younger ~60 ka post-Howiesons Poort industry lacking backed tools (Wadley 2006b). Stratigraphically, the Howiesons Poort succeeds the Still Bay, but it is not yet known whether there is a hiatus between these two industries at Sibudu Cave (Wadley 2006b).

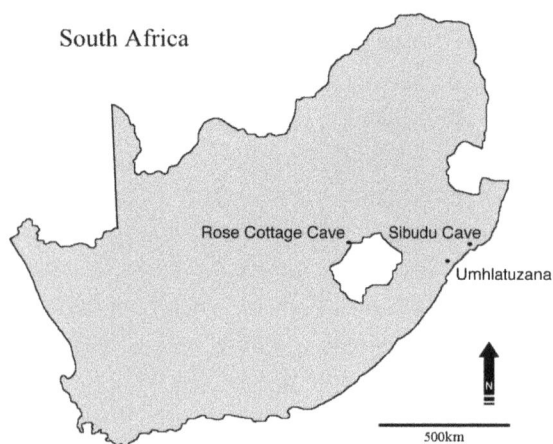

Figure 1: Map with deep-sequence KwaZulu-Natal sites and Rose Cottage Cave.

SELECTED METHODOLOGICAL DEVELOPMENTS

In 2004 we published the protocols and results of our first two blind tests (Wadley, Lombard and Williamson 2004). The original aim was to assess ML's ability to identify a variety of plant and animal residues using light microscopy. However, issues and problems that arose during the testing process made it clear that greater value might be gained from the lessons that we learnt about methodology and the direction for future micro-residue research (for discussions see Lombard and Wadley 2007a, 2007b and Wadley and Lombard 2007). Addressing problems identified during our first tests stimulated research and subsequently two more tests were conducted, and this time they were totally field-based (Lombard and Wadley 2007a).

The series of tests helped to evaluate and improve the quality of data that could be gained through micro-residue analysis using light microscopy. For example it was shown that, based on their microscopic morphological appearance, animal residues could have been previously mistaken for plant residues (Figure 3). This discovery has far-reaching implications for the interpretation of archaeological assemblages, site functions and associated behavioural hypotheses. The same tests also demonstrated that the rock types used to produce stone tools might influence the quality of functional interpretations based on micro-residue analysis (for example, highly reflective quartz makes recognition of residues challenging, and the smooth surfaces of quartz do not readily retain residues).

We introduced a multi-stranded approach (Lombard and Wadley 2007a, 2007b; Wadley and Lombard 2007). This requires that functional and hafting interpretations are based not only on the presence of single identified micro-residues, but also on their association with related residues, their frequency, distribution patterns, layering, orientation, and the way in which they adhere to the tool. We believe that this approach, in combination with use-wear and macrofracture analyses (see Lombard 2005, 2006b, 2007a for application of this approach), controlled curatorial circumstances and the microscopic study of associated sediment samples and potential contaminants, provides a cautious, but secure strategy for the detailed interpretation of archaeological residues (Lombard and Wadley 2007a, 2007b; Wadley and Lombard 2007). ML also developed a method for creating quantifiable, comparable data for the interpretation of micro-residue distribution patterns (Lombard 2004, 2005, 2006a, 2007a, 2007b, in press). Using this method it is shown that, with adequate tool samples and good organic residue preservation, sufficient data can be generated to compare results from different assemblages or micro-stratigraphic archaeological contexts. The results can also be statistically tested for the possibility of coincidental distribution patterns (Lombard 2005, 2007b). Interpretations based on such data have the potential to enhance considerably the resolution of our knowledge of human behaviour in the distant past.

IMPLICATIONS FOR OUR CURRENT KNOWLEDGE OF HUNTING AND HAFTING BEHAVIOURS

A brief background

Effective hunting with hafted weapons has long been part of the 'modernity' debate. Although evidence for active hunting or the presence of hafted stone tools was sometimes considered to imply modern behaviour (e.g. Ambrose 2001 and Klein 2000 and references in both), anatomically archaic humans also seem to have hafted tools and hunted (e.g. Rots and Van Peer 2006). This

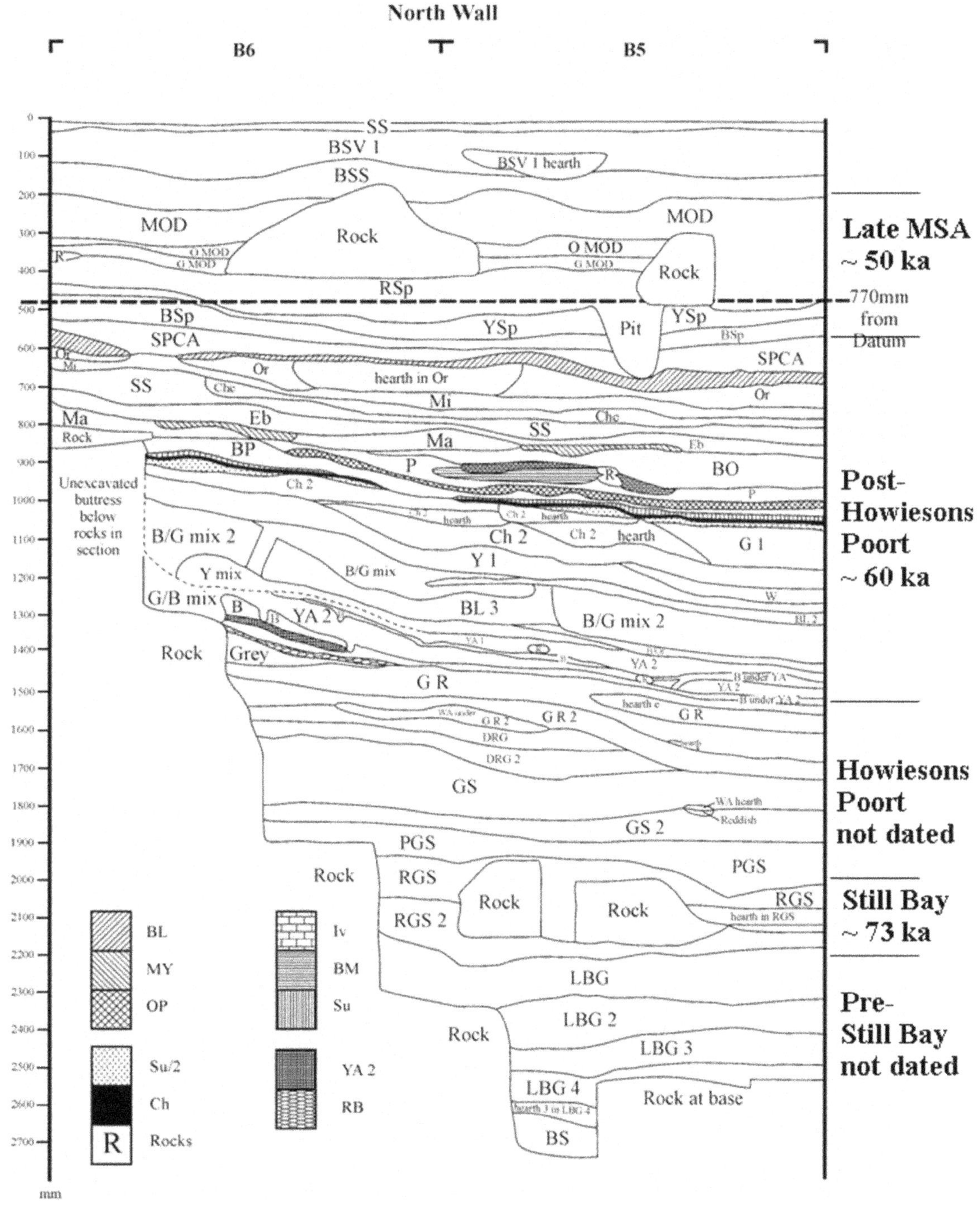

Figure 2: North wall stratigraphy at Sibudu Cave showing the main occupational phases and associated approximate ages.

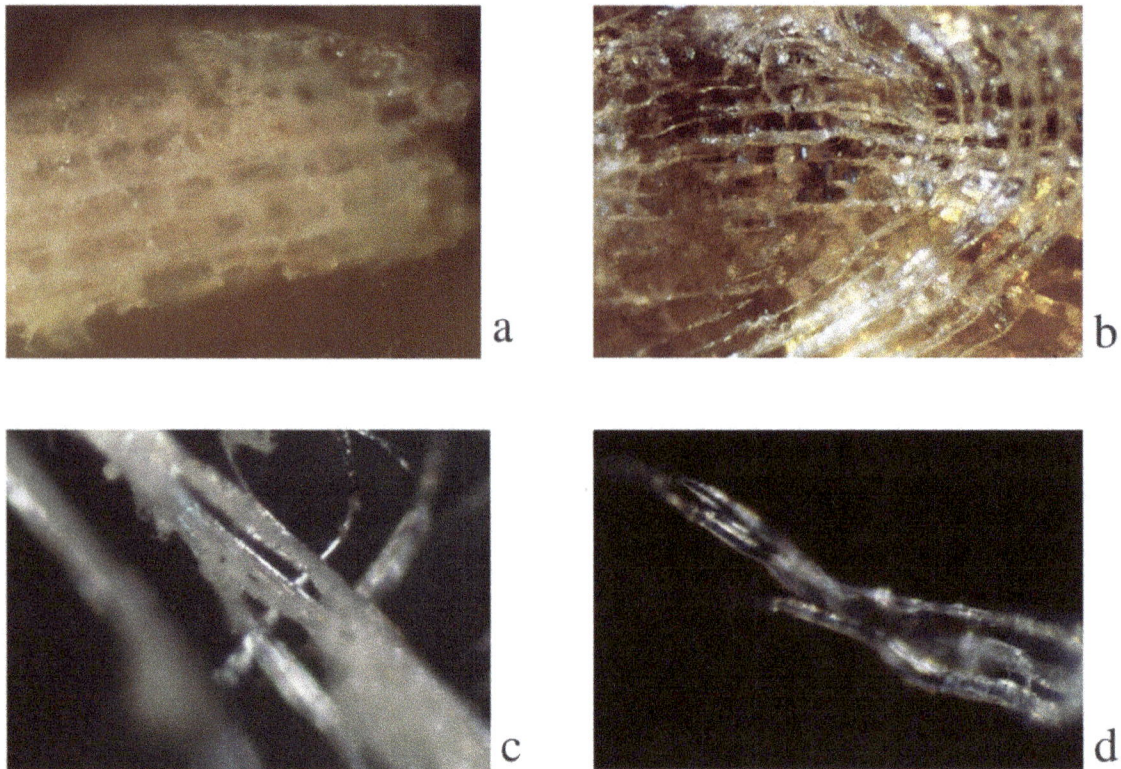

Figure 3: Selected images of replicated vegetal and faunal microresidues highlighting identification complications. (a) Inner epidermal cells of wet wood (Combretum zeheri) on a tool as a result of scraping bark, photographed at 200x. (b) Longitudinally orientated, striated muscle tissue on a tool as a result of cutting beef, photographed at 200x. (c) Plant fibres on the edge of a tool used to scrape fibrous leaf (Sanseviera pearsonii), photographed at 200x. (d) Collagen fibre on the edge of a replicated tool used to cut fatty cartilage of an Aepyceros melampus carcass, photographed at 200x. Figure originally published in Lombard and Wadley (2007a).

indicates a complex relationship between technological sophistication and behaviour that could be considered symbolic (and therefore modern). We know that symbolic behaviour can be traced with certainty to at least 82 ka in North Africa (Bouzouggar *et al.* 2007) and to a similar age in southern Africa (e.g. Henshilwood *et al.* 2002, 2004) because of personal ornaments that have been found at Taforalt (Morocco) and Blombos (South Africa). Thus we focus on hunting and hafting behaviour during the MSA not to provide further evidence for symbolic behaviour, but to increase our knowledge of past technologies, highlighting the complex, variable and multi-faceted nature of the archaeological record. Our brief overview of previously published results provides a summary of our contributions based on this approach. For more detailed and contextual discussions, please consult the primary publications.

Selected results for the post-Howiesons Poort
A multi-analytical functional study suggests that lithic points from the post-Howiesons Poort and late Middle Stone Age layers at Sibudu Cave, with ages between about 60 ka and 50 ka ago, were predominantly used as hafted spear tips. The quantification, plotting and chi-square statistical tests on the distribution patterns of 440 residue occurrences on 24 unbroken points show that the distribution of the residue types cannot be considered coincidental (Lombard 2004, 2005). The traces indicate the use of wooden shafts – revealing a hidden wood-working industry for which there is little additional evidence in the southern African Middle Stone Age (Figure 4). The stone points were probably glued to the shafts with an adhesive and then lashed with plant twine for added strength during use.

The evidence indicates that the spears were most likely used as thrusting or throwing spears. This impression is supported by technological and morphometric data that fall within the expected range of thrusting or throwing spears (Shea 2006; Villa and Lenoir 2006; Villa and Lenoir in press; Villa et al. 2005). Further indicators for effective hunting during this phase at Sibudu Cave come from the associated faunal assemblage (Cain 2006; Plug 2004; Wells 2006). The age profiles of the Sibudu Cave samples show that there are not many bones from juvenile or very old animals: most animals are adults, with some sub-adults. Carnivore-damaged bone is scarce in relation to the size of the sample. These factors indicate that the people using the cave between 50 and 60 ka actively hunted rather than scavenged, and were regularly targeting large animals in their prime (Plug 2004; Wells 2006).

Selected results for the Howiesons Poort
Based on their small size and apparent standardisation, it has often been hypothesized that backed tools from the Howiesons Poort were hafted (H.J. Deacon 1989, 1993; J. Deacon 1995; Wurz 1999). Initial micro-residue evidence for the hafting of such tools from the Howiesons Poort at Rose Cottage Cave (between about 68 and 60 ka old), was provided by Gibson et al. (2004). The spatial distributions of ochre, plant tissue, plant fibres and white starchy residue on 48 backed tools were seen as indications for hafting. The data were interpreted to indicate that backed blades could have been hafted laterally, segments (the type fossil of the Howiesons Poort Industry and also called crescents or lunates) might have been placed transversely into their hafts, while obliquely backed blades were possibly hafted with their short axis in the haft (Gibson et al. 2004).

A more recent study of the micro-residues and other associated use-traces on all the unwashed Howiesons Poort segments from Sibudu Cave provides further resolution for hafting technologies during this industry (Lombard in press). Here it is shown that most segments were indeed hafted, but that it might be difficult to distinguish whether they were hafted transversely, longitudinally or as pairs to form a point. When cumulative use-trace data (micro-residues, use-wear and macrofractures) are considered, however, it is shown that segments were probably hafted in a variety of positions and that there might have been differences in preferred hafting configurations during sequential phases of the Howiesons Poort at Sibudu Cave.

Interpretation of the micro-residues through the Howiesons Poort sequence also shows that there might have been time-related variability in haft materials; the oldest segments seem to have been hafted to bone and the youngest ones to wood, with a transitional phase where either bone or wood may have been used as shafts (Figure 5). These outcomes provide crucial evidence for relatively quick change and variability in technological behaviour during the Howiesons Poort, thereby contradicting the notion of slow change or even stasis during Middle Stone Age industries. Bone-hafting during the Howiesons Poort adds a new dimension to our understanding of hafting technologies and the composite tools of which segments were components (Lombard in press).

The micro-residues showed that segments were mostly used throughout the Howiesons Poort sequence to process animal material, and based on associated macrofracture and use-wear data, there are strong indications that they were predominantly used as inserts (tips, barbs or cutting inserts) for hunting tools (Lombard 2006b, in press) (Figure 6). Preliminary impressions of the Howiesons Poort faunal material indicate a wide range of species, with the little blue duiker well-represented (Clark and Plug submitted). Wadley (2006b) suggests that such tiny forest dwellers could have been trapped rather than hunted with spears or arrows. If we accept the interpretation for the principal hunting function of segments and Wadley's suggestion for the use of traps as reflecting real scenarios for meat procurement, it implies that people used highly variable and specialised hunting technologies more than 60 ka ago.

Selected results for the Still Bay
Recently a study was conducted on a sample of Still Bay pointed tools from Sibudu Cave (older than 70 ka). The sample size is still small, so that assemblage-level studies similar to those conducted on the Howiesons Poort segments are not yet possible. However, a detailed tool-by-

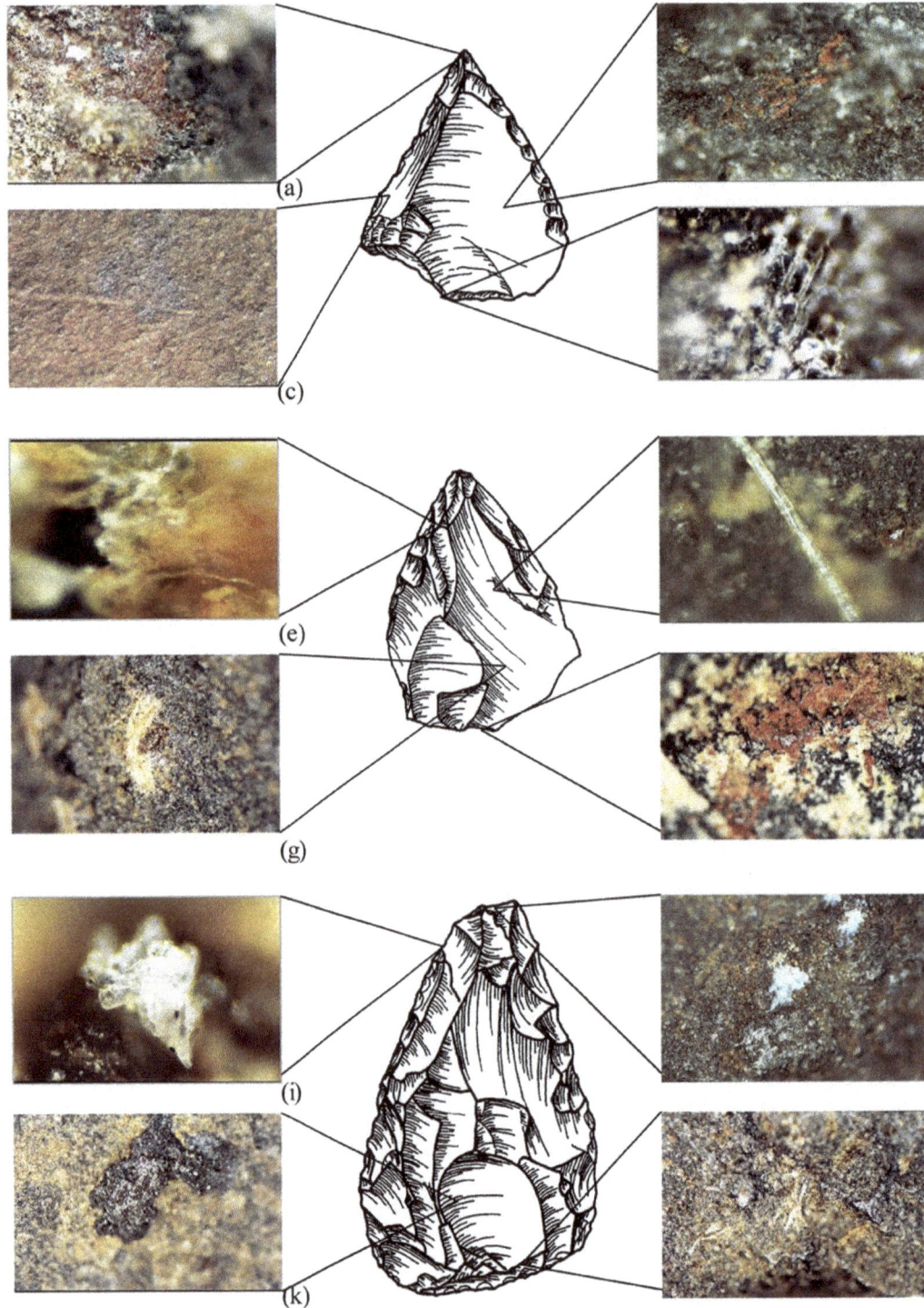

Figure 4: Selected tools and micrographs from the post-Howiesons Poort point sample, Sibudu Cave, indicating hunting and hafting. (a) A thick blood residue deposit, photographed at 200x. (b) A diagonally deposited ochre smear, photographed at 200x. (c) A transverse striation associated with ochre and plant exudate, photographed at 50x. (d) Bark cells or epidermal cell tissue, photographed at 500x. (e) Animal tissue, photographed at 500x. (f) Animal hair, photographed at 200x. (g) Woody residue trapped under a resin deposit, photographed at 100x. (h) A thick ochre and macerated wood deposit, associated with a diagonal striation and bright wood polish, photographed at 200x. (i) Bone and or sheet collagen, photographed at 500x. (j) Fatty residue, photographed at 100x. (k) Thick resinous deposit with wood cell imprint, photographed at 200x. (l) Woody fibres and resin associated with bright wood polish, photographed at 100x. Figure originally published in Lombard (2005).

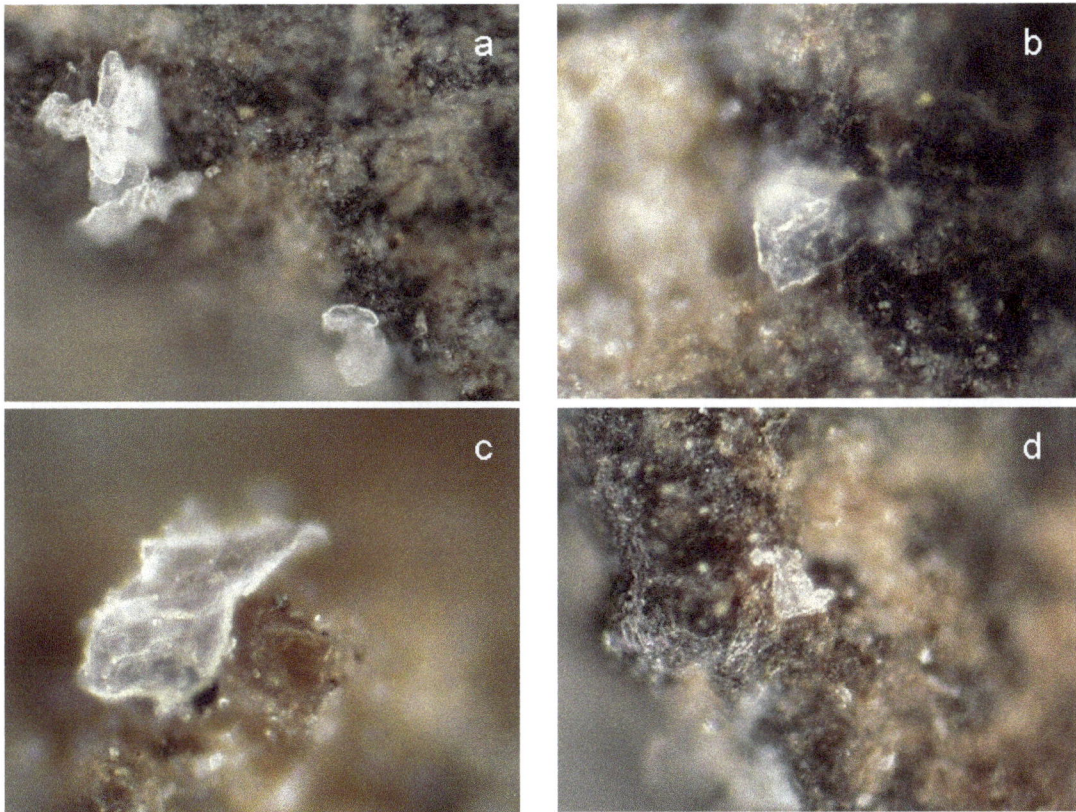

Figure 5: Bone residues superimposed on resinous residues on the backed portions of four different Howiesons Poort segments from Sibudu Cave; (a) photographed at 100x, (b) photographed at 200x, (c) photographed at 500x, (d) photographed at 200x. Figure originally published in Lombard (in press).

a

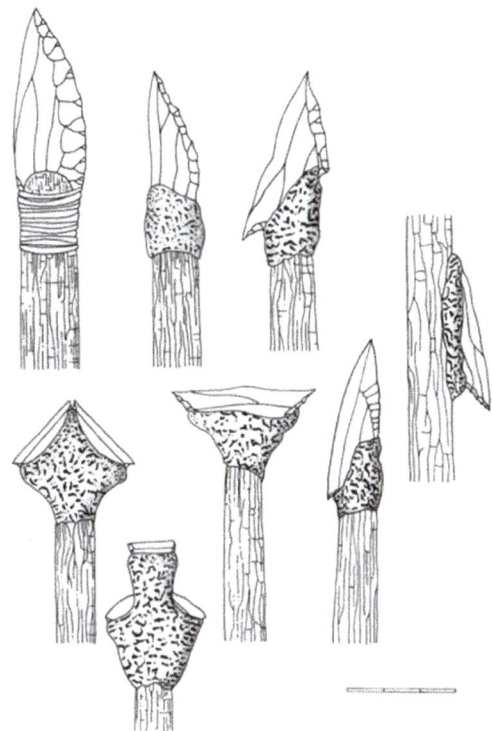

b

Figure 6: (a) Selected Howiesons Poort segments from the Umhlatuzana assemblage. (b) Potential hafting configurations re-drawn by L. Davis after Nuzhnyj (2002).

tool analysis made it possible to test existing hypotheses and generate new working hypotheses for the functions and hafting technologies of these tools (Lombard 2006b). For example, Wadley (2006b) suggested that Sibudu's double-pointed, bifacial points with asymmetrical bases were not intended to be reversible in their hafts, but that the bases were pointed to facilitate a type of hafting that was favoured at the time. The micro-residue distribution patterns and other use-traces on the two unbroken points of this type support this suggestion (Figure 7). Furthermore, both tools show signs of having been used as knives for butchering activities. It is thus possible that the asymmetrically-pointed bases were an adaptation to facilitate the effective hafting of the tools as knives (see Lombard 2006b) (Figure 8). This also supports Minichillo's (2005) and Shea's (2006) suggestions that some Still Bay points from the Cape were used as knives.

There is one triangular bifacial point In the Sibudu sample that is similar in morphology to post-Howiesons Poort points. Macro-fractures and the distribution of animal residues on this point, and also on distal fragments of other points, show that they could have been used as hunting weapons. Thus, a current working hypothesis is that the asymmetrical points with pointed bases were hafted as knives, while symmetrical, triangular points were possibly hafted as hunting weapons. Continued work at Sibudu Cave and other sites with Still Bay assemblages will test this hypothesis. The preliminary results imply, however, that Middle Stone Age points from Still Bay contexts might have been used and hafted differently from points from the younger Middle Stone Age phases.

The use of ochre during the Middle Stone Age
Ochre has been intensely discussed in the literature because its presence is sometimes regarded as evidence for early symbolism (e.g. Ambrose 1998; Barham 1998, 2002; Barton 2005; Conard 2005; d'Errico 2003; Hovers *et al.* 2003; Watts 1998, 2002; Wreschner 1980, 1982). The large quantities of ochre retrieved from Middle Stone Age sites (Watts 2002) and the engraved ochre fragments from Blombos Cave with an age of about 77 ka (Henshilwood *et al.* 2002) mean that the Middle Stone Age of South Africa is central to any debates about ochre. The research conducted on the tools from Sibudu Cave and Rose Cottage Cave and related experimental work has considerably augmented our understanding of the applications of pigmentatious materials such as iron hydroxides and iron oxides, casually referred to as ochre (see Gibson *et al.* 2004; Hodgskiss 2006; Lombard 2004, 2005; Wadley 2005a, 2005b, 2006a; Wadley, Williamson and Lombard 2004). For example, when micro-residue analysis was used to establish the relationship between resin and ochre on a sample of 53 Howiesons Poort segments from Sibudu Cave it was demonstrated that most of the ochre (80%, total n of occurrences = 502) and resin occurrences (87%, total n of occurrences = 585) are located on the backed portions (Lombard 2006a) (Figure 9). These portions of backed tools are generally associated with hafting based on ethnographic and archaeological examples from a variety of contexts (e.g. Becker and Wendorf 1993; Bocquentin and Bar-Yosef 2004; Clark 1977; Clark *et al.* 1974; Goodwin 1945; Phillipson 1976).

The results are interpreted as compelling, direct evidence that the tools were hafted, and that ground ochre was used in the adhesive recipe. It supports the hafting evidence for backed tools from the Howiesons Poort at Rose Cottage Cave and previous observations about the association of ochre with Middle Stone Age hafting technology (Gibson *et al.* 2004; Wadley, Williamson and Lombard 2004). A similar trend for the distribution of ochre and resin residues on the Sibudu segments was documented on a sample of 30 non-quartz Howiesons Poort segments from Umhlatuzana Rock Shelter, about 100 km south-west of Sibudu Cave in KwaZulu-Natal (Lombard 2007b). However, when a sample of 25 quartz segments from Umhlatuzana Rock Shelter was analysed using the same methodology, 269 resin occurrences and only 43 ochre occurrences were counted. Although both residues were concentrated on the backed edges, 68% of the quartz segments have resin but no ochre on them. The same is true for only 23% of the non-quartz sample from the shelter, and 9.5% of the Sibudu Cave sample (Lombard 2007b).

The quartz and crystal quartz segments are not only generally smaller than those made on hornfels and dolerite, but they are also less elongated. Based on these morphological attributes

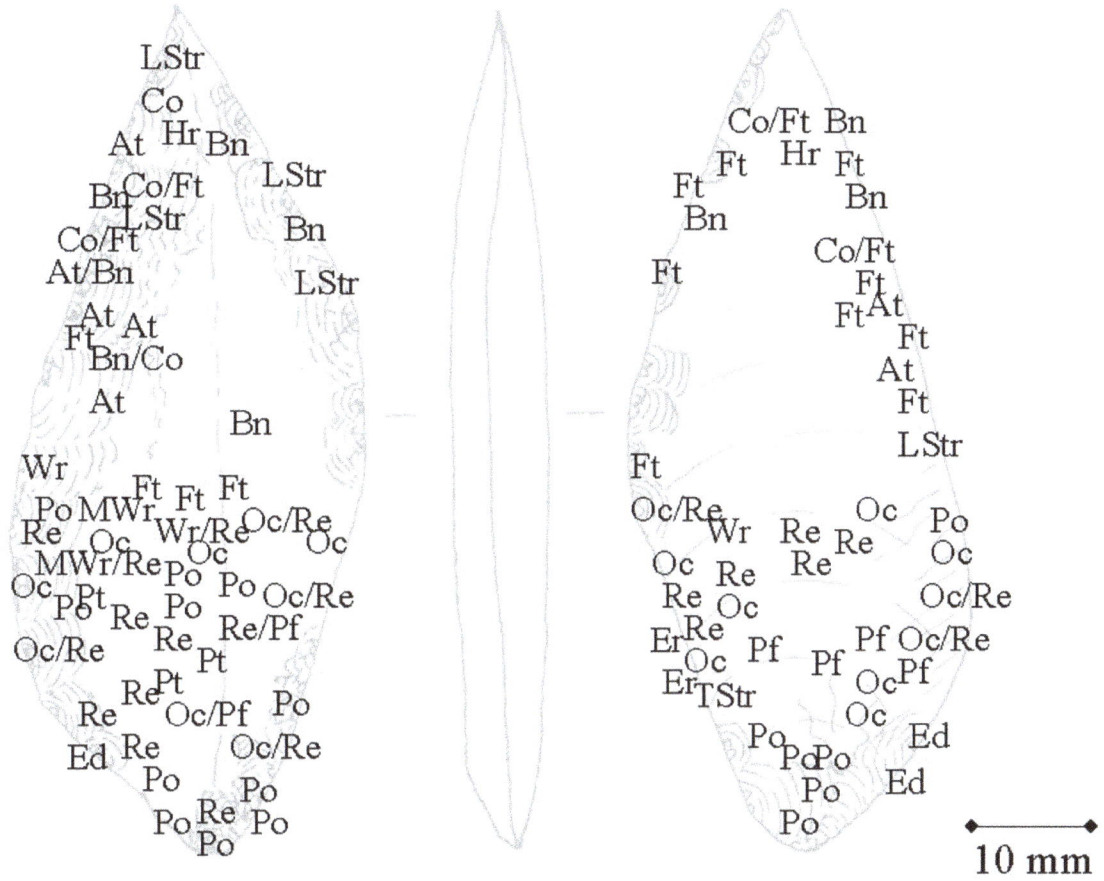

Figure 7: Example of the use-trace plots on a double-pointed, asymmetrical bifacial point from the Still Bay at Sibudu Cave. At = animal tissue, Bl = blood, Bn = bone, Bs = brown stain, Co = collagen, Ed = edge damage, Er = edge rounding, Ft = animal fat, Hr = hair, LStr = longitudinal striation, MWr = macerated woody residue, Oc = ochre, Pf = plant fibre, Po = polish, Pt = plant tissue, Re = resin, TStr = Transverse striations, Wr = woody residue. Figure originally published in Lombard (2006b).

Figure 8: Hypothetical reconstruction of a double pointed, asymmetrical bifacial Still Bay point as a butchering knife with a wooden haft based on use-trace analysis. Drawing by P. Letley.

Figure 9: (a – c) Howiesons Poort segments from Sibudu Cave with macroscopically visible ochre residues on their backed portions. (d) Ochre grains in a clear resin deposit recorded on the backed portion of a Howiesons Poort segment from Sibudu Cave, photographed at 200x. (e) Ochre grains in a clear resin deposit on a replicated stone tool that was hafted to a wooden haft with an ochre-loaded adhesive.

we have suggested that they could have been hafted differently from the larger, longer segments produced on other raw materials (Delagnes *et al.* 2006). Quartz is also very hard (Moh's scale 7; Bishop *et al.* 2001) and the surface is smooth and glass-like. During replication and blind testing (Lombard and Wadley 2007a) it was found that residues do not adhere to the hard, smooth surfaces of quartz to the same degree as they seem to adhere to more porous, coarser-grained raw materials. It is therefore feasible to consider that a different, possibly more 'sticky', adhesive recipe may have been used for the hafting of quartz tools. Our ochre research reported here and elsewhere does not necessarily conflict with interpretations of Middle Stone Age ochre-use as symbolic. Instead, our findings imply that at least from about 70 ka ago, and probably before, people had sophisticated knowledge of the properties of ochre, which made it a suitable aggregate for use in strong glues. Our findings enrich understanding of past technologies, which were by no means primitive.

CONCLUSION

The aim of this paper is to illustrate how stone tool micro-residue analysis in South Africa has evolved in close correlation with Middle Stone Age excavations, experimental research projects and international research trends. The methodology and reference collections, which were developed during replication for experimental projects and blind tests, enable us to generate quantitative data. Such data can be used for comparison with archaeologically recovered assemblages from various contexts at the same site, or even assemblages from different sites. One example is the comparison of the ochre and resin distribution patterns on quartz and non-quartz Howiesons Poort segments from two KwaZulu-Natal sites (Lombard 2007b).

Both the Howiesons Poort and Still Bay technocomplexes are central to the debate about the emergence of modern cognitive behaviour (Ambrose 2006; H.J. Deacon 1989, 1995, 2001; Deacon and Wurz 1996; Henshilwood *et al.* 2001a, 2001b; Lombard 2007a; Minichillo 2005; Wadley 2007; Wurz 1999; Wurz and Lombard in press). Micro-residue analyses conducted within the context of these bigger focus areas are starting to provide detailed, empirical evidence for hunting and butchery activities as well as insight into the complexities of hafting technologies practised during the Middle Stone Age.

Micro-residue analyses provide evidence to support other data that imply effective and innovative hunting (e.g. Marean and Assefa 1999). The results of these focused research projects inform on ancient technological skills and planning abilities. More than 70 ka ago, people understood the properties of various raw materials and tool shapes, sizes and weights, and they seem to have adapted their hunting, butchery and adhesive technologies accordingly. We have generated evidence for change and variability during the Middle Stone Age that might indicate cumulative advances in cognition from at least 70 ka ago. If we accept that people in South Africa behaved symbolically before 70 ka ago (e.g. Henshilwood and d'Errico 2005), we ought to expect that the evidence for change and complexity in behaviours will become increasingly evident in the archaeological record after about 70 ka.

It is understood here that the mere existence of, or evidence for, changes in technologies such as hunting weapons or hafting strategies do not provide evidence for or against symbolic behaviour. The discernible technological variability in the Middle Stone Age record helps, however, to dispel previous notions of simple, unchanging technologies and subsistence patterns. The closer that we get to understanding everyday life in the remote past, represented in tasks such as hunting and hafting, the more it seems that these ancient hunter-gatherers may have behaved similarly to their more recent counterparts.

ACKNOWLEDGEMENTS

We thank the Editorial Committee for the invitation to contribute to this volume and express our gratitude to the late Tom Loy who introduced residue studies to researchers in South Africa. We are indebted to all students and colleagues who partook in the Rose Cottage Cave and Sibudu Cave excavations or contributed to ACACIA (Ancient Cognition and Culture in Africa) research projects. Our appreciation also goes to the Department of Archaeology at the University of the Witwatersrand for the use of their microscope and digital micrograph equipment. The research of Marlize Lombard is funded by PAST (Palaeontological Scientific Trust) and supported by the Natal Museum. Lyn Wadley received funds from the National Research Foundation of South Africa and is supported by the University of the Witwatersrand. The opinions expressed here, or any oversights, are those of the authors and are not necessarily to be attributed to the funding agencies or supporting organisations.

REFERENCES

Ambrose, S. H. 1998. Chronology of the Later Stone Age and food production in East Africa. *Journal of Archaeological Science* 25:179-184.

Ambrose, S.H. 2001. Palaeolithic technology and human evolution. *Science* 291:1789-1753.

Ambrose, S.H. 2006. Howiesons Poort lithic raw material procurement patterns and the evolution of modern human behaviour. *Journal of Human Evolution* 50:356-369.

Barham, L. 1998. Possible early pigment use in south-central Africa. *Current Anthropology* 39:703-710.

Barham, L. 2002. Systematic pigment use in the Middle Pleistocene of south-central Africa. *Current Anthropology* 43: 181-190.

Barton, L. 2005. Origins of culture: functional and symbolic uses of ochre. *Current Anthropology* 46:499.

Becker, M. and F. Wendorf 1993. A microwear study of a Late Pleistocene Quadan assemblage from southern Egypt. *Journal of Field Archaeology* 20:389-398.

Bishop, A.C., A.R. Woolley, and W.R. Hamilton 2001. *Minerals, Rocks and Fossils*. London: George Philip's.

Bocquentin, F. and O. Bar-Yosef 2004. Early Natufian remains: evidence for physical conflict from Mt. Carmel, Israel. *Journal of Human Evolution* 47:19-23.

Bouzouggar, A., N. Barton, M. Vanhaeren, F. d'Errico, S. Collcutt, T. Higham, E. Hodge, S. Parfitt, E. Rhodes, J-L Schwenniger, C. Stringer, E. Turner, S. Ward, A Moutmir and A. Stambouli 2007. 82,000-year-old shell beads from North Africa and implications for the origins of modern human behaviour. *PNAS* 104:9964-9969.

Cain, C.R. 2006. Human activity suggested by the taphonomy of 60 ka and 50 ka faunal remains from Sibudu Cave. *Southern African Humanities* 18(1):241-260.

Clark, J.D. 1977. Interpretations of prehistoric technology from ancient Egyptian and other sources. Part II: prehistoric arrow forms in Africa as shown by surviving examples of the traditional arrows of the San Bushmen. *Paléorient* 3:127-150.

Clark, J.D., J.L. Phillips and P.S. Staley. 1974. Interpretations of prehistoric technology from ancient Egyptian and other sources. Part 1: ancient Egyptian arrows and their relevance for African prehistory. *Paléorient* 2:323-388.

Clark, J.L. and I. Plug submitted. Animal exploitation strategies during the South African Middle Stone Age: Howiesons Poort and post-Howiesons Poort fauna from Sibudu Cave.

Conard, N.J. 2005. An overview of the patterns of behavioural change in Africa and Eurasia during the Middle and Late Pleistocene. In F. d'Errico and L. Backwell (eds) *From Tools to Symbols: From Early Hominids to Modern Humans*, pp. 295-332. Johannesburg: Witwatersrand University Press.

Deacon, H.J. 1989. Late Pleistocene palaeoecology and archaeology in the southern Cape, South Africa. In P. Mellars and C.B. Stringer (eds) *The Human Revolution: Behavioural and Biological Perspectives of the Origins of Modern Humans*, pp. 547-564. Edinburgh: Edinburgh University Press.

Deacon, H.J. 1993. Southern Africa and modern human origins. In M.J. Aitken, C,B. Stringer and P. Mellars (eds*)* *The Origins of Modern Humans and Impact of Chronometric Dating*, pp. 104-117. Princeton: Princeton University Press.

Deacon, H.J. 1995. Two late Pleistocene-Holocene depositaries from the Southern Cape, South Africa. *South African Archaeological Bulletin* 50:121-131.

Deacon, H.J. 2001. Modern human emergence: an African archaeological perspective. In P.V. Tobias, M.A. Raath, J. Moggi-Cecci and G.A. Doyle (eds) *Humanity from African Naissance to Coming Millennia*, pp. 213-222. Johannesburg: University of the Witwatersrand Press.

Deacon, H.J. and S. Wurz 1996. Klasies River Main Site, Cave 2: a Howiesons Poort occurrence. In G. Pwiti and R. Soper (eds) *Aspects of African Archaeology*, pp. 213-218. Harare: University of Zimbabwe Publications.

Deacon, J. 1995. An unsolved mystery at the Howieson's Poort name site. *South African Archaeological Bulletin* 50:110-120.

Delagnes, A., L. Wadley, P. Villa and M. Lombard 2006. Crystal quartz backed tools from the Howiesons Poort at Sibudu Cave. *Southern African Humanities* 18(1):43-56.

d'Errico, F. 2003. The invisible frontier: a multiple species model for the origin of behavioural modernity. *Evolutionary Anthropology* 12:188-202.

d'Errico, F., C.S. Henshilwood, G. Lawson, M. Vanhaeren, A-M. Tillier, M. Soressi, F. Bresson, B. Maureille, A. Nowell, J. Lakarra, L. Backwell and M. Julien 2003. Archaeological evidence for the emergence of language, symbolism, and music – an alternative multidisciplinary perspective. *Journal of World Prehistory* 17:1-70.

Gibson, N. E., L. Wadley and B.S. Williamson 2004. Residue analysis of backed tools from the 60 000 to 68 000 year-old Howiesons Poort layers of Rose Cottage Cave, South Africa. *Southern African Humanities* 16:1-11.

Goodwin, A.J.H. 1945. Some historical Bushman arrows. *South African Journal of Science* 41:429-443.

Henshilwood, C.S. and F. d'Errico 2005. Being modern in the Middle Stone Age: individuals and innovation. In C. Gamble and M. Porr (eds) *The Hominid Individual in Context: Archaeological Investigations of Lower and Middle Palaeolithic Landscapes, Locales and Artefacts*, pp. 244-264. London: Routledge.

Henshilwood, C.S., F. d'Errico, C.W. Marean, R.G. Milo and R. Yates 2001a. An early bone tool industry from the Middle Stone Age at Blombos Cave, South Africa: implications for the origins of modern human behaviour, symbolism and language. *Journal of Human Evolution* 41:631-678.

Henshilwood, C.S., J.C. Sealy, R.J. Yates, K. Cruz-Uribe, P. Goldberg, F.E. Grine, R.G. Klein, C. Poggenpoel, K.L. van Niekerk and I. Watts 2001b. Blombos Cave, southern Cape: preliminary report on the 1992-1999 excavations of the Middle Stone Age levels. *Journal of Archaeological Science* 28:421-448.

Henshilwood, C.S., F. d'Errico, R. Yates, Z. Jacobs, C. Tribolo, G.A.T. Duller, N. Mercier, J.C. Sealy, H. Valladas, I. Watts and A.G. Wintle 2002. Emergence of modern human behaviour: Middle Stone Age engravings from South Africa. *Science* 295:1278-1280.

Henshilwood, C.S., F. d'Errico, M. Vanhaeren, K. van Niekerk and Z. Jacobs 2004. Middle Stone Age shell beads from South Africa. *Science* 304:404.

Henshilwood, C.S. and C.W. Marean 2003. The origin of modern human behaviour: critique of the models and their test implications. *Current Anthropology* 44:627-651.

Henshilwood, C.S. and C.W. Marean 2006. Remodelling the origins of modern human behaviour. In H. Soodyall (ed.) *The Prehistory of Africa: Tracing the Lineage of Modern Man*, pp. 31-48. Johannesburg: Jonathan Ball Publishers.

Hodgkiss, T. 2006. In the mix: replication studies to test the effectiveness of ochre in adhesives for hafting. Unpublished MSc Dissertation. Johannesburg: School of Geography, Archaeology and Environmental Sciences, University of the Witwatersrand.

Hovers, E., S. Ilani, O. Bar-Yosef and B. Vandermeersch 2003. An early use of colour symbolism: ochre use by modern humans in Qafzeh Cave. *Current Anthropology* 44:491-522.

Jacobs, Z., L. Wadley, A.G. Wintle and G.A.T. Duller submitted. New ages for the post-Howiesons Poort, late and final Middle Stone Age at Sibudu Cave, South Africa. *Journal of Archaeological Science.*

Klein, R.G. 2000. Archaeology and the evolution of human behaviour. *Evolutionary Anthropology* 9:17-36.

Kuhn, S.L. and E. Hovers 2006. General Introduction. In E. Hovers and S.L. Kuhn (eds) *Transitions Before the Transition: Evolution and Stability in the Middle Palaeolithic and Middle Stone Age*, pp. 1-11. New York: Springer.

Lombard, M. 2004. Distribution patterns of organic residues on Middle Stone Age points from Sibudu Cave, KwaZulu-Natal, South Africa. *South African Archaeological Bulletin* 59:37-44.

Lombard, M. 2005. Evidence for hunting and hafting during the Middle Stone Age at Sibudu Cave, KwaZulu-Natal, South Africa: a multianalytical approach. *Journal of Human Evolution* 48:279-300.

Lombard, M. 2006a. Direct evidence for the use of ochre in the hafting technology of Middle Stone Age tools from Sibudu Cave, KwaZulu-Natal. *Southern African Humanities* 18(1):57-67.

Lombard, M. 2006b. First impressions on the functions and hafting technology of Still Bay pointed artefacts from Sibudu Cave. *Southern African Humanities* 18(1):27-41.

Lombard, M. 2007a. Archaeological use-trace analyses of stone tools from South Africa. Unpublished PhD Thesis. Johannesburg: School of Geography, Archaeology and Environmental Sciences, University of the Witwatersrand.

Lombard, M. 2007b. The gripping nature of ochre: the association of ochre with Howiesons Poort adhesives and Later Stone Age mastics from South Africa. *Journal of Human Evolution* 53:406-419.

Lombard, M. in press. Finding resolution for the Howiesons Poort through the microscope: micro-residue analysis of segments from Sibudu Cave, South Africa. *Journal of Archaeological Science* (2007) doi:10.1016/j.jas.2007.02.021.

Lombard, M., I. Parsons and M.M. Van der Ryst 2004. Middle Stone Age lithic point experimentation for macro-fracture and residue analyses: the first set of experiments and preliminary results with reference to Sibudu Cave points. *South African Journal of Science* 100:159-166.

Lombard, M. and L. Wadley 2007a. The morphological identification of micro-residues on stone tools using light microscopy: progress and difficulties based on blind tests. *Journal of Archaeological Science* 34:155-165.

Lombard, M. and L. Wadley 2007b. Micro-residues on stone tools: the bigger picture from a South African Middle Stone Age perspective. In H. Barnard and J.W. Eerkens (eds) Theory and practice of archaeological residue analysis, pp. 18-28. *BAR International Series* 1650: Oxford.

Marean, C.W. and Z. Assefa 1999. Zooarcheological evidence for the faunal exploitation behavior of Neandertals and Early Modern Humans. *Evolutionary Anthropology* 8:22-37.

Marean, C.W. and Z. Assefa 2005. The Middle and Upper Pleistocene African record for the biological and behavioural origins of modern humans. In A.B. Stahl (ed.) *African Archaeology: a Critical Introduction*, pp. 93-129. Oxford: Blackwell Publishing.

McBrearty, S. and A.S. Brooks 2000. The revolution that wasn't: a new interpretation of the origin of modern human behaviour. *Journal of Human Evolution* 39:453-563.

Mellars, P. 1973. The character of the middle-upper Palaeolithic transition in south-west France. In C. Renfrew (ed.) *The Explanation of Cultural Change*, pp. 255-276. London; Duckworth.

Mellars, P. 1995. Symbolism, language and the Neanderthal mind. In P. Mellars and K.R. Gibson (eds) *Modelling the Human Mind*, pp. 15-32. Cambridge: Macdonald Institute for Archaeological Research.

Mellars, P. 2005. The impossible coincidence: a single-species model for the origins of modern human behaviour in Europe. *Evolutionary Anthropology* 14:12-27.

Mellars, P. 2006. Why did modern populations disperse from Africa ca. 60,000 years ago? A new model. *Proceedings of the National Academy of Science* 103:9381-9386.

Minichillo, T.J. 2005. Middle Stone Age lithic study, South Africa: an examination of modern human origins. Ph.D Thesis: University of Washington.

Pargeter, J. 2007. To tip or not to tip: a multi-analytical approach to the hafting and use of Howiesons Poort segments. Unpublished Honours Dissertation. Johannesburg: School of Geography, Archaeology and Environmental Studies, University of the Witwatersrand.

Phillipson, D.W. 1976. The prehistory of eastern Zambia. *Memoir Number Six of the British Institute in Eastern Africa*, Nairobi.

Pienaar, M. 2006. Dating the Stone Age at Rose Cottage Cave, South Africa. Unpublished M.A. Dissertation. Pretoria: Department of Anthropology and Archaeology, University of Pretoria.

Plug, I. 2004. Resource exploitation: animal use during the Middle Stone Age at Sibudu Cave, KwaZulu-Natal. *South African Journal of Science* 100:151-158.

Rots, V. and P. van Peer. 2006. Early evidence of complexity in lithic economy: core-axe production, hafting and use at Late Middle Pleistocene site 8-B-11, Sai Island (Sudan). *Journal of Archaeological Science* 31:1287-1299.

Rots, V. and B.S. Williamson 2004. Microwear and residue analysis in perspective: the contribution of ethnographical evidence. *Journal of Archaeological Science* 31:1287-1299.

Shea, J.J. 2003. Close encounters: Neanderthals and modern humans in the Middle Palaeolithic Levant. *The Review of Archaeology* 24:42-56.

Shea, J.J. 2006. The origins of lithic projectile point technology: evidence from Africa, the Levant, and Europe. *Journal of Archaeological Science* 33:823-846.

Soriano, S., P. Villa and L. Wadley 2007. Blade technology and tool forms in the Middle Stone Age of South Africa: the Howiesons Poort and post-Howiesons Poort at Rose Cottage Cave. *Journal of Archaeological Science* 34:681-703.

Valladas, H., L. Wadley, M. Mercier, C. Tribolo, J.L. Reyss and J.L. Joron 2005. Thermoluminescence dating on burnt lithics from Middle Stone Age layers at Rose Cottage Cave. *South African Journal of Science* 101:169-174.

Villa, P. and M. Lenoir 2006. Hunting weapons of the Middle Stone Age and the Middle Palaeolithic: spear points from Sibudu, Rose Cottage and Bouheben. *Southern African Humanities* 18(1):89-122.

Villa, P. and M. Lenoir in press. Hunting and hunting weapons of the Lower and Middle Paleolithic of Europe. In M. Richards and J.J. Hublin (eds) *The Evolution of Hominid Diet*. Springer, Leipzig.

Villa, P., A. Delagnes and L. Wadley 2005. A late Middle Stone Age artefact assemblage from Sibudu (KwaZulu-Natal): comparisons with the Middle Palaeolithic. *Journal of Archaeological Science* 32:399-422.

Wadley, L. 2001. What is cultural modernity? A general view and a South African perspective from Rose Cottage Cave. *Cambridge Archaeological Journal* 11:201-221.

Wadley, L. 2005a. Putting ochre to the test: replication studies of adhesives that may have been used for hafting tools in the Middle Stone Age. *Journal of Human Evolution* 49:587-601.

Wadley, L. 2005b. Ochre crayons or waste products? Replications compared with MSA 'crayons' from Sibudu Cave, South Africa. *Before Farming* 2005/3:1-12.

Wadley, L. 2005c. A typological study of the final Middle Stone Age tools from Sibudu Cave, KwaZulu-Natal. *South African Archaeological Bulletin* 60:51-63.

Wadley, L. 2006a. Revisiting cultural modernity and the role of ochre in the Middle Stone Age. In H. Soodyall (ed.) The Prehistory of Africa: Tracing the Lineage of Modern Man, pp. 49-63. Johannesburg: Jonathan Ball Publishers.

Wadley, L. 2006b. Partners in grime: results of multi-disciplinary archaeology at Sibudu Cave. *Southern African Humanities* 18(1):315-341.

Wadley, L. 2007. Announcing a Still Bay Industry at Sibudu Cave. *Journal of Human Evolution* 52:681-689.

Wadley, L. and Z. Jacobs 2006. Sibudu Cave: background to the excavations, stratigraphy and dating. *Southern African Humanities* 18(1):1-26.

Wadley, L. and M. Lombard 2007. Small things in perspective: the contribution of our blind tests to micro-residue studies on archaeological stone tools. *Journal of Archaeological Science* 34:1001-1010.

Wadley, L., M. Lombard and B.S. Williamson 2004. The first residue analysis blind tests: results and lessons learnt. *Journal of Archaeological Science* 31:1491-1450.

Wadley, L., B.S. Williamson and M. Lombard 2004. Ochre in hafting in Middle Stone Age southern Africa: a practical role. *Antiquity* 78:661-675.

Watts, I. 1998. The origin of symbolic culture: the Middle Stone Age of Southern Africa and Khoisan ethnography. Unpublished PhD Thesis. London: College University of London.

Watts, I. 2002. Ochre in the Middle Stone Age of southern Africa: ritualised display or hide preservative? *South African Archaeological Bulletin* 31:5-11.

Wells, C.R. 2006. A sample integrity analysis of faunal remains from the RSp layer at Sibudu Cave, *Southern African Humanities* 18(1):315-341.

Williamson, B.S. 1996. Preliminary stone tool residue analysis from Rose Cottage Cave. *Southern African Field Archaeology* 5:36-44.

Williamson, B.S. 1997. Down the microscope and beyond: microscopy and molecular studies of stone tool residues and bone implements from Rose Cottage Cave. *South African Journal of Science* 93:458-464.

Williamson, B.S. 2000a. Prehistoric stone tool residue analysis from Rose Cottage Cave and other southern African sites. Unpublished Ph.D. Thesis. Johannesburg: School of Geography, Archaeology and Environmental Studies, University of the Witwatersrand.

Williamson, B.S. 2000b. Direct testing of rock painting pigments for traces of haemoglobin at Rose Cottage Cave, South Africa. *Journal of Archaeological Science* 27:755-762.

Williamson, B.S. 2004. Middle Stone Age tool function from residue analysis at Sibudu Cave. *South African Journal of Science* 100:174-178.

Williamson, B.S. 2005. Subsistence strategies in the Middle Stone Age at Sibudu Cave: the microscopic evidence from stone tool residues. In F. d'Errico and L. Backwell (eds) *From Tools to Symbols: From Early Hominids to Modern Humans*, pp. 513-524. Johannesburg: University of the Witwatersrand Press.

Wreschner, E.E. 1980. Red ochre and human evolution: a case for discussion. *Current Anthropology* 21:631-644.

Wreschner, E.E. 1982. Red ochre, the transition between Lower and Middle Palaeolithic and the origin of modern man. In A. Ronen (ed.) *The Transition From Lower to Middle Palaeolithic and the Origin of Modern Man*, pp. 35-39. BAR International Series 151: University of Haifa.

Wurz, S. 1999. The Howiesons Poort backed artefacts from Klasies River: an argument for symbolic behaviour. *South African Archaeological Bulletin* 54:38-50.

Wurz, S. and M. Lombard in press. 70 000-year-old geometric backed tools from the Howiesons Poort at Klasies River, South Africa: were they used for hunting? In K. Seetah and B. Gravina (eds) *Bones for Tools, Tools for Bones: the Interrelationship of Lithic and Bone Raw Materials*. McDonald Institute for Archaeological Research monographs.

5

A MICROSTRATIGRAPHIC INVESTIGATION INTO THE LONGEVITY OF ARCHAEOLOGICAL RESIDUES, STERKFONTEIN, SOUTH AFRICA

Peta Jane Jones
School of Social Science
The University of Queensland
St Lucia QLD 4072 Australia

ABSTRACT

Controversial claims that proteinaceous residues have been detected on two-million–year-old stone tools from the stratigraphic unit Member five (M5) of the Sterkfontein site, South Africa, are examined through analysis of the microstratigraphy of M5's cemented breccia infill. This study was undertaken to determine the composition of the burial matrix and to understand the post-depositional processes at the site, as a means of examining the feasibility of proteinaceous residue preservation. Petrographic analysis of the samples revealed the main constituents of the M5 breccia to be calcite and clay. Two calcite structural types were observed, which may indicate two or more phases of cementation through calcification. The clay minerals appear to have bonded with the residues to create a fixed and stable environment. Chemical experiments found the M5 breccia had a relatively high potential for cation exchange and both rock types (breccia and dolomitic bedrock) were basic.

KEYWORDS

proteinaceous archaeological residues, clay, calcite, Sterkfontein

INTRODUCTION

Archaeological residues are microscopic remains of prehistoric lifeways, preserving minute traces of plants (such as cellulosic matter, starch and pollen) and animals (such as blood, plasma, feathers and bone) over considerable time periods (Briuer 1976; Dominguez-Rodrigo *et al.* 2001; Fullagar 2006; Fullagar and Field 1997; Fullagar and Jones 2004; Fullager *et al.* 1996; Fullagar *et al.* 2006; Garling 1998; Gerlach *et al.* 1996; Hardy 2004; Hardy and Rogers 2001; Hardy *et al.* 1997; Haslam 2003; Hyland *et al.* 1990; Kooyman *et al.* 1992; Loy 1983, 1998; Loy and Dixon 1998; Loy and Hardy 1992; Newman and Julig 1989; Richards 1989; Shanks *et al.* 2005; Tuross and Dillehay 1995; Williamson 1997). Residue recovery is a major addition to archaeological investigation, allowing for another level of inference: 'The ability to identify residues on stone artefacts will allow researchers to link tool use with animal species to support inferences about human cultural practices' (Shanks *et al.* 2005:36).

In late 1997, Loy claimed the detection of blood residues on Oldowan stone tools from the Sterkfontein cave site in the Republic of South Africa (Loy 1998). The stratigraphic unit from which the stone tools were recovered, the Oldowan Infill from Member 5 (M5), has been dated to approximately two million years (Kuman and Clarke 2000). The potential validation of these claims would demonstrate remarkable post-depositional survival of these organic residues (Figure 1) over a multi-million-year timeframe, regardless of the pre-burial factors that lead to the introduction of the residues to the tools.

From burial to excavation archaeological material is part of the taphonomic system, which includes all processes acting upon the materials in situ (Hanson 1980:157). The archaeological environment involves complex interactions between soil organic matter, soil micro-organisms and soil structure and properties (Haslam 2004:1718). Initial degradation of organic material is rapid, before it plateaus out and stabilises, leaving low molecular weight fragments which have the potential to form durable residues (Haslam 2004; Kimura *et al.* 2001; Sensabaugh *et al.* 1971a; see also Barton this volume). A small number of studies report that preservation of residues over significant timeframes is unlikely (Eisele *et al.* 1995; Tuross *et al.* 1996). However, the majority of investigations indicate that proteinaceous residues retain sufficient structure through time and in varying environmental and depositional conditions to allow for analysis (Cattaneo *et al.* 1990, 1993; Gurfinkel and Franklin 1988; Hyland *et al.* 1990; Loy 1983; Sensabaugh *et al.* 1971a, 1971b).

Physical, biochemical and molecular interactions have the potential to affect blood residue preservation (Brown 1988; Craig and Collins 2000; Heydari 2007; Wadley *et al.* 2004). The pre-burial processes that occur in the biostratinomic interval and contribute to residue preservation involve a complex interplay of the amount of blood deposited, UV exposure time and drying conditions of the blood residues (which allow the air to reduce the oxygenated blood to a more stable condition) (Cattaneo *et al.* 1993; Hortolà 2002; Loy 1983; Tuross *et al.* 1996). The cultural use of a tool in its systemic context may also contribute to blood residue preservation (Cattaneo *et al.* 1993; Hardy and Rogers 2001; Hortolà 1992; Shanks *et al.* 2001). For example, tool use can create minute indentations, such as microcracks, on an artefact's surface that can sequester ancient biological residues away from degradational conditions (Cattaneo *et al.* 1993; Newman and Julig 1989; Shanks *et al.* 1999, 2001, 2005; Tuross and Dillehay 1995).

Archaeological site formation processes (Enloe 2006; Heydari 2007; Latham and Herries 2004; Morin 2006; Piló *et al.* 2005; Rick *et al.* 2006; Schiffer 1996; Ward *et al.* 2006;) and diagenetic conditions of the host site are primary controls on residue survival. A stable soil matrix protects residues from percolating rainwater and chemical, physical, and biological processes (Cattaneo, *et al.* 1993; Kimura *et al.* 2001; Loy 1983). Alkaline soils favour residue recoverability as low pH may deprotonate basic functional groups that displace positively charged residues from the surface. Additionally, Loy (1987) highlighted the potential significance of high clay/silt content, which can contribute to stabilisation of proteinaceous residues.

Interactions between the dynamic depositional, burial and diagenetic processes that affect an archaeological site produce a highly specific set of conditions. Therefore, understanding the reasons for the preservation of residues on stone tools is beyond the scope of experimental studies (for example Wadley *et al.* 2004), which cannot sufficiently replicate these processes or the timeframes over which they occur. In contrast, the rocks that comprise the site preserve a signature of these processes, providing a record of site conditions. Determining the potential preservation mechanisms of the Sterkfontein blood residues requires a taphonomic investigation of the geological processes affecting site formation and how these create and affect the microenvironment of the cultural deposits.

This study reports a microstratigraphic investigation into the M5 geological matrix that enveloped the stone tools on which Loy's (1998) purported blood residues were detected, focusing on the key elements that may be conducive to residue preservation. This includes defining the composition of the burial matrix through petrographic analysis and determining cation exchange capacity (CEC) and pH. From this study the depositional and post-depositional processes that affected M5 can be identified to better constrain the processes that might lead to ancient residue preservation, and therefore provides a basis for conducting future research.

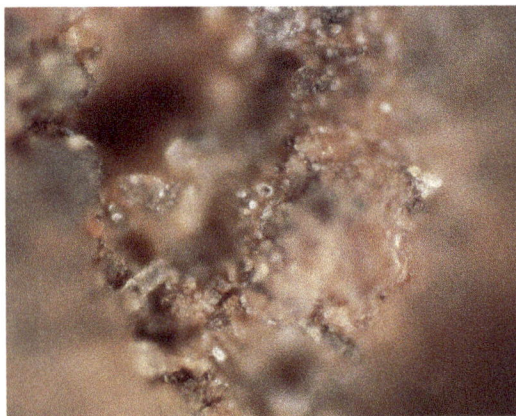

Figure 1. Proteinaceous blood residues found on M5 Oldowan stone tool no. StK2580 mag. 50x (micrograph and interpretation courtesy of Paul Kajewski, University of Queensland).

THE STERKFONTEIN SITE

The Sterkfontein site is situated within the republic of South Africa (Figure 2), 60 km northwest of Johannesburg near Krugersdorp, in the Gauteng region. The region is 1000 to 1800 m above sea level and is covered with grassland similar to savannah, with undulating hills, scattered trees and bushes. The geomorphic setting describes Sterkfontein as a small hill (1491 m), with cave entrances on top, approximately 45 m above a valley that supports the northeasterly flowing Blauuwbank River (Martini *et al.* 2004). The Blauuwbank River is situated 500 m downhill from the Sterkfontein deposits, and has its source in the Witswatersrand mountain range southwest of the Sterkfontein site. The rainy season occurs between October and April, maintaining green eastern lowlands and grasslands on the higher plateau regions of the Highveld and Bushveld. Evidence for this seasonal rain is recorded in the palaeoenvironmental data from Sterkfontein (Vrba 1975).

The Sterkfontein caves are part of a limestone karst deposit comprising several large connecting caves carved out of the Malmani dolomite of the Transvaal sequence (SACS 1980). The dolomite outcrops in the south and dips to the north, with roof collapse occurring in east-west lines (Kuman 1994b). The karsts developed through the intermittently falling water table, which leached calcium carbonate from the dolomitic bedrock (Butterick *et al.* 1993; Wilkinson 1983). Speleological survey of Sterkfontein undertaken by Martini *et al.* (2004) confirms the cave systems formed from a single speleogenetic event, with cavity infills introduced through a pit entrance, mainly in the vadose setting (above the water table), whereas surrounding phreatic passages (situated below the water table) were only partially filled or were left open. Further leaching formed joint-determined cavities, passage networks and wall pockets in the phreatic zone, which were subsequently enlarged through cave-ins above the water table. As a result of the developing karst landscape the surface became unstable in areas with the production of narrow slots (or avens) and eroded surficial sections, allowing the influx of fragmented rocks, fauna, clays and other detrital material. The total exposure of the Sterkfontein caves is approximately 65 x 35 m (Wilkinson 1983), and the opening of the Sterkfontein cave is convoluted, with large dolomite overhangs and treacherous secreted holes (Figure 3). More than 500 hominid-associated fossils have been recovered from the brecciated infills.

Member 5: artefacts and residues

Member 5 was formed between 2.8 and 1.5 million years ago (Partridge and Watt 1991). Kuman and Clarke (2000:827) describe the paleoenvironment for M5 as representing more open woodland and grassland than M4, however the Oldowan infill in the eastern section of M5 suggests a wet localised paleoclimate. The member comprises a pinkish to reddish-brown, well calcified breccia with abundant rock debris, which was replaced by an orange, sandy breccia during periods of collapse, erosion and infilling (Clarke 1994; Partridge and Watt 1991) (Figure 4).

Evidence of Australopithecus and Homo species, early Pleistocene fauna and archaeological artefacts (totalling 9000 pieces ranging from Oldowan to Early Acheulean technologies; Kuman 1994a) have been found within this deposit. The tools found in M5 include chopper-like cores, discoid cores, and protobifaces (Figure 5). All flakes are smaller than 10 cm, have limited retouch, and lack variety throughout the assemblage (Kuman 1994b). The Oldowan industry comprises 2800 tools dated to 1.7–2.0 million years on the basis of associated fossils. The Oldowan assemblage associated with M5 is the first systematic evidence for hominid tool use in South Africa, and

Figure 2. Location map of Sterkfontein, Swartkrans, Kromdraai, and other archaeologically significant sites (after Gauteng Department of Agriculture, Conservation and Environment South Africa: 1998).

Figure 3. Surficial exposure of Sterkfontein.

Figure 4. Orange, sandy debris-rich breccia, comprising Sterkfontein M5 (pencil is 13 cm long).

M5 has been interpreted as representing the first major hominid occupation of caves in the Sterkfontein Valley (Vrba 1980:270)

Loy (1998) analysed residues on 35 stone tools from M5. Of these tools, 43% had animal residues and 17% had plant residues; the rest contained a mixture of both. A bird feather (possibly of the Falconiform order), starch grains, plant fibres, wood, cells and collagen were observed. Other residues included anucleate circular red blood cells (5-8 µm in diameter), nucleated cells, proteinaceous films mixed with clay (Figure 6), degraded and fragmented hairs, feather barbule fragments, collagen fibrils, collagen sheets and collagen powder. Of the residues found by Loy, those applicable to this study are blood proteins.

MICROSTRATIGRAPHIC ANALYSIS

The microstratigraphic analysis for this study involved petrography, pH determination and cation exchange capacity (CEC) experimentation to investigate factors contributing to residue preservation. Petrographic and pH analyses were used to understand the composition of the matrix surrounding the residues, and the CEC experiments indicated how readily the surrounding matrix would exchange cations. The geological samples used in this study were Sterkfontein dolomitic bedrock and breccia from the Oldowan Infill of M5. Dolomite is a chemical sedimentary rock that develops in situ whereas breccia is a clastic sedimentary rock formed from cemented rock fragments.

Petrography

The petrographic analysis involved microstratigraphic study of thin and polished sections, including the shape, form and colour of individual minerals, crystals, and grains, from which the depositional history of a rock can be inferred (Berry and Mason 1959; Kerr 1959). The optical mineralogy of each section was studied and distinguishing features such as colour, pleochroism, birefringence, cleavage, twinning, extinction, relief, crystal form and alteration were noted. Analysis points were taken at random coordinates from areas of general interest. To determine rock volume percentages, 4500 coordinates were observed on each of the thin and polished sections, with the variables at each coordinate entered into a database.

Bloombaum's (1970) 'mapping sentence analysis', adapted by Loy (1972), was used to classify the information in the petrographic database. Observed petrographic attributes were used to build unique 'facets', which consisted of an accumulation of diagnostic traits for each unidentified constituent. Standard definitions were also formulated into facets for comparative purposes (Berry and Mason 1959; Ehlers 1987). Crystals were identified by comparing the breccia and dolomite constituents with these 'ideal' text-book facets. Mineral crystals rarely display the perfect features demonstrated in text-book examples because perfect crystal forms seldom occur under natural conditions (Watson 1972). Using facets as a categorical tool enables relationships between variables to be recognised and analysed, leading to the identification of constituents. Patterns identified included alteration to clay, rock volume percentage and structure.

pH

pH (a symbol for hydrogen-ion concentration) is a measure of the acidity or alkalinity of a solution. The method for soil pH determination used in this study was taken from Cope and Evans (1985). Two gram fragments from each of the breccia and dolomite samples (broken up

Figure 5. Oldowan stone tools recovered from Sterkfontein M5.

Figure 6. SEM micrograph of proteinaceous residue film mixed with clay found on Oldowan stone tool no. StK3312 (scale bar = 10 μm; micrograph and interpretation courtesy of Tom Loy).

as a result of section production and the CEC experiment) were ground to a coarse powder with a mortar and pestle. Milli-Q water was used to mix the ground samples. Before the samples were added, the pH of the water was calibrated using an electronic pH meter. The pH of the samples mixed into the water was then subtracted from the pH of the water to determine the breccia and dolomite pH values.

Cation exchange capacity

This study describes the first use of cation exchange capacity (CEC) field work to investigate the preservation of ancient residues (Jones 1998). Orlov (1992) described cation exchange as a reversible chemical reaction and the CEC of a material as the sum of exchangeable cations it contains. Soil matrices possess an exchange complex comprising various cations, including Ca^{2+}, Mg^{2+}, K^+, and Na^+, and these exchangeable cations are attracted by negative charges (Marshall *et al.* 1996). The capacity for cations to be exchanged varies with the type of clay, crystal size, isomorphous substitution and pH (Marshall *et al.* 1996).

This study used a simplified methylene blue method for rapid determination of CEC in the field (Savant 1993). The determination of CEC is defined as milligram-equivalents (me.) in 100 g of soil, and the range of exchange capacity is from <1.0 to >100 me./100 g. Two of the geological samples (one breccia and one dolomite) were ground to a fine powder using a mortar and pestle. Two grams of each ground sample were measured using electronic scales. Each amount was then transferred to 250 ml capacity Erlenmeyer flasks, and 50 ml of 1% sodium carbonate solution (Na_2CO_3) were added to each. After 10–15 minutes of shaking, the suspensions were titrated: 1 ml of 0.01 M methylene blue solution was released into the Na_2CO_3 suspension. The sample was then gently swirled by hand for 60 seconds. A glass rod was dipped into the solution and then dot-blotted onto Whatman No. 1 filter paper.

Observations were recorded and then the flask was placed under a burette where 1 ml methylene blue solution was released again. The procedure was continued until a halo could be observed around the deposited solution. The volume of methylene blue required to create a halo was multiplied by a constant (0.535) to determine a number that was ascribed a CEC value.

RESULTS

Petrography

The results of the facet analyses of the thin sections showed that the breccia samples had seven components (Figure 7). These components with their relative proportions were: calcite (49%); clay (38%); dolomite (7%); quartz (5%); and chlorite, biotite, and bone (together totalling approximately 1%). The key elements were calcite and clays.

Variations in crystal structure of the breccia samples showed that both anhedral and euhedral crystal growths occurred in the calcite (Figure 8). Approximately 91% of the calcite groups displayed anhedral crystals and 51% of the calcite groups displayed euhedral crystal types. Rhombohedral crystal morphology was apparent on 85% of the calcite studied and was attributed to both anhedral and euhedral crystal types. The euhedral crystal types surround air-filled voids and were large compared to the anhedral crystals. The anhedral crystals formed a flow-like distribution.

The breccia contained a high proportion (38%) of clay minerals. The clay crystal type was massive with no identifiable structure. There was no deviation from this form; it comprised 100% of the observed clay's structure. Chlorite, biotite, bone fragments and dolomite were the minerals displaying the greatest amount of clay formation. Quartz also had some clay-forming crystals; however, quartz is a strong, dense mineral that does not break down as easily as other minerals (Hamblin 1992).

The dolomite rock comprised clasts within the breccia and was formed from dolomite, calcium, magnesium and trace amounts of manganiferous minerals. The dolomite in the breccia sections displayed the largest variation in structure but was of no real consequence to this study because the dolomite formed before the archaeological deposits. The total amount of anhedral crystal morphology of the observed quartz grains was 100%, whereas chlorite displayed 87.5% lathlike crystal structure. The amount of mineral biotite observed was 100% lathlike in structure, with 10% of the crystals held within the boundaries of the bone structures also displaying lathlike morphology.

The breccia constituents were considered immature because they were not well rounded or well sorted. The very poor sorting, massive fabric, fine to medium size and subangular roundness of the breccia grains (Figure 9) is indicative of the depositional environment. The breccia was initially deposited in a high-energy environment, as is evident in the extremely coarse nature of the clasts (larger clasts require more depositional energy). The modifying energy was low, as reflected in the angularity and sorting of the breccia grains. This interpretation is consistent with the palaeoenvironmental reconstructions of Scott and Partridge (1994), whose research into manifestations of Pliocene warming in southern Africa showed that episodic sheetfloods were responsible for the shift from fine clastic sediments to coarser colluvial sediments.

Total Rock Volume Percentages for M5 Breccia

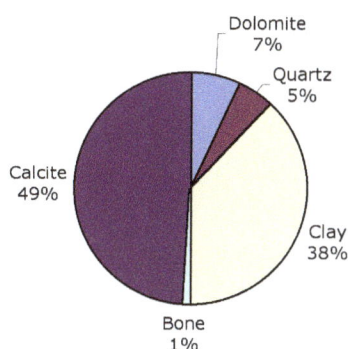

Figure 7. Total rock volume percentages for breccia samples.

pH and cation exchange capacity

The pH results ranged from 8.34–9.86 for the dolomite and 8.34–10.38 for the breccia sample, indicating alkaline conditions. The cation exchange capacity analysis of the Sterkfontein samples produced ambiguous results. The initial halo (the determining factor) was faint so the next definite halo was also recorded to give a range of CEC values. This was not unexpected – results from Savant (1993) also gave a range of values. The 2 g of ground dolomite taken from the surface of the Sterkfontein site had a cation exchange capacity of 2–5%. The 2 g of ground breccia taken from M5 of the Sterkfontein site displayed a cation exchange capacity of 15–40%.

DISCUSSION

This investigation has identified clay and calcite as the major constituents of the breccia matrix surrounding the M5 Oldowan stone tools, with a relatively high pH. The CEC results show the calcite samples to have a higher exchange capacity than the dolomite. As discussed below, these findings are broadly supportive of the potential for long-term blood residue survival at Sterkfontein as claimed by Loy (1998) and are further enhanced by the cave setting. Compared to open-air sites, rockshelters and caves are considered major contributors to influencing the preservation of archaeological material, particularly if stone tools are buried quickly (Tuross *et al.* 1996; Ward *et al.* 2006; Heydari 2007). In addition Kimura *et al.* (2001) suggest that tools protected from the outside environment, for example in caves and similar structures, would have an increased chance of preserving analysable DNA.

Clay and minor constituents

The petrographic examination of the M5 samples found clay and calcite to be the major constituents, with other minerals and fossilised bone also observed in minor amounts. Intermittently wet conditions are ideal for clay formation, because water accelerates the chemical breakdown of rock minerals to clays. Therefore, the high clay content (and the pH values) of the M5 deposits indicate a well-watered environment at the time that the stone tools containing residues were buried. However, the degree of clay formation in the breccia minerals is not significant enough to account for the high clay content recorded. Therefore, it is suggested that the majority of the clays were introduced into the system through primary deposition. The nearby river and the Malmani dolomite may have been a major source of the other breccia minerals. The clay and a proportion of the less abundant breccia minerals were likely introduced into the cave deposits through fluid dynamics and rockfall. The more distantly transported constituents, including quartz, chlorite and biotite, would have entered the cave via rain, floods or rock fall as a result of disturbances at the cave's entrances.

The bone found in the breccia may have been deposited in a similar manner to the quartz, chlorite and biotite; through gravitation and sheet flooding. An additional factor that may have contributed to bone deposition is hominid interaction. The Sterkfontein caves may have been used as a disposal centre for the deceased; unsuspecting hominids may have stumbled and fallen into the treacherous caves; or predators may have dropped bones from over hanging trees, such as with the myriad of small bones found in the breccia thin sections that likely represent the remains of an owl's prey, including tiny lizards or mice (Brain *et al.* 1988). The Sterkfontein Oldowan tools

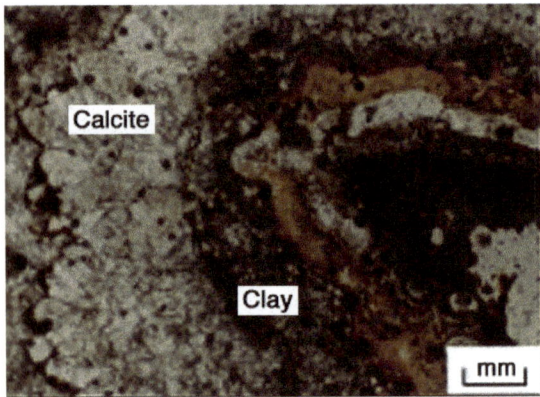

Figure 9. Photograph of M5 breccia showing poor sorting, massive fabric and subangular grains, indicative of a high-energy depositional environment.

Figure 8. Micrograph of calcite crystals tightly packed around breccia minerals of Sterkfontein's M5 deposit (scale in mm).

would have entered the site in a similar manner to the bones, for example, as part of surface debris dropping in, or through hominid use of the cave for shelter and food procurement.

Upon burial, the clay would have come into contact with any blood residues. Pinck and Allison's (1951) investigations describe the preservation of proteins bound to clay, and Gurfinkel and Franklin (1988) suggest that blood binds to soil and clay (see also Loy 1983; Vettori *et al.* 1996). Clay complexes have been reported to protect bound DNA (including proteinaceous constituents) (Alvarez *et al.* 1998; Crecchio and Stotzky 1998; Gallori *et al.* 1994; Stotzky 1986). Residual blood molecules on a stone tool have a highly reactive surface that contains a large number of positive and negative charges, which bind to ionic clay. The ionic bonding creates a linkage formed by the transfer of electrons from one atom (residue) to another atom (clay). Features to be considered in this exchange include cation size and valence, also known as the polarizing power; calcium cations are strongly polarizing (Oades 1989). The nature of the bonds can make recovering proteinaceous residues from ancient mineral surfaces extremely difficult (Craig and Collins 2000).

As organic material and clay minerals create persistent bonds, this suggests that clays bound to proteins (adsorbed) may act as a shield protecting archaeological proteinaceous blood residues from the detrimental effects of fluids, either ground water or surficial water (Jones 1998). The large surface area of some clay species allows for water to be adsorbed, swelling the clay structure and forming a diffuse layer of exchangeable cations extending out from the negatively charged surface to which reactive blood surfaces bind (Loy 1987). Cross-linking of covalent bonds between protein molecules may produce a mass of proteins through aggregation. The resultant organic-bonded materials act as a shield protecting the residues from the detrimental effects of diagenesis through microbial attack and water washing (Gurfinkel and Franklin 1988).

Calcite

The second major constituent of the M5 breccia was calcite. The potential importance to residue longevity of a cemented stable matrix developed through calcite precipitation has not previously been considered, although Thackeray (1997) reported the Sts 5 (Mrs Ples) fragments excavated from M5 were covered by a thin veneer of calcite, which would have formed prior to the cranial bone being buried in sediment.

Calcite has cemented the M5 cave deposits containing the stone tools, clasts, and clay. The calcite content of the breccia displayed two structural morphologies: anhedral crystals, which were small, amorphous crystals in a flow-like distribution, and euhedral crystals, which were more evenly shaped, with larger, well-formed crystals. The majority of the euhedral crystals were distributed in circular patches surrounding an air-filled void. These two crystal morphologies most likely indicate two phases of calcification (Jones 1998). The flow-like distribution of the small anhedral calcite crystals indicates a period of rapid deposition in an area of little space for growth. The euhedral or equant calcite growths surrounding air-filled voids may have formed after the unconsolidated breccia was cemented by the first calcite deposition (James and Choquette 1988; Moore 1989).

Analysis of the site suggests that the two phases of calcification were the result of either a climatic event (Jones 1998) or water table variation. The anhedral calcite may have been associated with increased water flux producing a significant amount of calcite in solution that was deposited rapidly as small, amorphous crystals. The second phase of calcification could then have occurred over a longer period of time, during which there was relatively little rainfall with smaller amounts of calcite solution released at any one time. This environment would have allowed gradual growth of large, well-formed crystals. An alternative explanation may be that the anhedral crystals grew in the phreatic zone and the well-formed euhedral crystals grew in the vadose zone. The use of staining procedures to determine which cementation took place first (Scholle et al. 1989) would be a worthwhile avenue of future research.

It is suggested here that calcite plays a role in the preservation mechanics of proteinaceous archaeological residues. The CEC experiment found that the M5 breccia had a higher capacity to exchange cations than the dolomite. The relatively high CEC value of the breccia could be attributed to the presence of the calcite, which possesses a large amount of cations with high mobility (Press and Siever 1986). These exchangeable cations are available in solution and are attracted to the negatively charged surface of the clay, creating a weak ionic bond (Press and Siever 1986). Calcium is a divalent cation with a higher charge and is therefore more strongly attracted to the clay surface then other free cations (Marshall et al. 1996).

Mitterer and Cunningham (1985:17) state that organic material interacts with calcium carbonate systems in two ways: (i) adsorption of the organic material to calcium carbonate surfaces, and (ii) complexation or chelation of free cations by dissolved or adsorbed organic material. If clay bonds to blood residues and calcite bonds weakly to clay, a protective shield for the blood residues may result. Furthermore the crystallised form of calcite can help protect the clay and blood from degradative processes through the tight packing of crystals and continual growth from solution and further sequestration of the proteins away from the active soil environment (Jones 1998). Upon lithification the clay and calcite are cemented together and the clay sediment dries, which would sequester water away within clay minerals. These clay minerals are capable of extracting and absorbing water from most other materials. Dried clay sediments are strong and stable, particularly with the dominant presence of the alkaline clay mineral montmorillonite. The calcification of the M5 deposits therefore cemented the immediate environment of the blood-clay bond and further protected them from detrimental processes.

Other Sites

The reported detection of two million year old blood residues on Oldowan stone tools from Sterkfontein (Loy 1998) is supported by the taphonomic history of the site as recorded in the microstratigraphy. This study suggests that a cave environment and a well cemented, clay-rich burial matrix with an alkaline pH and high CEC favours residue preservation over extreme timeframes. This hypothesis can be indirectly tested by reviewing sites with similar and differing taphonomic contexts that both preserve and do not preserve proteinaceous residues.

Cave and rockshelter sites that have been found to preserve blood residues include the Enclosed Chamber of Rocky Cape South, Tasmania (Fullagar 2004) and La Quina, southwest France (Hardy et al. 1997). In these cases it is assumed that the physical barrier of the enclosed or covered sites was a critical contributing factor in residue preservation, as neither site has a significant clay component in the burial matrix. While the preservation timeframes are significantly less than at Sterkfontein (6,000 and 40,000-48,000 years old, respectively), if the site morphology persists through time then the preserving physical barrier will not be diminished, even over millions of years.

Blood residues 30,000 to 36,000 years old have been discovered in clay-rich burial matrices in the Australian megafauna site of Cuddie Springs (Field *et al.* 2006; Garling 1998) and the wetland Monte Verde site in Chile (Tuross and Dillehay 1995). The Cuddie Springs and Monte Verde sites formed in low energy environments at the surface (i.e. not in a cave). The clay content of the burial matrix may have been the primary contributing factor in the preservation of the residues; however, the host stone tools were recovered beneath strata that may have provided a physical barrier to detrimental surficial processes (Cuddie Springs – compacted clay layer; Monte Verde – peat layer rich in silica gel). These strata may have played a similar role to the lithifiying calcite in the Sterkfontein site.

Archaeological sites of comparable age and taphonomic history to Sterkfontein that do not preserve blood residues include the Peninj site in Tanzania and the Okote Member of the Koobi Fora Formation in East Africa. In Peninj, tools dated to 1.5 Ma have been recovered from consolidated sands and clays in a muddy sandstone matrix that include carbonates and calcite (Dominguez-Rodrigo *et al.* 2001). The site maintained a good preservation environment, but there is no evidence of animal processing activities. It is likely therefore that the absence of blood residues is a function of tool use rather than non-preservation. The geology of the upper part of the surficial Koobi Fora site is defined as a caliche (calcium carbonate block) with sands, gravels, sandy silt with calcareous concretions, and volcanic tuff comprising the other layers (Rogers and Harris 1992:43). Hardy and Rogers (2001) report the detection of residues on stone tools dated to 1.5-1.6 Ma. These residues included woody and non-woody plant tissue, starch grains, and hair fragments. Hardy and Rogers (2001) indicate that the Okote artefacts were used to process plant and animals, which is consistent with cutmarks found on bone from the site (Bunn 1981). These tool use reconstructions suggest that in the case of the Koobi Fora site, the absence of blood residues on the stone tools is due to non-preservation despite the presence of clay in the lithified burial matrix.

A brief review of international case studies therefore supports some of the key findings from this study. The detection of proteinaceous archaeological residues between 6,000 and 48,000 years old in Australia, Chile and France is due to varying combinations of clay fraction in the burial matrix, the sealing of archaeological deposits through lithification or a similar process, and the protected environment of cave systems and rockshelters. The absence of blood residues from stone tools used to process animals in the lithified, calcareous and partly silty burial matrix of the Koobi Fora site in East Africa suggests that the physical barrier of a cave system may be the primary requisite factor in the possible preservation of proteinaceous residues for millions of years, as in the Sterkfontein site. However, this is likely to be an over-simplification, with the potential for the preservation of blood residues for greater than two million years likely requiring a specific combination of the factors described in this study, and for these to have influenced the taphonomy of the site in a particular series of events.

CONCLUSIONS

From this investigation it can be stated that the burial matrix of the Oldowan infill from M5 has provided and maintained a stable and non-detrimental environment to archaeological remains, attested to by the exceptional preservation of hominid remains and the presence of a large number of stone artefacts. Significant factors in this context included both the burial matrix and the environmental setting (dolomitic karst caves that provided ingress points for the stone tools and covering sediments, and that may have attracted hominins in the first place). Loy's (1998) claim of blood residue preservation on M5 Oldowan tools, combined with the findings of this study, therefore suggests a specific scenario as follows:

1. Hominids use stone tools multiple times in the processing of animals which leaves behind a thick layer of blood residues;

2. Blood dries onto the stone tool surface, residues are denatured and/or sequestered into microcracks;
3. Stone tools are transported, residues come into contact with clay;
4. Some degree of binding occurs which creates hydrophobic organic-protein complexes;
5. Archaeological remains and sediment are transported;
6. Deposition from entrances i.e. side of hills, avens or slopes;
7. Clay sediment provides a large surface area, swelling capability, diffuse layers of exchangeable cations and a negatively charged surface which all encourage organic-mineral complexes;
8. The aggregation of proteins with clay makes them more resistant to degradation;
9. Rainfall releases calcium ions from the overlying dolomitic bedrock (leaching);
10. The leached solution filters through the underlying gravity-derived sediments and precipitates as calcium carbonate (calcite) that subsequently cements the deposits into brecciated infill;
11. Calcium ions are adsorbed to organic material;
12. Weak bonds form between clay and calcite surfaces;
13. Precipitated calcite creates a well-cemented clay/calcite matrix enclosing and protecting the hominid-bearing deposits, including the blood residues;
14. The breccia matrix maintains a regulated environment that is not directly exposed to surficial water and was possibly buried with further compaction from overlying brecciated deposits, protected from detrimental conditions.

ACKNOWLEDGEMENTS

I am indebted to Meg Heaslop, Andrew Jones and Michael Haslam for their input into the various manuscripts that came their way. This study would not be possible without the aid of Ron Clarke, Kathy Kuman, and Francis Thackeray, who patiently showed me around the Sterkfontein region and dedicated their time to collecting rock samples. Thank you to Peter Colls whose geotechnical skills helped create the thin sections that were integral to this study. The comments offered by two reviewers were also greatly appreciated. Funds were supplied in part by an Australian Postgraduate Award. Thank you also to my mentor Dr T.H. Loy for having faith in me in the first place and for offering continual support, I miss you dearly.

REFERENCES

Alvarez, A.J., M. Khanna, G.A. Toranzos, and G. Stotzky 1998. Amplification of DNA bound on clay minerals. *Molecular Ecology* 7:775–778.

Berger, L. 1998. The dawn of humans: redrawing our human tree? *National Geographic* 198:90–99.

Berry, L.G. and B. Mason 1959. *Mineralogy: Concepts, Descriptions, Determinations*. San Francisco: W. H. Freeman and Company.

Bloombaum, M. 1970. Doing Smallest Space Analysis. *Journal of Conflict Resolution* 14:409–416.

Brain, C.K., C.S. Churcher, J.D. Clark, F.E. Grine, P. Shipman, R.L. Susman, A. Turner and V. Watson 1988. New evidence of early hominids, their culture and environment from the Swartkrans cave, South Africa. *South African Journal of Science* 84:828–835.

Briuer, F.L. 1976. New clues to stone tool function: plant and animal residues. *American Antiquity* 41:478–484.

Brown, P. 1988. Residue analysis of stone artefacts from Yambon, West New Britain. Unpublished BA (Honours) thesis. Sydney: University of Sydney.

Bunn, H.T. 1981. Archaeological evidence for meat-eating by Plio-Pleistocene hominids from Koobi Fora and Olduvai Gorge. *Nature* 291:574–577.

Butterick, D.B., J.L. van Rooy, J.L. and R. Ligthelm, R. (1993). Environmental geological aspects of the dolomites of South Africa. *Journal of African Earth Sciences* 16:53–61.

Cattaneo C., K. Gelsthorpe, P. Phillips and R.K. Sokol 1990. Blood in ancient human bone. *Nature* 347:339.

Cattaneo C., K. Gelsthorpe, P. Phillips and R.K. Sokol 1993. Blood residues on stone tools: indoor and outdoor experiments. *World Archaeology* 25:29–43.

Chafetz, H.S., B.H. Wilkinson and K.M. Love 1985. Morphology and composition of non-marine carbonate cements in near-surface settings. In N. Schneidermann and P.M. Harris (eds) *Carbonate Cements* (No. 36), pp. 337–347. Oklahoma: Society of Economic Paleontologists and Mineralogists.

Clarke, R.J. 1994. On some new interpretations of Sterkfontein stratigraphy. *South African Journal of Science* 90:211–214.

Cope, J.T. and C.E. Evans 1985. Soil Testing. *Advances in Soil Science* 1:201–228.

Craig, O.E. and M.J. Collins 2000. An improved method for the immunological detection of mineral bound protein using hydrofluoric acid and direct capture. *Journal of Immunological Methods in Enzymology* 236:89–97.

Crecchio, C. and G. Stotzky 1998. Binding of DNA on humic acids: effect on transformation of *Bacillus subtilis* and resistance to DNase. *Soil Biology and Biochemistry* 30:1061–1067.

Dashman, T. and G. Stotzky 1982. Adsorption and binding of amino acids on momoinic montmorillonite and kaolinite. *Soil Biology and Biochemistry* 14:447–456.

Dominguez-Rodrigo, M., J. Serralonga, J. Juan-Tresserras, L. Alcala and L. Luque 2001. Woodworking activities by early humans: a plant residue analysis on Acheulian stone tools from Peninj (Tanzania). *Journal of Human Evolution* 40:289–299.

Ehlers, E.G. 1987. *Optical mineralogy* Vol. 2. London: Blackwell Scientific Publications.

Eisele, J.A., D.D. Fowler, G. Haynes and R.A. Lewis 1995. Survival and detection of blood residues on stone tools. *Antiquity* 69:36–46.

Enloe, J.G. 2006. Geological processes and site structure: Assessing integrity at a Late Paleolithic open-air site in northern France. *Geoarchaeology* 21:523–540.

Field, J., S. Wroe and R. Fullager 2006. Blitzkrieg: fact and fiction at Cuddie Springs. *Australasian Science* 27:28–29.

Fullagar, R. 2006. Residues and usewear. In J. Blame and A. Paterson (eds) *Archaeology in Practice: A student guide to archaeological analysis,* pp. 207–234. Malden: Blackwell Publishing.

Fullagar, R. and J. Field 1997. Pleistocene seed-grinding implements from the Australian arid zone. *Antiquity* 71:300–307.

Fullagar, R.J. and R. Jones 2004. Usewear and residue analysis of stone artefacts from the Enclosed Chamber, Rocky Cape, Tasmania. *Archaeology in Oceania* 39:79–93.

Fullagar, R., J. Furby and B. Hardy 1996. Residues on stone artifacts: state of a scientific art. *Antiquity* 70:74–745.

Fullagar, R.J., J. Field, T. Denham and C. Lentfer 2006. Early and mid Holocene tool use and processing of taro (*Colocasia esculenta*), yam (*Dioscorea* sp.) and other plants at Kuk Swamp in the highlands of Papua New Guinea. *Journal of Archaeological Science* 33:595–614.

Gallori, E., M. Bazzicalupo, L. Dal Canto, R. Fani, P. Nannipieri, C. Vettori, and G. Stotzky 1994. Transformation of *Bacillus subtilis* by DNA bound on clay in non-sterile soil. *Federation of European Microbiological Societies Microbiology Ecology* 15:119–126.

Garling, S.J. 1998. Megafauna on the menu? Haemoglobin crystallization of blood residues from stone artefacts at Cuddie Springs. In R. Fullagar (ed.) *A Closer Look*, pp. 29–48. Sydney: Sydney University.

Gerlach, S.C., M. Newman and E.J. Knell Jr. 1996. Blood protein residues on lithic artifacts from two archaeological sites in the De Long Mountains, northwestern Alaska. *Arctic* 49:1–10.

Grimshaw, R.W. 1971. *The Chemistry and Physics of Clays and Allied Ceramic Materials.* London: Ernest Benn Limited.

Gurfinkel, D.M. and U.M. Franklin 1988. A study of the feasibility of detecting blood residue on artifacts. *Journal of Archaeological Science* 15:83–97.

Hamblin, W.K. 1992. *Earth's Dynamic Systems* (6th ed). New York: Macmillan Publishing Company.

Hanson, C.B. 1980. Fluvial taphonomic processes: models and experiments. In A.K. Behrensmeyer and A.P. Hill (eds) *Fossils in the Making: Vertebrate taphonomy and paleoecology,* pp. 156–181. Chicago: University of Chicago Press.

Hardy, B.L. 2004. Neanderthal behaviour and stone tool function at the Middle Palaeolithic site of La Quina, France. *Antiquity* 78:547–565.

Hardy, B.L., and M.J. Rogers 2001. Microscopic investigation of stone tool function from Okote Member sites, Koobi Fora, Kenya [abstract]. *Journal of Human Evolution* 40:A9.

Hardy, B.L., V. Raman and R.A. Raff 1997. Recovery of mammalian DNA from Middle Palaeolithic stone tools. *Journal of Archaeological Science* 24:601–611.

Harter, R.D. and G. Stotzky 1971. Formation of clay-protein complexes. *Soil Science Society of America Proceedings* 35:383–389.

Haslam, M. 2003. Evidence for maize processing on 2000-year-old obsidian artefacts from Copan, Honduras. In D. M. Hart and L. A. Wallis (eds) *Phytolith and Starch Research in the Australian-Pacific-Asian regions: The state of the art*, pp. 153–161. Canberra: Pandanus Books.

Haslam, M. 2004. The decomposition of starch grains in soils: implications for archaeological residue analyses. *Journal of Archaeological Science* 31:1715–1734.

Heydari, S. 2007. The impact of geology and geomorphology on cave and rockshelter archaeological site formation, preservation, and distribution in the Zagros mountains of Iran. *Geoarchaeology* 22:653–669.

Hiscock, P. 1985. The need for a taphonomic perspective in stone artefact analysis. *Queensland Archaeological Research* 2:82–95.

Hortolà, P. 1992. SEM characterization of blood stains on stone tools. *The Microscope* 40:111–113.

Hortolà, P. 2002. Red blood cell haemotaphonomy of experimental human bloodstains on techno-prehistoric lithic raw materials. *Journal of Archaeological Science* 29:733–739.

Hyland, D., J. Tersak, J. Adovasio and M. Siegel 1990. Identification of the species of origin of residual blood on lithic material. *American Antiquity* 55:104–112.

James, N. P. and P.W. Choquette 1988. Introduction. In N.P. James and P.W. Choquette (eds) *Paleokarst*, pp. 1–21. New York: Springer-Verlag.

Jones, P. 1998. *A Microstratigraphic Investigation into the Longevity of Archaeological Blood Residues, Sterkfontein, South Africa*, Unpublished BA (Honours) thesis. St Lucia: Department of Anthropology and Sociology, The University of Queensland.

Kerr, P.F. 1959. *Optical Mineralogy.* 3rd ed. New York: McGraw-Hill Book Company.

Kimura, B., S.Brandt, B. Hardy and W. Hauswirth 2001. Analysis of DNA from ethnoarchaeological stone scrapers. *Journal of Archaeological Science* 28:45–53.

Kooyman, B., M.E. Newman and H. Ceri 1992. Verifying the reliability of blood residue analysis on archaeological tools. *Journal of Archaeological Science* 19:265–269.

Kuman, K. 1994a. The archaeology of Sterkfontein — past and present. *Journal of Human Evolution* 27:471–495.

Kuman, K. 1994b. The archaeology of Sterkfontein: preliminary findings on site formation and cultural change. *South African Journal of Science* 90:215–219.

Kuman, K. and R.J. Clarke 2000. Stratigraphy, artefact industries and hominid associations for Sterkfontein, Member 5. *Journal of Human Evolution* 38:827–847.

Latham, A.G. and. A.I.R. Herries 2004. The formation and sedimentary infilling of the cave of hearths and historic cave complex, Makapansgat, South Africa. *Geoarchaeology* 19:323–342.

Loy, T.H. 1972. *Archaeological Survey of Yoho National Park: 1971.* Report submitted to Department of Indian and Northern Affairs: National Historic Parks and Sites Branch.

Loy, T.H. 1983. Prehistoric blood residues: detection on tool surfaces and identification of species of origin. *Science* 220:1269–1270.

Loy, T.H. 1987. Recent advances in residue analysis. In W. Ambrose and J. M. J. Mummery (eds) *Archaeometry: Further Australasian Studies*, pp. 57–65. Australian National University, Canberra.

Loy, T.H. 1998. Organic residues on Oldowan tools from Sterkfontein Cave, South Africa [abstract]. In H.S.M.A. Raath, K.L.K.D. Barkhan and P.V. Tobias (eds) *Dual Congress of the International Association for the Study of Human Paleontology, and International Association of Human Biologists*, pp. 74–75, Johannesburg: Department of Anatomical Sciences, University of the Witswatersrand Medical School.

Loy, T.H. and E.J. Dixon 1998. Blood residues on fluted points from eastern Beringia. *American Antiquity* 63:21–46.

Loy, T.H. and B.L. Hardy 1992. Blood residue analysis of 90 000 year old stone tools from Tabun Cave, Israel. *Antiquity* 66:24–35.

Marshall, T.J., J.W. Holmes and C.W. Rose 1996. *Soil Physics.* 3rd ed. Cambridge: Cambridge University Press.

Martini, J.E.J., P.E.Wipplinger, H.F.G. Moen and A. Keyser 2004. Contribution to the speleology of Sterkfontein Cave, Gauteng province, South Africa. *Speleogenesis and Evolution of Karst Aquifers* 2:1–20.

Mitterer, R.M. and R.C. Cunningham Jr. 1985. The interaction of natural organic matter with grain surfaces: Implications for calcium carbonate precipitation. In N. Schneidermann and P.M. Harris (eds) *Carbonate Cements* (No.36), pp. 17–31. Oklahoma: Society of Economic Paleontologists and Mineralogists.

Moore, C.H. 1989. *Carbonte Diagenesis and Porosity.* New York: Elsevier Publishing Company.

Morin, E. 2006. Beyond stratigraphic noise: unraveling the evolution of stratified assemblages in faunalturbated sites. *Geoarchaeology* 21:541–565.

Newman, M. and P. Julig 1989. The identification of protein residues on lithic artifacts from a stratified boreal forest site. *Canadian Journal of Archaeology* 13:119–132.

Oades, J.M. 1989. An introduction to organic matter in mineral soils. In J. B. Dixon and S. B. Weed (eds) *Minerals in soil environments.* 2nd ed., pp. 1244. Wisconsin: Soil Science Society of America.

Orlov, D.S. 1992. *Soil Chemistry.* Rotterdam: A.A. Balkema.

Partridge, T.C. and I.B. Watt 1991. The stratigraphy of the Sterkfontein hominid deposit and its relationship to the underground cave system. *Palaeontologia Africana* 28:35–40.

Piló, L.B., A.S.Auler, W.A.Neves, X.Wang, H. Cheng, and R. Lawrence Edwards 2005. Geochronology, sediment provenance, and fossil emplacement at Sumidouro Cave, a classic Late Pleistocene/Early Holocene paleoanthropological site in Eastern Brazil. *Geoarchaeology* 20:751–764.

Pinck, L.A. and F.E. Allison 1951. Resistance of a protein montmorillonite complex to decomposition by soil microorganisms. *Science* 114:130–131.

Press, F. and R. Siever 1986. *Earth.* 4th ed. New York: W. H. Freeman and Company.

Richards, T. 1989. Initial results of a blood residue analysis of lithic artefacts from Thorpe Common rockshelter, South Yorkshire. In I. P. Brooks and P. Philips (eds) Breaking the stony silence: Papers from the Sheffield lithics conference 1988, pp. 73–90. British Series 123. Oxford: British Archaeological Reports.

Rick, T.C., J.M. Erlandson and R.L. Vellanoweth 2006. Taphonomy and site formation processes on California's Channel Islands. *Geoarchaeology* 21:567–589.

Rogers, M.J. and J.W.K. Harris 1992. Recent investigations in landscape archaeology at East Turkana. *Nyame Akuma* 38:41–47.

South African Committee for Stratigraphy (SACS) 1980. *Stratigraphy of South Africa.* Part 1: Lithostratigraphy of the Republic of South Africa, South West Africa/Namibia and the Republics of Bophuthatswana, Transkei and Venda Handbook 8. Pretoria: Department of Mineral and Energy Affairs.

Savant, N.K. 1993 Simplified methylene blue method for rapid determination of cation exchange capacity of mineral soils. *Communications in Soil Science and Plant Analysis* 23:3357–3364.

Schiffer, M.B. 1996. *Formation Processes of the Archaeological Record.* Salt Lake City: University of Utah Press.

Scholle, P.A., N.P. James and J.F. Read (eds) 1989. *Carbonate Sedimentology and Petrology.* Vol 4. Washington: American Geophysical Union.

Scoffin, T. P. 1987. *An Introduction to Carbonate Sediments and Rocks.* New York: Chapman and Hall.

Scott, L. and T.C. Partridge 1994. Some manifestations of Pliocene warming in southern Africa. In R.S. Thompson (ed.) *Pliocene Terrestrial Environments and Data/Model Comparisons.* U.S. Geological Survey Open-File Report 94-23: 54-55.

Sensabaugh, G.F., A.C. Wilson and P.L. Kirk 1971a. Protein stability in preserved biological remains I. Survival of biologically active proteins in an 8-year-old sample of dried blood. *International Journal of Biochemistry* 2:545–557.

Sensabaugh, G.F., A.C. Wilson and P.L. Kirk 1971b. Protein stability in preserved biological remains II. Modification and aggregation of proteins in an 8-year-old sample of dried blood. *International Journal of Biochemistry* 2:558–568.

Shanks, O.C., R.Bonnichsen, A.T. Vella and W. Ream 2001. Recovery of protein and DNA trapped in stone tools microcracks. *Journal of Archaeological Science* 28:965–972.

Shanks, O.C., M. Kornfeld and D.D. Hawk 1999. Protein analysis of Bugas-Holding tools: new trends in immunological studies. *Journal of Archaeological Science* 26:1183–1191.

Shanks, O.C., L. Hodges, L. Tilley, M. Kornfeld, M.L. Larson and W. Ream 2005. DNA from ancient stone tools and bones excavated at Bugas-Holding, Wyoming. *Journal of Archaeological Science* 32:27–38.

Stotzky, G. 1986. Influence of soil mineral colloids on metabolic processes, growth, adhesion, and ecology of microbes and viruses. In P.M. Huang and M. Schnitzer (eds) *Interactions of Soil Minerals with Natural Organics and Microbes*, no.17. Madison: Soil Science Society of America.

Thackeray, J.F. 1997. Cranial bone of 'Mrs Ples'; fragments adhering to matrix. *South African Journal of Science* 93:169–170.

Tuross, N. and T.D. Dillehay 1995. The mechanism of organic preservation at Monte Verde, Chile, and one use of biomolecules in archaeological interpretation. *Journal of Field Archaeology* 22:97–110.

Tuross, N., I. Barnes and R. Potts 1996. Protein identification of blood residues on experimental stone tools. *Journal of Archaeological Science* 23:289–296.

Vettori, C., D.Paffetti, G. Pietramellara, G. Stotzky and E. Gallori 1996. Amplification of bacterial DNA bound on clay minerals by the random amplified polymorphic DNA (RAPD) technique. *Federation of European Microbiological Societies Microbiology Ecology* 20:251–260.

Vrba, E. S. 1975. Some evidence of chronology and palaeoecology of Sterkfontein, Swartkrans and Kromdraai from the fossil Bovidae. *Nature* 254:301–304.

Vrba, E.S. 1980. The significance of bovid remains as indicators of environment and predation patterns. In A.K. Behrensmeyer and A.P. Hill (eds) *Fossils in the Making: Vertebrate taphonomy and paleoecology,* pp. 247–271. Chicago: University of Chicago Press.

Wadley, L., M. Lombard and B. Williamson 2004. The first residue analysis blind tests: results and lessons learnt. *Journal of Archaeological Science* 31:1491–1501.

Ward, I.A.K., R. Fullagar, T. Boer Mah, L.M. Head, P.S.C. Taçon and K. Mulvaney 2006. Comparison of sedimentation and occupation histories inside and outside rock shelters, Keep-River region, northwestern Australia. *Geoarchaeology* 21:1–27.

Watson, J. 1972. Rocks and Minerals. In J.A.G. Thomas (ed.) *Introducing Geology: 4. Rocks and Minerals,* pp. 65. London: George Allen and Unwin Limited.

Wilkinson, M.J. 1983. Geomorphic Perspectives on the Sterkfontein Australopithecine Breccias. *Journal of Archaeological Science* 10:515–529.

Williamson, B. 1997. Down the microscope and beyond: microscopy and molecular studies of stone tool residues and bone samples from Rose Cottage Cave. *South African Journal of Science* 93:458–464.

6

MOUNTAINS AND MOLEHILLS: SAMPLE SIZE IN ARCHAEOLOGICAL MICROSCOPIC STONE-TOOL RESIDUE ANALYSIS

Michael Haslam

Leverhulme Centre for Human Evolutionary Studies
University of Cambridge
Cambridge CB2 1QH, United Kingdom
Email: mah66@cam.ac.uk

ABSTRACT

The passing of Thomas H. Loy in late 2005 presents an opportunity to reflect on the field of archaeological stone-tool microscopic residue analysis, of which Loy was a pioneer. This paper discusses one important and neglected aspect of residue studies from an historical perspective: reliance by residue analysts on small sample sizes. A review of stone-tool microscopic residue studies published over the first 30 years of the discipline (1976-2006) demonstrates that half of these examined 25 or fewer artefacts, with sample sizes of three or fewer artefacts being most common. In addition, sample sizes are shown to be decreasing in recent years. These findings have implications for the applicability of the novel data provided by residues to broader archaeological debates, and are discussed in relation to the specialisation of archaeological microscopic residue analyses. The review provides a caution against over-interpretation and emphasises the need for a more explicit link between sample sizes and research questions, while recognising the validity and usefulness of appropriately theorised studies of small numbers of artefacts.

KEYWORDS

residue analysis, microscopy, sample size, stone tools, Thomas H. Loy

INTRODUCTION

The passing in October 2005 of a pioneer of archaeological microscopic residue analysis, Thomas H. Loy, offers an opportunity to reflect on the historical trajectory of the field. Although building on earlier traceological work exemplified by Semenov (1964), the traditional origin date of the stone-tool microscopic residue technique is 1976, with the publication by Briuer of 'New clues to stone tool function: plant and animal residues' in American Antiquity. This was followed by the initiation of rigorous and multi-stranded investigation of blood residues by Loy (1983) and subsequent increased emphasis on starch grain residues and other plant microfossils (e.g. Loy *et al.* 1992). The residue analysis field is presently characterised by a growing diversity of approaches and geographic foci and extends to a time depth of well over 1,000,000 years (Dominguez-Rodrigo *et al.* 2001; Hardy and Rogers 2001), with this success owing in no small measure to Loy's influence.

This paper does not intend to provide an exhaustive historical overview, however, and instead takes this opportunity to examine a neglected issue of significance to current practice, namely the sample sizes employed by stone-tool residue analysts. This issue is discussed in the context of the role of novel scientific data within archaeological reconstructions, using a review of the archaeological microscopic stone-tool residue literature from 1976-2006. The aims of this review are twofold: first to ascertain the range of variation and any trends in the numbers of artefacts examined by analysts; and second to ask whether or not sample size has or should have a constraining influence on the conclusions drawn from a stone-tool residue study.

SAMPLE SIZE IN MICROSCOPIC STONE-TOOL RESIDUE ANALYSIS

Sampling issues have a long tradition in archaeological investigation (see outlines in Baxter 2003; Mueller 1975; Redman 1974; also Orton 2000). To assess residue study sample sizes for this study, an annotated list of stone-tool microscopic residue analyses published between 1976 and 2006 was assembled (see Appendix). Certain parameters were followed for inclusion, namely that the residue data must be from a published account of original research on archaeological stone artefacts using light microscopy. This excludes therefore publications that discuss residue findings but report no new results (e.g. Denham and Barton 2006; Villa *et al.* 2005; Wood 1998), as well as unpublished theses and manuscripts (e.g. Broderick 1982; Fullagar 1982; Furby and Loy 1994; Gorman 2000; Higgins 1988; Loy *et al.* 1999; Paull 1984; Petraglia *et al.* 2002) and residue studies not employing, or ambiguous about the use of, light microscopy (e.g. Anderson 1980; Anderson *et al.* 2004; Anderson-Gerfaud 1990; Cosgrove 1985; Loy and Nelson 1986; McBryde 1974; Morwood 1981; Nelson *et al.* 1986; Potter 1994; Robertson 1996). Care has also been taken to avoid double listing of separate publications that report the same results (e.g. Dodson *et al.* 1993 and Furby *et al.* 1993; Fullagar 1988 and 1989; Fullagar 1992 and Fullagar *et al.* 1998; Pearsall 2003, Pearsall *et al.* 2004 and Chandler-Ezell *et al.* 2006; Perry 2002 and 2005) and in these situations the more detailed study is listed.

The studies reviewed were from the English-language literature, as the vast majority of residue results have been published in English, although a limited number of studies exist in other languages (e.g. Babot 2001; Pagan Jimenez *et al.* 2005). The Appendix to this paper updates that provided in Haslam (2006b: Appendix 1), and the literature surveyed covers only those publications available up to the time of data compilation in mid-2006. A small number of publications appearing around or after this time are therefore not included. These restrictions are not expected to affect the conclusions of the survey to any significant degree, however the dynamic nature of the field emphasises the need for periodic review.

The key concern of most researchers using sampling theory is obtaining a representative sample that will allow quantification and extrapolation of results from a limited dataset to a larger (often unknown) population. A relevant distinction here is between target and sampled populations (Orton 2000:41), the former being the set of artefacts about which we wish to make statements, and the latter being the available artefacts from which we select pieces for analysis. The very notion of sampling implies a target population under investigation, although the myriad approaches for analysing archaeological objects may require a narrowing or broadening of this target depending on the interests of the archaeologist. If the sampled population contains biases (e.g. all analysed stone tools are from one site), then it may not be possible to make meaningful statements about a wider target population (e.g. the use of stone tools across all sites in a region). It is also difficult to assess the validity of statements about any population when sampling procedures are not explicitly discussed in published studies, as is the case for the large majority of those reviewed for this paper. For example, while the artefacts in Haslam (2003) were randomly selected, this fact was not reported in the published study, which simply lists the number of artefacts examined. Non-random sampling introduces further potential biases.

Data from the studies reviewed in the Appendix suggest that stone-tool microscopic residue analyses have yet to adequately address or account for the influence of sampling biases, or establish representative samples appropriate for the desired scale of analysis. Clearly, there have been both large (>2000 artefacts, e.g. Boot 1993; Briuer 1976) and very small (single artefact, e.g. Babot and Apella 2003; Fullagar 1993b; Loy 1985; Piperno, *et al.* 2004) samples investigated.

Table 1. Summary statistics for the reviewed studies

All studies			
Total no. of studies	96	Standard deviation	420.34
Total no. of analysed artefacts	13025	Median	26
Mean no. of artefacts per study	135.68	Mode	3

24 largest studies		24 smallest studies	
Total no. of analysed artefacts	11514	Total no. of analysed artefacts	62
Mean no. of artefacts per study	479.75	Mean no. of artefacts per study	2.58
Standard deviation	750.93	Standard deviation	1.41
Median	166.5	Median	3
Mode	150	Mode	3

Statistical analyses and graphical presentation of these numbers provide an initial perspective on the representativeness of the conclusions drawn from these studies. Table 1 outlines some basic descriptive statistics; as it examines sample sizes those studies not reporting such data are excluded. Where multiple studies are reported in a single paper they are treated separately, as indicated in the Appendix, giving a total of 96 sites in 91 publications. The table includes information on the entire selection as well as the 24 smallest and largest studies (comprising 25% of the total in each case) as a means of further clarifying the nature of the data.

The most obvious feature of the data is that the sample sizes are very heavily positively skewed (that is, towards small samples), as seen in the sequence of mean>median>mode. This is true of the sample as a whole as well as for the largest 25% of studies. Data for the smallest 25% of studies show a very slight negative skew in the sequence of mean<median=mode, but as the largest sample size in this data subset is five artefacts, this does not detract from the observed trend towards analyses of small samples. The most common sample size in the entire dataset (the mode), and therefore the most commonly reported microscopic residue artefact sample size, is three artefacts. The second most common sample size is a single artefact, and the third most common is two artefacts. Similarly, while the overall average is 136 artefacts per study, this figure is deceptive, as the standard deviation of 420 (more than three times the mean), and the median value of 26 clearly show. The influence of the higher outliers is also seen when the largest 25% of studies are considered, with one standard deviation (751) far exceeding the mean (480). Half of all studies have examined 25 or fewer artefacts.

The largest 25% of studies contains every residue study of over 70 artefacts and the chief impact on this dataset is made by the three studies (Boot 1993; Briuer 1976; Fullagar 1988) that exceed 1000 artefacts. Because of the strong positive skew, a more realistic view of past practice is provided by the smallest 25% of studies, with a mean sample size of 2.6 and standard deviation of 1.4. By way of comparison, the smallest 50% of studies have an average of 8.6 artefacts per study, again with a relatively large standard deviation of 7.2 (data not shown in Table 1). Additionally, the total number of stone artefacts examined microscopically for residues over the past three decades may be estimated at less than 15,000 worldwide, even allowing for those studies that are not included in Table 1 due to lack of published data. The three largest studies contribute over 7000 of these artefacts. Compared with other archaeological analytical methods and other lithic analysis techniques, the low total number of artefacts analysed and published globally to this point marks the field as a niche provider of archaeological information, and the focus on small samples warrants further investigation for its impact on the field.

Graphic presentations of numerical trends are more accessible than raw statistics alone, so these are combined in the remainder of this section. Figure 1 shows the spread of published sample sizes, with the x-axis presenting two different scales divided at the 400 artefact mark for practical reasons. Studies smaller than 400 artefacts are mapped within 20-artefact bands, beginning with 1–20 artefacts (represented as '20' on the x-axis), then 21–40 artefacts, and so on. After the 381–400 artefact bracket (represented as '400' on the x-axis), the studies are shown in 200-artefact bands in order to incorporate the few very large studies. The indication '600' on the x-axis therefore represents the 401–600 artefact bracket, and so on up to 2601–2800 artefacts. Figure 2 breaks down the dominant '1–20 artefact' category to reveal the emphasis on samples of three artefacts or fewer.

In addition to reviewing sample sizes by individual study, it is also beneficial to consider trends over time. In the following it should be stressed that phrases such as 'artefacts analysed per year' are used interchangeably with 'artefacts published per year', as it is rarely possible

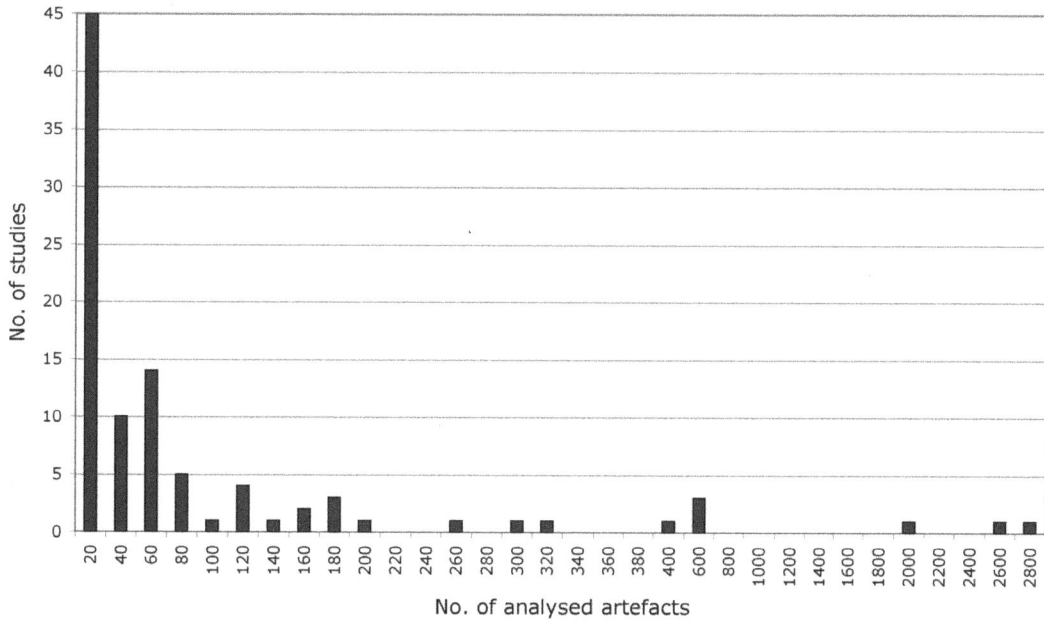

Figure 1. Number of stone artefacts analysed per stone-tool microscopic residue study, 1976-2006.

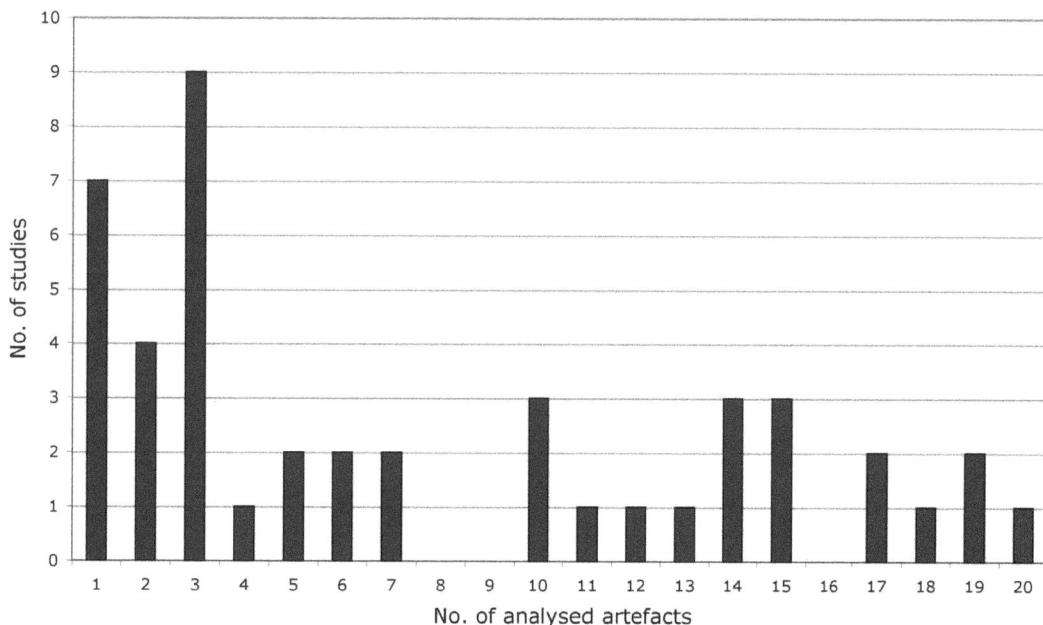

Figure 2. Number of stone artefacts analysed per microscopic residue study, 1976-2006 (1–20 artefact sample sizes only).

to differentiate year of analysis from year of publication. The issues considered here, then, include the number of publications per year reporting original stone-tool microscopic residue results (Figure 3), the cumulative number of authors contributing to these publications (Figure 4), the number of artefacts examined per year worldwide (Figure 5), and the median sample sizes for analyses published each year (Figure 6). The former two graphs support the contention that the field is expanding and, while not all the authors included in Figure 4 are themselves residue analysts, this trend does show the rapidly increasing number of archaeologists willing to incorporate microscopic residue data into their work. Figure 4 is conservative in that publications discussing residues but not presenting new data are not included in the surveyed literature for this paper; however trends toward multiple authorship counter this conservatism somewhat.

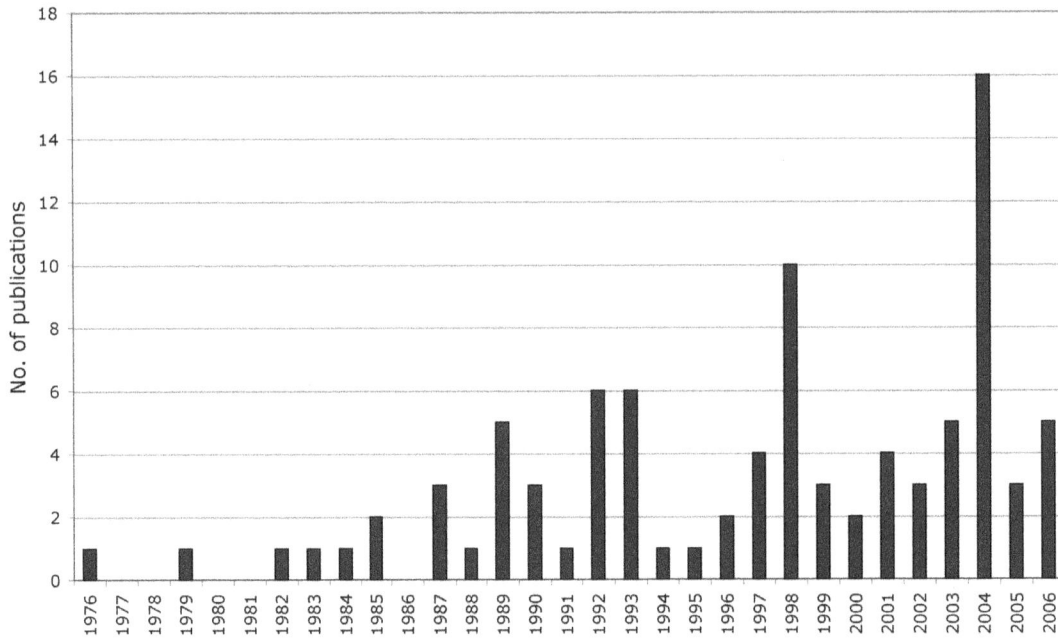

Figure 3. Number of publications reporting original results of microscopic stone-tool residue analyses per year, 1976-2006.

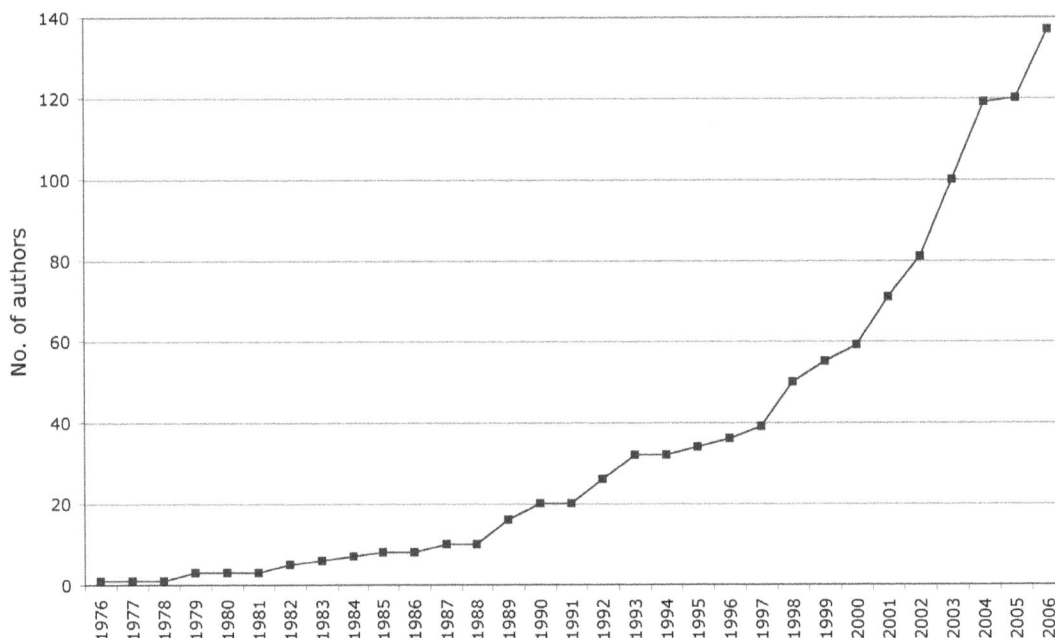

Figure 4. Cumulative total number of authors who have published original microscopic stone-tool residue results, 1976-2006.

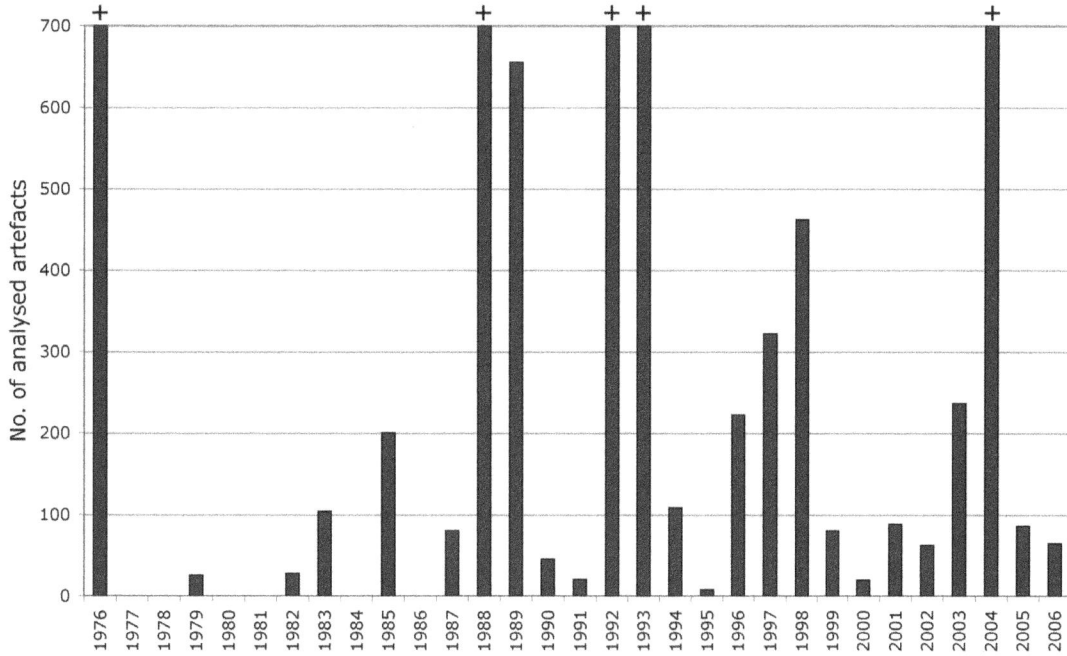

Figure 5. Number of stone artefacts analysed for microscopic residues per year, 1976-2006. Years marked with '+' have values beyond the range of the graph (1976 n=2551; 1988 n=1814; 1992 n=1263; 1993 n=2798; 2004 n=1687).

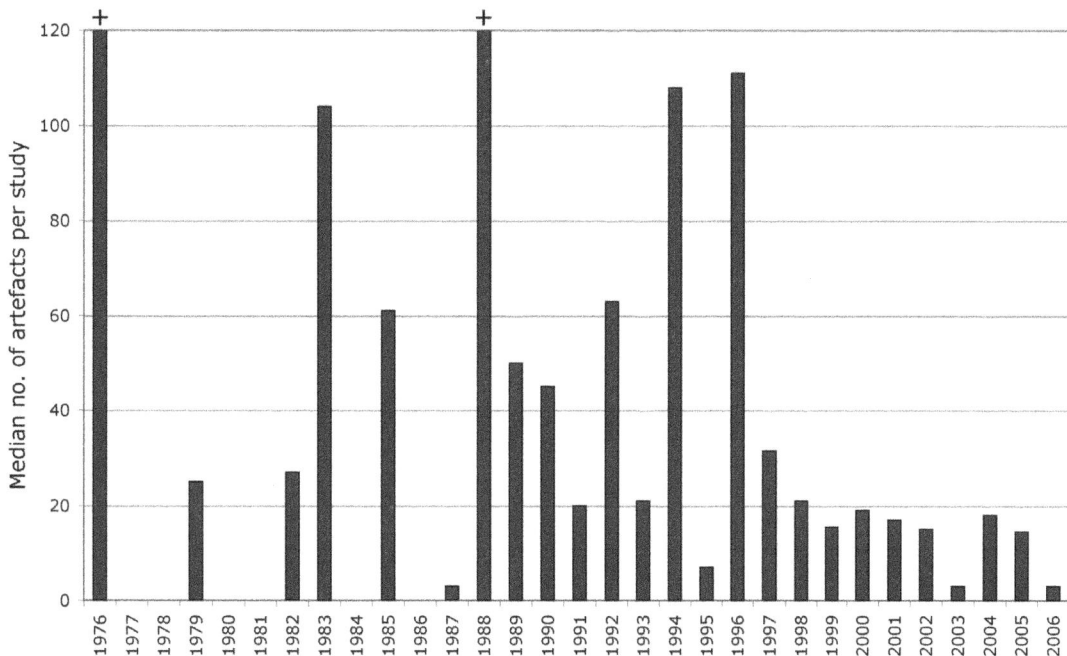

Figure 6. Median number of analysed artefacts per stone-tool microscopic residue study, 1976-2006. Years marked with '+' have values beyond the range of the graph (1976 n=2551; 1988 n=1814).

Figures 5 and 6 present two different perspectives on chronological quantification. Since 1976, five years have seen more than 700 artefacts analysed (Figure 5), largely attributable to the actions of individual researchers (Boot 1993; Briuer 1976; Fullagar 1988, 1992; Hardy 2004; and Williamson 2004, Wadley *et al.* 2004). Other than these exceptional occurrences, global residue research output is measured in the hundreds, and in many recent years the tens, of artefacts annually. There are however no striking trends in this highly variable data, even if a speculative 'pulse' of activity on a roughly five-year cycle is discerned over the past 20 years or so. Additionally, the figure for 2006 in this chart would be expected to increase with the inclusion of data from the latter part of that year. Figure 6 provides a valuable alternate viewpoint, displaying median annual sample sizes rather than means (which simply tend to follow a scaled-down version of the data in Figure 5) in an attempt to avoid the skew created by extreme outliers at the higher end of the scale. Two years (1976 and 1988) are again off the chart as they saw only one large published study each. Interestingly, once the field moved beyond its introductory phase and established itself in the late 1980s (the last year without at least one publication reporting an original set of light microscopy residue results was 1986; see Figure 3), there has been a slow and uneven but nevertheless relatively steady decline in median sample sizes. Note that unlike the totals in Figure 5, median values for the first half of 2006 are valid as they are independent of the number of studies. With only one exception, each year from 1988 to 1998 inclusive the median sample was above 20 artefacts per study; since 1998 the median has not exceeded this number. Importantly in this regard, 41 of the 91 publications (45%) surveyed for this review were published since 1998, forming a significant component of the residue literature currently contributing to archaeological debate. Implications of declining sample sizes for archaeological interpretation are discussed further below.

A final approach to the data (Figures 7 and 8) is based on geography rather than sample size, and again illustrates the strong influence of a few individuals in the growth of the residue field. Microscopic stone-tool residue analyses have to date been completed by researchers working in a number of countries, however these may be divided into four broad geographic locations for the purposes of initial assessment – Oceania (principally Australia and Papua New Guinea), the Americas (especially North and Central America and northern South America), Africa (chiefly South Africa) and Eurasia. In reporting the numbers of artefacts analysed for each of these regions, I have broken the 30-year period since 1976 into five-year blocks (note that for the sake of inclusiveness the few studies from the first part of 2006 have been included in the final block). It is apparent from Figure 7 that initial application of microscopic residue analysis occurred in the Americas, and was followed subsequently by rapid growth in Oceania, and more recently South Africa. Studies of Eurasian material (with a focus on western Europe and the Middle East) have not seen the same expansion in terms of studied artefacts but have nonetheless continued steadily since their inception in the late 1980s.

As seen earlier the presence of large outlier studies is a strong influence and in this instance is directly related to the geographic focus of key analysts: Briuer initially in North America, Fullagar and Boot in Australia/Papua New Guinea, Hardy and Loy in Europe and Williamson in South Africa. It is worth noting in the context of this volume that Loy was responsible for training and/or early collaboration with many of these researchers, indicating the high impact (quantitatively and pedagogically) of individuals in shaping a field with relatively few practitioners. Interestingly, after 1990 a decline or plateau in numbers of analysed artefacts occurs in Oceania and the Americas alongside growth in Eurasia and Africa, despite increases in the numbers of studies published in all regions over the same period (Figure 8). A pattern of initial examination of large samples, perhaps as a form of method validation or training, followed by fragmentation into a variety of much smaller analyses is therefore apparent for the Americas, Oceania and to a lesser extent Eurasia. Whether this pattern will hold for Africa remains to be seen.

REPRESENTATIVENESS AND STATISTICAL SIGNIFICANCE

Having briefly considered the trends in typical sample sizes employed in stone-tool microscopic residue analysis, we may now consider the relevance of this information for archaeological interpretation. In general, for any project the impact of sample size on the inclusiveness or representativeness of results is dependent on the question asked, the target population, and the qualitative or quantitative nature of the enquiry. Piperno *et al.* (2004) for example examined a single grindstone, yet the provenance (Upper Palaeolithic Israel) and observed residues (starches indicative of barley processing) warranted publication in a prestigious scientific journal. There is no

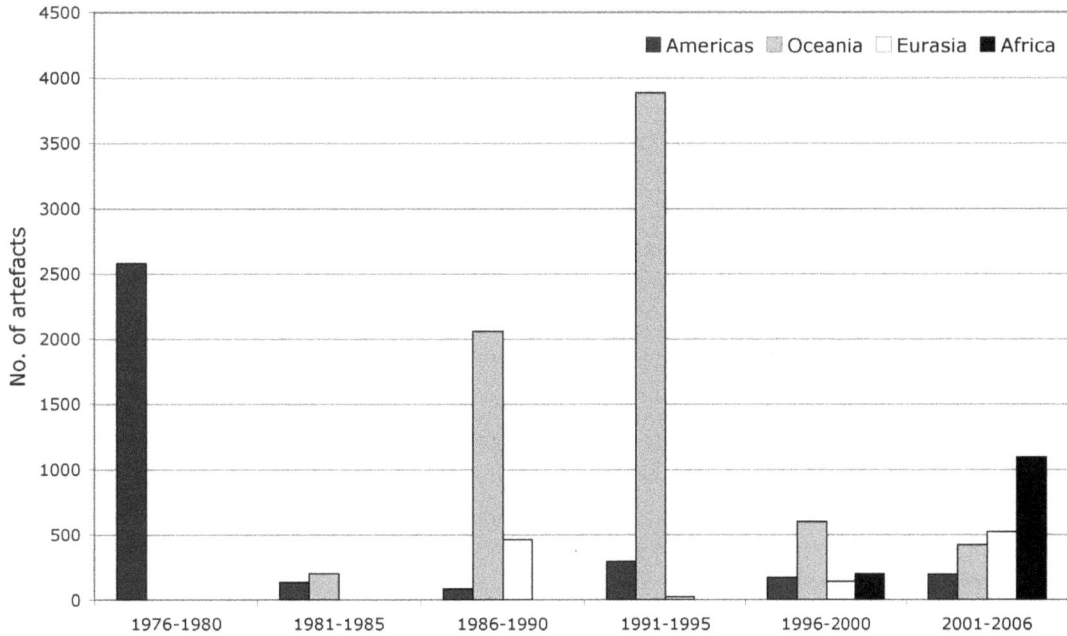

Figure 7. Number of stone artefacts analysed for microscopic residues per 5-year period, by geographic region.

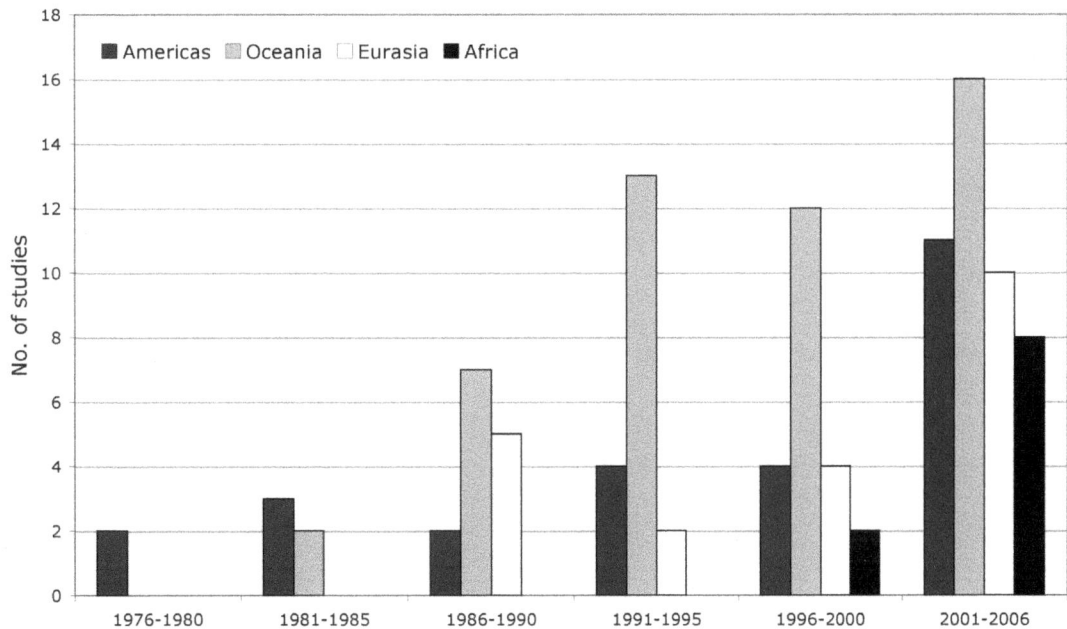

Figure 8. Number of published stone-tool microscopic residue studies per 5-year period, by geographic region.

need, and in fact it is somewhat counter-intuitive, to extrapolate the results of the earliest evidence for grass-seed processing to other sites or artefacts. The unspoken, but correct, assumption in this situation is that the implications for our understanding of past social organisation, dietary shifts and economic factors behind the use of a single artefact are profound enough to warrant publication. The routine residue analysis of non-descript artefacts that do not have a proxy importance gained through age, location or rarity may require justification in other terms, especially when choices are made as to which sites will be analysed and which cannot due to financial and time constraints. At the same time, it is usually the non-descript, 'everyday' artefacts that are most prevalent and available for study, even though these may appear essentially identical in potential information content from site to site in a given region. Factors other than perceived importance may also influence sampling decisions – for example artefacts of limited typological or display value may escape rigorous cleaning and therefore provide more appropriate candidates for residue study. The key is to ensure that the questions asked and the conclusions drawn are both relevant for the actual pieces of stone examined.

At the heart of the matter is a simple question: is it possible or meaningful to talk of a 'valid microscopic residue analysis sample size'? In this sense statistics may provide some guidance. The point at which a randomly selected sample from an unknown population size (the case for the vast majority of archaeological material) becomes statistically significant can be calculated for a number of confidence levels. The most common confidence level employed by archaeologists is 95%, although at its base this is an arbitrary choice (Cowgill 1977). The data generated by residue analysts can be considered categorical (non-numerical), in that there is no justifiable basis yet established for treating, for example, the presence of 15 starch granules or collagen fibres on a used tool edge as being meaningfully different to finding 12 or 30 granules or fibres on that edge, thanks largely to taphonomic factors. In addition, as studies often report the proportions of various residue types and inferred uses of analysed artefacts (e.g. Fullagar 1992; Lombard 2008), the following equation can be used to calculate the number of artefacts that must be examined to make statements of statistical significance for a particular confidence level (Drennan 1996):

$$n = \left(\frac{\sigma t}{ER} \right)^2$$

where   n = the required sample size.
        σ = the standard deviation of the sample.
        t = the Student's t distribution value for the desired confidence level.
        ER = the error range (or confidence interval) for the sample.

Using proportional data, the standard deviation is found by:

$$\sigma = s = \sqrt{pq}$$

where p = the proportion of artefacts expressed as a decimal fraction; q = 1-p.

If the proportions of artefacts used for various tasks are unknown to begin with (again typical in residue studies) then the most conservative guess of 50% can be used (Drennan 1996:143). This guess means that if there is any variation from 50%, the resulting error ranges will be smaller than those calculated — in other words the result will only become more precise. This information can be used to calculate the number of artefacts that must be examined to make a statistically valid statement about the inferred tasks represented by those artefacts. Using the t value (1.96) associated with the standard 95% confidence level for a large unknown population size, an error range of ±5%, and a proportion of 50%, then:

$$\sigma = \sqrt{pq} = \sqrt{(0.5)(1-0.5)} = 0.5$$

and

$$n = \left(\frac{(0.5)(1.96)}{0.05} \right)^2 = 384.16$$

To summarise, an analyst is required to examine 385 artefacts randomly selected from a given context to be able to speculate on the representativeness of their (categorically-described) residue findings for that context at a 95% confidence level. Of the 91 studies listing sample sizes in the Appendix, only eight exceed this figure. If an analyst wishes to divide their sample, for example into different temporal periods or spatial areas, each of those divisions would require 385 artefacts to make assertions at this confidence level, provided of course that the requisite number of artefacts were recovered. As a point of comparison, 25 of the studies in the Appendix analysed sufficient artefacts to assert statistical significance at the 90% confidence level (requiring 68 artefacts with a ±10% error range), and 66 had a sufficient sample to assert statistical significance at the 80% confidence level (requiring 11 artefacts with a ±20% error range). None of these studies contain statistical calculations of significance for the findings they report. In fairness, while heuristically valuable, confidence levels are not fixed boundaries that decisively assign a given study into a box labelled either 'true' or 'false' (Ringrose 1993), and as noted even one artefact may supply a wealth of information. The important point is for residue analysts to apply appropriate statistical reasoning, at appropriate times, to support and strengthen their conclusions.

The misuse of or blind faith in statistical analyses has been identified by Cowgill (1993:552-553) as one of the major shortcomings of archaeological processualism, while at the same time commenting that 'Quantitative techniques can be treated lightly only if one believes that remains of the past themselves have little bearing on our judgements about the merits of different stories'. To muddy the statistical waters further, there is also no necessary correlation between the number of artefacts analysed in a given study and the number with identifiable and interpretable residues. For example, Briuer's (1976) analysis, the first reported stone-tool microscopic residue study and still the second largest, found identifiable organic use residues on 37 (or 1.5%) of 2551 analysed flakes. With residue studies in their infancy, however, Briuer may have missed or been unable to identify residues that would now be routinely recorded. The largest reported study to date (Boot 1993) found 165 hafting or use residues on 2722 artefacts (6%); the third largest (Fullagar 1988) found residues on 13 of 1814 artefacts (0.7%). Much higher rates have been reported from other studies, including the early work of Loy (1983) who reported blood residues on 90 of 104 artefacts (87%). These results demonstrate that there is no sure ratio of analysed artefacts to artefacts with residues, a fact that may be intuitive but could cause problems if attempts at quantitative comparisons are made. These problems are compounded by variations in sampling rationale: artefacts for residue studies have been drawn randomly from the available assemblage (e.g. Haslam 2003); chosen selectively by the archaeologist as likely to have been used (e.g. Veth *et al.* 1997); chosen to cover a range of temporal periods and raw materials (e.g. Barton 1990); or the sample is the complete assemblage from a pit or site (e.g. Fullagar and David 1997). Attempting to compare inter-study percentage results of, for example, plant and animal processing tools (e.g. Boot 1999) may be inappropriate in light of these diverse sampling procedures.

The restrictions that small sample sizes impose on the broader applicability of results were recognised early on by Shafer and Holloway (1979:398), who noted that for their sample of 25 artefacts they 'fully realize the limited application of the overall findings regarding the use of tools in the assemblages being studied'. Fullagar (1987:26) likewise observed that in determining overall site function, 'for sites with very small numbers of tools, the proportion of materials worked may be meaningless', a sentiment which has obvious correlation with analyses of small samples, regardless of the assemblage size from which they are drawn. The majority of studies in the Appendix make no mention of a carefully reasoned sampling strategy based on a well-defined target population, which is either an omission in reporting protocols or evidence that the impact of sampling on the relevance and reliability of results has to this point largely been overlooked. It is also possible that the lack of explicit published sampling rationales correlates with indications that many stone-tool microscopic residue analyses may begin with an aim of simply 'finding out what's there' (Haslam 2006b), and only subsequently are the results broadened into discussions of, for example, regional subsistence practices (see Hardy and Svoboda [this volume] for an alternative to this scenario). The desirability of examining entire assemblages to produce defensible results was highlighted recently by Smith (2004:174) in a study of starch on grindstones, although the logistics of such a task often may be prohibitive. As microscopic residue analyses have increasingly taken a central role in key world archaeological debates, especially concerning early plant domestication and food processing (Denham, *et al.* 2003; Fullagar and Field 1997; Fullagar, *et al.* 2006; Perry, *et al.* 2006; Piperno, *et al.* 2000; Piperno, *et al.* 2004), it is likely that sampling issues will play a correspondingly important role in further progressing the value of the field.

TARGETS AND SAMPLES, MOUNTAINS AND MOLEHILLS

Microscopic stone-tool residue analysis is time-consuming and expensive, and with current median values of less than 20 artefacts per study, there are clearly benefits in openly discussing the kinds of questions we want microscopic residues to address. If total-assemblage projects were to become the norm, then there would be few restrictions on research targets – everything from broad studies of non-lithic artefact production, prehension and cognitive processes to classification of artefacts by use rather than form (and much more) would be worth exploring. In the short-term, however, more may be gained from questions that maximise the value of small samples, and selectively target larger numbers of artefacts when required. For example, in recent years researchers in the Americas in particular (e.g. Perry *et al.* 2006) have established a solid research agenda concentrated on the variety of human-plant interaction, plant processing and settlement information to be gained from focused studies of starch residues, with an emphasis on ground stone artefacts. Alternatively, for non-descript stone pieces we may profitably consider the social roles of individual artefacts (Haslam 2006a; Haslam and Liston 2008). This approach relies on contextual data for one or a few artefacts (demonstrated above to be the most commonly analysed sample sizes) as an alternative to extrapolating results over a site or region. Recent investigations of South African hafting and hunting technologies (e.g. Gibson *et al.* 2004; Lombard 2005; Wadley *et al.* 2004) also demonstrate the interpretive value of a clear research focus, building a broad perspective from successive related studies.

Perhaps the most obvious way to bring sampling into a more prominent position within residue analysis is to re-examine existing studies, first determining the scale (target population) of the presented discussion, then comparing this with the quantity and contexts of the analysed artefacts. This task would take more space than is available here and is by no means straightforward; as a start however, questions arising from a few brief examples may be considered to introduce the concept. Consider first the recent work of Dominguez-Rodrigo *et al.* (2001) on East African Acheulian handaxes, a significant study reporting the oldest microscopic use-residues to date and rigorously documenting clear associations between tools and their residues over a very large time period. From a sampling perspective, then, we may begin by asking how many Acheulian handaxes need to be analysed before any generalisations about early human use of these tools can be made? Are the three analysed by Dominguez-Rodrigo *et al.* (2001:292), from one site dated to 1.4-1.7 million years ago, sufficient? What about thirty or three hundred, and across how many sites? The authors identified traces of woodworking residues on two of the three handaxes; to what extent can this finding be interpreted as showing 'that humans, at a very early stage of their evolution, were producing wooden implements' (Dominguez-Rodrigo *et al.* 2001:297-298), with implications for complex hominid intelligence? In broader contextual perspective, what role should wooden spears from Germany or England made more than one million years later have in strengthening hypotheses about early East African handaxe use for spear-making (Dominguez-Rodrigo *et al.* 2001:289, 298)?

A discussion of potential handaxe use for the continents and hundreds of millennia across which these artefacts are found is well beyond the scope of this paper, and the requisite functional analyses of representative samples from well-dated Acheulian assemblages divided into different environmental, geographical, hypothesised site-function and temporal zones is non-existent. Nevertheless, given the large numbers of handaxes currently in museum collections and a long history of functional speculation and investigation (e.g. Keeley and Toth 1981; Kohn and Mithen 1999; Shick and Toth 1993; Whittaker and McCall 2001) this artefact type would appear to be an ideal target for a series of residue studies aimed at determining both basic population-wide use-data and the role of sampling in potentially biasing interpretations. We may ask, for example, if the proportion of plant to animal working tools seen in a given study varies independently of sample size, or is asymptotic for larger samples? Are there site- or region-specific patterns suggesting environmental and/or social influences on use? Were these artefacts even made for use as tools (Davidson 2002)? Dominguez-Rodrigo *et al.* have commendably established a necessary base-line for beginning to address such questions, but well-defined target populations are clearly required before broadening results beyond the site and time period examined.

Sampling concerns do not only affect such spatiotemporally vast constructs as the Acheulian, of course, and with this in mind we may examine a second important recent study. Perry (2005) successfully gathered both botanical and artefact functional data from starch analysis of five microlithic 'manioc' grater flakes from a Venezuelan site, providing excellent contextual information for the location some 1200-1500 years ago and demonstrating the range of activities for which these particular tools were employed. At the same time, Perry (2005:414) uses the results to evaluate 'the validity of the direct historical approach in the interpretation of archaeological artifacts from the middle Orinoco valley' by comparing the ethnographically-derived assumption of manioc grating to the actual tool functions observed microscopically. Again from a sampling perspective, the question at hand is whether the reported starch analysis of five artefacts, from one half of one level of a 1 m x 1 m test unit at one site, is sufficient for such an evaluation.

The Orinoco case study has relevance for all functional projects that attempt to test ethnographically-derived functional assumptions, and as the stated target is in this instance quite broad the sample ideally should be equally comprehensive. It is likely therefore that artefacts from a number of sites in the middle Orinoco and across a number of time periods leading up to the ethnographic present would require examination to judge the validity of the direct historical approach in this region. In the absence of such data it is difficult to judge the representativeness of the analysed pieces for the middle Orinoco assemblage. As a test, it seems reasonable to assume that a representative sample suitable for addressing the target question would be able to aid in identifying the point at which direct historical approach becomes valid, that is, when ethnographically-observed activities become the dominant ones for these microliths. The sample selected does not permit such an assessment. Perry has certainly demonstrated the inapplicability of the direct historical approach to interpreting her sample of five artefacts, and the extrapolation from this to the whole middle Orinoco may appear a minor one, but it is precisely the feasibility of such extrapolations that sampling designs make explicit. Tangentially, it could be asked whether narrowing the study's discussion solely to statements relevant to the analysed artefacts and their immediate context may diminish its value, and therefore restrict the potential publishing venues. In other words, there is a tension between on the one hand the restrictions imposed in microscopic residue analysis by a time-consuming practical component, and on the other a requirement to make the results as useful to others as possible.

In these examples a case may be argued for both large and small-scale interpretation of limited sample studies, depending on the reader's willingness to follow the author's leap from sample to target. Similar issues, including extrapolation from single site analyses to regional patterns, and from precise reconstruction of limited individual tool use to broad generalisations about short or long term human activities are worth discussing for many of the studies conducted to date. That said, all projects progress sequentially over time, and small sample sizes addressing questions aimed at small, well-defined target populations are entirely valid. The key point is that archaeologists engaged in microscopic stone-tool residue studies have not been as stringent in our identification of target populations as we could be, and explicit demonstrations of how sampling strategies (of however many artefacts) tie back to a given target are often conspicuously absent. In a related issue facing all archaeologists, we may ask whether the use of qualifier terms such as 'preliminary', or phrases such as 'hints at' or 'suggests' rather than 'demonstrates', justify making claims beyond the significance warranted by the sample size? Use of such terms may be tempting, particularly in a field such as microscopic residue analysis where time and financial constraints act to restrict the number of artefacts that may be examined, while as in any field publication pressures demand a high level of impact for the results. Equally important is the fact that residue analyses often produce striking results of specific tool-use that are likely to excite the imagination, perhaps contributing to an overstatement of significance. If care is not taken, however, results first reported as preliminary or speculative may after repetition come to be cited as reliable facts. Overall, the contribution of sampling theory to such issues is, as Orton (2000:206) points out, 'in providing a language and a frame of reference within which such problems can be discussed, rather than definitive answers'.

What then are the prospects for augmenting current practices that emphasise many studies of few artefacts rather than vice versa? Apart from a need to improve the reporting of sampling strategies and to explicitly tie the scale of conclusions to the scale of analysis, several additional options are available. First, as mentioned earlier, a theoretical turn towards highly contextualised interpretation of individual actions offers an alternate path for residue results from a limited number of artefacts (Haslam 2006a). This approach is designed to provide a humanised and agent-centred past as a further line of investigation complementary to current research agenda, rather than to directly address the 'big picture' questions of archaeology, and it therefore provides a potentially valuable but only partial solution. A second approach is to conduct a series of dove-tailed research projects at the one site or region, which may then by accretion answer questions at successively larger theoretical and geographical scales. Excellent examples of this approach include Fullagar's work in West New Britain, Williamson and more recently Lombard's studies of Rose Cottage Cave and Sibudu Cave in South Africa, and Hardy's work in Crimea.

One of the chief benefits of such long-term investment in particular sites and regions is that additional small-sample research may be justifiably incorporated into an established framework. Researchers working on well-studied sites, especially where lithic technological information is also available, are in the best position to identify relevant large-scale issues with which residue analysis may engage. The contributions to this volume by each of the researchers mentioned demonstrate that the necessary vitality exists in the field to generate productive discourse on this issue. As a natural corollary to this last point, where discussions of broad trends are a desired outcome then initiation of further projects that aim from the outset to examine significant proportions of a site or regional assemblage will be essential. And finally, a third path for increasing the scale of analyses without incurring significantly greater costs is to conduct initial rapid screening of artefact assemblages for residue-bearing artefacts. Subsequent sample selection from among those artefacts with clear residues avoids some of the problems noted earlier but introduces new challenges for relating the residue-bearing assemblage to the lithic assemblage and target population as a whole.

CONCLUSION

With a trend over the past decade toward smaller sample sizes, and typical samples measured in the tens of artefacts or less, sampling theory suggests that analysts should be implementing an accompanying constriction in the scale of conclusions drawn from microscopic stone-tool residues. In practice, such constrictions may be alleviated or overcome through explicit recognition of the limitations built into small-scale analyses. In this regard it would be insufficient, for example, to label a study 'preliminary' and then to disregard this fact in either the conclusions of that same study or in future publications. In comparison with lithic technological research in which many thousands of artefacts or complete assemblages are routinely recorded for a given site, it is perhaps surprising that microscopic lithic residue analyses have yet to come under greater scrutiny for their sampling practices, and in this regard the success of the technique may obscure its limitations somewhat. There are logistical impediments to the widespread use of a whole-assemblage approach to residues, however it is only through such projects that we gain the necessary baseline data against which the reliability of small-sample conclusions can be evaluated. Caution in interpreting small-sample studies and an explicit focus on the relative merits of different sample sizes can only result in greater explanatory power and wider acceptance and integration for all microscopic stone-tool residue analyses.

ACKNOWLEDGEMENTS

Throughout this paper I have referred to 'Loy', as academic convention dictates. To me he will however always be Tom, an inspiring mentor and excellent storyteller. My thanks go to Tom for introducing me to residues, and to my fellow archaeologists over the years in the Archaeological Science Laboratory and School of Social Science at the University of Queensland for their encouragement. Thanks also to the growing band of microscopic residue analysts who make a review like this one possible, and who are slowly but surely changing the way we see the past. Valuable constructive comments and support were received from Richard Fullagar, Gail Robertson and Chris Clarkson. Parts of this paper were presented in the UQ Working Papers in Archaeology seminar series in March 2007. Thank you to everyone who provided feedback on that occasion. Portions of the research contained in this paper were funded by an Australian Postgraduate Award.

REFERENCES

Akerman, K., R. Fullagar and A. van Gijn 2002. Weapons and wunan: production, function and exchange of Kimberley points. *Australian Aboriginal Studies* 2002/1:13-42.

Anderson, P. 1980. A testimony of prehistoric tasks: diagnostic residues on stone tool working edges. *World Archaeology* 12(2):181-193.

Anderson, P., J. Chabot and A. van Gijn 2004. The functional riddle of 'glossy' Canaanean blades and the Near Eastern threshing sledge. *Journal of Mediterranean Archaeology* 17(1):87-130.

Anderson-Gerfaud, P. 1990. Aspects of behaviour in the Middle Palaeolithic: functional analysis of stone tools from southwest France. In P. Mellars (ed.) *The Emergence of Modern Humans: An archaeological perspective*, pp. 389-418. New York: Cornell University Press, Ithaca.

Atchison, J. and R. Fullagar 1998. Starch residues on pounding implements from Jinmium rock-shelter. In R. Fullagar (ed.) *A Closer Look: Recent Australian studies of stone tools*, pp. 109-125. Sydney: Archaeological Computing Laboratory, University of Sydney.

Attenbrow, V., R. Fullagar and C. Szpak 1998. Stone files and shell fish-hooks in southeastern Australia. In R. Fullagar (ed.) *A Closer Look: Recent Australian studies of stone tools*, pp. 127-148. Sydney: Archaeological Computing Laboratory, University of Sydney.

Babot, M. d. P. 2001. La molienda de vegetales almidonosos en el noroeste Argentino prehispanico. *Publicatcion Especial de la Asociacion Paleontologica Argentina* 8:59-60.

Babot, M. d. P. and M.C. Apella 2003. Maize and bone: residues of grinding in Northwestern Argentina. *Archaeometry* 45(1):121-132.

Balme, J., G. Garbin and R.A. Gould 2001. *Residue analysis and palaeodiet in arid Australia. Australian Archaeology* 53:1-6.

Barton, H. 1990. Raw Material and Tool Function: A residue and use-wear analysis of artefacts from a Melanesian rockshelter. Unpublished BA (Hons) thesis. Sydney: University of Sydney.

Barton, H., R. Torrence and R. Fullagar 1998. Clues to stone tool function re-examined: comparing starch grain frequencies on used and unused obsidian artefacts. *Journal of Archaeological Science* 25:1231-1238.

Barton, H. and J.P. White 1993. Use of stone and shell artefacts at Balof 2, New Ireland, Papua New Guinea. *Asian Perspectives* 32(2):169-181.

Baxter, M. 2003. *Statistics in Archaeology*. Arnold, London.

Boot, P.G. 1993. Analysis of resins and other plant residues on stone artefacts from Graman, New South Wales. In B.L. Fankhauser and J.R. Bird (eds) *Archaeometry: Current Australasian Research*, pp. 3-12. Occasional Papers in Prehistory, No. 22. Canberra: Department of Prehistory, Research School of Pacific Studies, The Australian National University.

Boot, P.G. 1999. Preservation of use related residues on stone artefacts from Graman. In by M.-J. Mountain and D. Bowdery (eds) *Taphonomy: The analysis of processes from phytoliths to megafauna*, pp. 35-40. Canberra: ANH Publications.

Brass, L. 1998. Modern stone tool use as a guide to prehistory in the New Guinea Highlands. In R. Fullagar (ed.) *A Closer Look: Recent Australian studies of stone tools*, pp. 19-28. Sydney: Archaeological Computing Laboratory, University of Sydney.

Briuer, F. L. 1976. New clues to stone tool function: plant and animal residues. *American Antiquity* 41(4):478-484.

Broderick, M. 1982. Residue analysis of artifacts and flakes recovered from the Telep Site DhRp 35. Appendix 2. In W.R.B. Peacock (ed.) *The Telep Site: A late autumn fish camp of the Locarno Beach Culture type*. Unpublished report on file, Heritage Conservation Branch, Victoria, British Columbia, Canada.

Cattaneo, C., K. Gelsthorpe, P. Phillips and R. J. Sokol 1993. Blood residues on stone tools: indoor and outdoor experiments. *World Archaeology* 25(1):29-43.

Chandler-Ezell, K., D. Pearsall and J. A. Zeidler 2006. Root and tuber phytoliths and starch grains document manioc (*Manihot esculenta*), arrowroot (*Maranta arundinacea*), and Lleren (*Calathea sp.*) at the Real Alto site, Ecuador. *Economic Botany* 60(2):103-120.

Cosgrove, R. 1985. New evidence for early Holocene aboriginal occupation in northeast Tasmania. *Australian Archaeology* 21:19-36.

Coughlin, E. and C. Claassen 1982. Siliceous and microfossil residues on stone tools: a new methodology for identification and analysis. In R.M. Gramly (ed.) *The Vail Site: A Palaeo-Indian encampment in Maine*. Buffalo, NY: Bulletin of the Buffalo Society of Natural Sciences.

Cowgill, G.L. 1977. The trouble with significance tests and what we can do about it. *American Antiquity* 42(3):350-368.

Cowgill, G.L. 1993. Distinguished Lecture in Archeology: beyond criticizing New Archeology. *American Anthropologist* 95(3):551-573.

Davidson, I. 2002. The finished artefact fallacy: Acheulean hand-axes and language origins. In A. Wray (ed.) *The Transition to Language*, pp. 180-203. Oxford: Oxford University Press.

Davis, K. 1999. Preliminary analysis of starch residues on ground stone tools from Mleiha. In M. Mouton (ed.) *Mleiha, Vol. 1: Environnement, Stratégies de Subsistance et Artisanats*, pp. 89-96. Lyon: Maison de l'Orient méditerranéen.

Denham, T.P., S.G. Haberle, C. Lentfer, R. Fullagar, J. Field, M. Therin, N. Porch and B. Winsborough 2003. Origins of agriculture at Kuk Swamp in the Highlands of New Guinea. *Science* 301:189-193.

Denham, T.P. and H. Barton 2006. The emergence of agriculture in New Guinea: a model of continuity from pre-existing foraging practices. In D.J. Kennett and B. Winterhalder (eds) *Behavioral Ecology and the Transition to Agriculture*, pp. 237-264. Berkeley: University of California Press.

Dodson, J., R. Fullagar, J. Furby, R. Jones and I. Prosser 1993. Humans and megafauna in a late Pleistocene environment from Cuddie Springs, north western New South Wales. *Archaeology in Oceania* 28:94-99.

Dominguez-Rodrigo, M., J. Serralonga, J. Juan-Tresserras, L. Alcala and L. Luque 2001. Woodworking activities by early humans: a plant residue analysis on Acheulian stone tools from Peninj (Tanzania). *Journal of Human Evolution* 40:289-299.

Downs, E.F. and J.M. Lowenstein 1995. Identification of archaeological blood proteins: a cautionary note. *Journal of Archaeological Science* 22:11-16.

Drennan, R.D. 1996. *Statistics for Archaeologists: A commonsense approach*. New York: Plenum Press.

Frederickson, C. 1985. The detection of blood on prehistoric flake tools. *New Zealand Archaeological Association Newsletter* 28(3):155-164.

Fullagar, R. 1982. What's the Use? An analysis of Aire Shelter II, Glenaire, Victoria. Unpublished MA (Prelim) thesis. La Trobe University.

Fullagar, R. 1987. Use-wear and residue analysis of Birrigai stone artefacts. In J. Flood, B. David, J. Magee and B. English, Birrigai: a Pleistocene site in the south-eastern highlands. *Archaeology in Oceania* 22(1):25-26.

Fullagar, R. 1988. Recent developments in Australian use-wear and residue studies. In S. Beyries (ed.) *Industries Lithique: Traceologie et technologie*, pp. 133-145. British Archaeological Reports International Series 411.

Fullagar, R. 1989. The potential of lithic use-wear and residue studies for determining stone tool functions. In P. Gorecki and D. Gillieson (ed.) *A Crack in the Spine: Prehistory and ecology of the Jimi-Yuat Valley, Papua New Guinea*, pp. 209-223. Townsville: James Cook University of North Queensland.

Fullagar, R. 1992. Lithically Lapita: functional analysis of flaked stone assemblages from West New Britain Province, Papua New Guinea. In J.-C. Galipaud (ed.) *Poterie Lapita et Peuplement*, pp. 135-143. Noumea: ORSTOM.

Fullagar, R. 1993a. Flaked stone tools and plant food production: a preliminary report on obsidian tools from Talasea, West New Britain, PNG. In P.C. Anderson, S. Beyries, M. Otte and H. Plisson (eds) *Traces et Fonction: Les gestes retrouvés*, pp. 331-337. vol. 50. Liège: ERAUL.

Fullagar, R. 1993b. Taphonomy and tool use: a role for phytoliths in use-wear and residue analyses. In B.L. Fankhauser and J.R. Bird (eds) *Archaeometry: Current Australasian research*, pp. 21-27. Occasional Papers in Prehistory, No. 22. Canberra: Department of Prehistory, Research School of Pacific Studies, The Australian National University.

Fullagar, R. and B. David 1997. Investigating changing attitudes towards an Australian Aboriginal dreaming mountain over >37,000 years of occupation via residue and use wear analyses of stone artefacts. *Cambridge Archaeological Journal* 7(1):139-144.

Fullagar, R. and J. Field 1997. Pleistocene seed-grinding implements from the Australian arid zone. *Antiquity* 71(272):300-307.

Fullagar, R., J. Field, T.P. Denham and C. Lentfer 2006. Early and mid Holocene tool-use and processing of taro (Colocasia esculenta), yam (Dioscorea sp.) and other plants at Kuk Swamp in the highlands of Papua New Guinea. *Journal of Archaeological Science* 33:595-614.

Fullagar, R. and R. Jones 2004. Usewear and residue analysis of stone artefacts from the Enclosed Chamber, Rocky Cape, Tasmania. *Archaeology in Oceania* 39(2):79-93.

Fullagar, R., T. Loy and S. Cox 1998. Starch grains, sediments and stone tool function: evidence from Bitokara, Papua New Guinea. In R. Fullagar (ed.) *A Closer Look: Recent Australian studies of stone tools*, pp. 49-60. Sydney: Archaeological Computing Laboratory, University of Sydney.

Fullagar, R., B. Meehan and R. Jones 1992. Residue analysis of ethnographic plant-working and other tools from Northern Australia. In P. C. Anderson (ed.) *Prehistoire de l'Agriculture: Nouvelles Approches Experimentales et Ethnographiques*, pp. 39-53. Paris: Editions du Centre National de la Recherche Scientifique.

Furby, J., R. Fullagar, J. Dodson and I. Prosser 1993. The Cuddie Springs bone bed revisited, 1991. In M.A. Smith, M. Spriggs and B.L. Fankhauser (eds) *Sahul in Review: Pleistocene archaeology in Australia, New Guinea and Island Melanesia*, pp. 204-210. Canberra: Department of Prehistory, Research School of Pacific Studies, The Australian National University.

Furby, J. and T.H. Loy 1994. Analysis of blood residues on stone artefacts from Cuddie Springs, northwestern New South Wales. Paper presented at the Australasian Archaeometry Conference, University of New England, Armidale.

Garling, S.J. 1998. Megafauna on the menu? Haemoglobin crystallisation of blood residues from stone artefacts at Cuddie Springs. In R. Fullagar (ed.) *A Closer Look: Recent Australian studies of stone tools*, pp. 29-48. Sydney: Archaeological Computing Laboratory, University of Sydney.

Gibson, N.E., L. Wadley and B.S. Williamson 2004. Microscopic residues as evidence of hafting on backed tools from the 60000 to 68000 year-old Howiesons Poort layers of Rose Cottage Cave, South Africa. *Southern African Humanities* 16:1-11.

Gorman, A. 2000. *The Archaeology of Body Modification: The identification of symbolic behaviour through usewear and residues on flaked stone tools*. Unpublished PhD thesis, University of New England.

Grace, R. 1996. Use-wear analysis: the state of the art. *Archaeometry* 38(2):209-229.

Green, R.C. and D. Anson 2000. Excavations at Kainapirina (SAC), Watom Island, Papua New Guinea. *New Zealand Journal of Archaeology* 20:29-94.

Hall, J., S. Higgins and R. Fullagar 1989. Plant residues on stone tools. In W. Beck, A. Clarke and L. Head (eds) *Plants in Australian Archaeology*, pp. 136-160. Tempus. vol. 1. Brisbane: Anthropology Museum, University of Queensland.

Hanslip, M. 1999. *Expedient Technologies? Obsidian artefacts in Island Melanesia*. Unpublished PhD thesis, Australian National University.

Hardy, B L. 1998. Microscopic residue analysis of stone tools from the Middle Paleolithic site of Starosele. In K. Monigal and V. Chabai (eds) *The Middle Paleolithic of the Western Crimea, Vol. 2*, pp. 179-196. Liege: Etudes et Recherches Archeologiques de l'Universitie de Liege.

Hardy, B L. 2004. Neanderthal behaviour and stone tool function at the Middle Palaeolithic site of La Quina, France. Antiquity 78(301):547-565.

Hardy, B.L. and M. Kay 1998. Stone tool function at Starosele: combining residue and use-wear evidence. In K. Monigal and V. Chabai (eds) *The Middle Paleolithic of the Western Crimea, Vol. 2*, pp. 197-209. Liege: Etudes et Recherches Archeologiques de l'Universitie de Liege.

Hardy, B.L., M. Kay, A.E. Marks and K. Monigal 2001. Stone tool function at the paleolithic sites of Starosele and Buran Kaya III, Crimea: behavioral implications. *Proceedings of the National Academy of Sciences* 98(19):10972-10977.

Hardy, B.L. and M.J. Rogers 2001. Microscopic investigation of stone tool function from Okote Member sites, Koobi Fora, Kenya [abstract only]. *Journal of Human Evolution* 40(3):A9.

Haslam, M. 2003. Evidence for maize processing on 2000 year old obsidian artefacts from Copan, Honduras. In D.M. Hart and L.A. Wallis (eds) *Phytolith and Starch Research in the Australian-Pacific-Asian Regions: The state of the art*, pp. 153-161. Canberra: Pandanus Books.

Haslam, M. 2006a. An archaeology of the instant? Action and narrative in archaeological residue analyses. *Journal of Social Archaeology* 6(3):402-424.

Haslam, M. 2006b. *Archaeological Residue and Starch Analysis: Interpretation and taphonomy*. Unpublished PhD thesis. St Lucia: School of Social Science, The University of Queensland.

Haslam, M. 2006c. Potential misidentification of in situ archaeological tool-residues: starch and conidia. *Journal of Archaeological Science* 33(1):114-121.

Haslam, M. and J. Liston 2008. The use of flaked stone artifacts in Palau, Western Micronesia. *Asian Perspectives* 47(2):405-428.

Higgins, S. 1988. *Starch Grain Differentiation on Archaeological Residues: A feasibility study*. Unpublished BA (Hons) thesis. St Lucia: Department of Anthropology and Sociology, The University of Queensland.

Hodder, I. 1991. *Reading the Past: Current approaches to interpretation in archaeology*. 2nd ed. Cambridge: Cambridge University Press.

Hurcombe, L. M. 1992. *Use Wear Analysis and Obsidian: Theory, experiments and results*. Sheffield: University of Sheffield.

Hyland, D. C., J. M. Tersak, J. M. Adovasio and M. I. Siegel 1990. Identification of the species of origin of residual blood on lithic material. *American Antiquity* 55(1):104-112.

Iriarte, J., I. Holst, O. Marozzi, C. Listopad, E. Alonso, A. Rinderknecht and J. Montana 2004. Evidence for cultivar adoption and emerging complexity during the mid-Holocene in the La Plata basin. *Nature* 432:614-617.

Jones, R. 1987. Ice-age hunters of the Tasmanian wilderness. *Australian Geographic* 8:26-45.

Jones, R. 1987. 1990. From Kakadu to Kutikina: the southern continent at 18000 years ago. In C. Gamble and O. Soffer (eds) *The World at 18000BP. Low Latitudes*, pp. 264-295. London: Unwin Hyman.

Kealhofer, L., R. Torrence and R. Fullagar 1999. Integrating phytoliths within use-wear/residue studies of stone tools. *Journal of Archaeological Science* 26:527-546.

Keeley, L. and N. Toth 1981. Microwear polishes on early stone tools from Koobi Fora, Kenya. *Nature* 293:464-465.

Kohn, M. and S. Mithen 1999. Handaxes: products of sexual selection? *Antiquity* 73:518-526.

Lombard, M. 2004. Distribution patterns of organic residues on Middle Stone Age points from Sibudu Cave, Kwazulu-Natal, South Africa. *South African Archaeological Bulletin* 59:37-44.

Lombard, M. 2005. Evidence of hunting and hafting during the Middle Stone Age at Sibidu Cave, KwaZulu-Natal, South Africa: a multianalytical approach. *Journal of Human Evolution* 48:279-300.

Lombard, M. 2008. Finding resolution for the Howiesons Poort through the microscope: micro-residue analysis of segments from Sibudu Cave, South Africa. *Journal of Archaeological Science* 35(1):26-41.

Loy, T. H. 1983. Prehistoric blood residues: detection on tool surfaces and identification of species of origin. *Science* 220:1269-1271.

Loy, T. H. 1985. Preliminary residue analysis: AMNH specimen 20.4/509. In D.H. Thomas (ed.) *The Archaeology of Hidden Cave*, pp. 224-225. New York: Museum of Natural History Press.

Loy, T. H. 1987. Recent advances in blood residue analysis. In W.R. Ambrose and J.M.J. Mummery (eds) *Archaeometry: Further Australasian Studies*, pp. 57-65. Canberra: Australian National University.

Loy, T. H. 1990. Getting blood from a stone. *Australian Natural History* 23(6):470-479.

Loy, T. H. 1991. Prehistoric organic residues: recent advances in identification, dating and their antiquity. In W. Wagner and A. Pernicka (eds) *Archaeometry '90: Proceedings of the 27th International Symposium on Archaeometry*, pp. 645-656. Boston: Birkhauser Verlag.

Loy, T. H. 1993. The artefact as site: an example of the biomolecular analysis of organic residues on prehistoric tools. *World Archaeology* 25(1):44-63.

Loy, T. H. 1994. Residue analysis of artifacts and burned rock from the Mustang Branch and Barton sites (41HY209 and 41HY202). In R.A. Ricklis and M B. Collins (eds) *Archaic and Late Prehistoric Human Ecology in the Middle Onion Creek Valley, Hays County, Texas*, pp. 607-627. vol. 2. Austin, Texas: Texas Archaeological Research Laboratory, The University of Texas at Austin.

Loy, T. H. 1998. Blood on the axe. *New Scientist* 159(2151):40-43.

Loy, T.H. and E.J. Dixon 1998. Blood residues on fluted points from eastern Beringia. *American Antiquity* 63(1):21-46.

Loy, T H. and B.L. Hardy 1992. Blood residue analysis of 90,000-year-old stone tools from Tabun Cave, Israel. *Antiquity* 66:24-35.

Loy, T.H., K. Kuman, J. Halliday and P. Unwins 1999. Organic residues on 2Myr Oldowan tools from Sterkfontein Cave, South Africa. In *Abstracts of the 64th Annual Meeting of the Society for American Archaeology*. Chicago: Society for American Archaeology.

Loy, T. H. and D. E. Nelson 1986. Potential applications of the organic residues on ancient tools. In J.S. Olin and M.J. Blackman (eds) *Proceedings of the 24th International Archaeometry Symposium*, pp. 179-185. Smithsonian Institution Press, Washington, D.C.

Loy, T. H., M. Spriggs and S. Wickler 1992. Direct evidence for human use of plants 28,000 years ago: starch residues on stone artefacts from the northern Solomon Islands. *Antiquity* 66:898-912.

Loy, T. H. and A. R. Wood 1989. Blood residue analysis at Cayönü Tepesi, Turkey. *Journal of Field Archaeology* 16:451-60.

Lu, T. 2003. Starch residue analysis of the Zengpiyan tools. In F. Xiangou (ed.) *Zengpiyan-A Prehistoric Cave in South China*, pp. 646-651. Beijing: Cultural Relics Publishing House.

Mazza, P.P.A., F. Martini, B. Sala, M. Magi, M.P. Colombini, G. Giachi, F. Landucci, C. Lemorini, F. Modungo and E. Ribechini 2006. A new Palaeolithic discovery: tar-hafted stone tools in a European Mid-Pleistocene bone-bearing bed. *Journal of Archaeological Science* 33(9):1310-1318.

McBryde, I. 1974. *Aboriginal Prehistory in New England.* Sydney: Sydney University Press.

McBryde, I. 1984. Backed blade industries from the Grama rock shelters, New South Wales: some evidence on function. In V.N. Misra and P. Bellwood (eds) *Recent Advances in Indo-Pacific Prehistory*, pp. 231-249. New Delhi: Oxford & IBH Publishing Co..

Morwood, M. 1981. Archaeology of the central Queensland highlands: the stone component. *Archaeology in Oceania* 16(1):1-52.

Mueller, J.W. (ed.) 1975. *Sampling in Archaeology.* Tuscon: University of Arizona Press.

Nelson, D E., T.H. Loy, J. Vogel and J. Southon 1986. Radiocarbon dating blood residues on prehistoric stone tools. *Radiocarbon* 28(1):170-174.

Newman, M. and P. Julig 1989. The identification of protein residues on artifacts from a stratified boreal forest site. *Canadian Journal of Archaeology* 13:119-132.

Orton, C. 2000. *Sampling in Archaeology.* Cambridge: Cambridge University Press.

Pagan Jimenez, J.R., M.A. Rodriguez Lopez, L.A. Chanlatte Baik and Y. Narganes Storde 2005. La temprana introducción y uso de algunas plantas domésticas, silvestres y cultivos en Las Antillas precolombinas: una primera revaloración desde la perspectiva del "arcaico" de Vieques y Puerto Rico. *Diálogo Antropológico* 3(10):7-33.

Paull, W. 1984. Organic residue analysis of stone and bone tools. In M.P.R. Magne and R.G. Matson (eds) *Athapaskan and Earlier Archaeology at Big Eagle Lake, British Columbia*, pp. 239-266. Unpublished report to the Social Sciences and Humanities Research Council of Canada.

Pavlides, C. 2004. From Misisil Cave to Eliva Hamlet: rediscovering the Pleistocene in interior West New Britain. In V. Attenbrow and R. Fullagar (ed.) *A Pacific Odyssey: Archaeology and Anthropology in the Western Pacific. Papers in Honour of Jim Specht*, pp. 97-108. Sydney: Australian Museum.

Pawlik, A.F. 2004a. An Early Bronze Age pocket lighter. In E.A. Walker, F. Wenban-Smith and F. Healy (eds) *Lithics in Action*, pp. 149-151. Oxford: Lithic Studies Society Occasional Paper No. 8. Oxbow Books.

Pawlik, A.F. 2004b. Identification of hafting traces and residues by scanning electron microscopy and energy-dispersive analysis of X-rays. In E.A. Walker, F. Wenban-Smith and F. Healy (eds) *Lithics in Action*, pp. 169-179. Oxford: Lithic Studies Society Occasional Paper No. 8. Oxbow Books.

Pearsall, D. 2003. Integrating biological data: phytoliths and starch grains, health and diet, at Real Alto, Ecuador. In D.M. Hart and L.A. Wallis (eds) *Phytolith and Starch Research in the Australian-Pacific-Asian Regions: The state of the art*. pp. 187-200. Canberra: Pandanus Books.

Pearsall, D., K. Chandler-Ezell and J.A. Zeidler 2004. Maize in ancient Ecuador: results of residue analysis of stone tools from the Real Alto site. *Journal of Archaeological Science* 31:423-442.

Perry, L. 2002. Starch analyses reveal multiple functions of quartz "manioc" grater flakes from the Orinoco Basin, Venezuela. *Interciencia* 27(11):635-639.

Perry, L. 2004. Starch analyses reveal the relationship between tool type and function: an example from the Orinoco valley of Venezuela. *Journal of Archaeological Science* 31:1069-1081.

Perry, L. 2005. Reassessing the traditional interpretation of "manioc" artifacts in the Orinoco Valley of Venezuela. *Latin American Antiquity* 16(4):409-426.

Perry, L., D.H. Sandweiss, D.R. Piperno, K. Rademaker, M.A. Malpass, A. Umire and P. de la Vera 2006. Early maize agriculture and interzonal interaction in southern Peru. *Nature* 440:76-79.

Petraglia, M., S.L. Bupp, S.P. Fitzell and K.W. Cunningham (eds) 2002. *Hickory Bluff: Changing perceptions of Delmarva archaeology*. Delaware Department of Transportation Archaeology Series No. 175.

Piperno, D.R. and I. Holst 1998. The presence of starch grains on prehistoric stone tools from the humid neotropics: indications of early tuber use and agriculture in Panama. *Journal of Archaeological Science* 25:765-776.

Piperno, D.R., A.J. Ranere, I. Holst and P. Hansell 2000. Starch grains reveal early root crop horticulture in Panamanian tropical forest. *Nature* 407:894-897.

Piperno, D R., E. Weiss, I. Holst and D. Nadel 2004. Procesing of wild cereal grains in the Upper Palaeolithic revealed by starch grain analysis. *Nature* 430:670-673.

Potter, D. 1994. Strat 55, Operation 2012, and comments on lowland Maya blood ritual. In T.R. Hester, H.J. Shafer and J.D. Eaton (eds) *Continuing Archeology at Colha, Belize*, pp. 31-37. Texas Archaeological Research Laboratory, The University of Texas at Austin, Austin.

Redman, C.L. 1974. *Archaeological Sampling Strategies*. Reading, Mass: Addison-Wesley.

Richards, T. 1989. Initial results of a blood residue analysis of lithic artefacts from Thorpe Common rockshelter, South Yorkshire. In I.P. Brooks and P. Phillips (eds) *Breaking the Stony Silence: Papers from the Sheffield Lithics Conference 1988*, pp. 73-90. Oxford: British Archaeological Reports British Series 213.

Ringrose, T J. 1993. Bone counts and statistics: a critique. *Journal of Archaeological Science* 29:121-157.

Risberg, J., L. Bengtsson, B. Kihlstedt, C. Lidstrom Holmberg, M. Olausson, E. Olsson and C. Tingvall 2002. Siliceous microfossils, especially phytoliths, as recorded in five prehistoric sites in Eastern Middle Sweden. *Journal of Nordic Archaeological Science* 13:11-26.

Robertson, G. 1996. An application of environmental scanning electron microscopy and image analysis to starch grain differentiation. *Tempus* 6:169-182.

Robertson, G. 2002. Birds of a feather stick: microscopic feather residues on stone artefacts from Deep Creek Shelter, New South Wales. *Tempus* 7:175-182.

Robertson, G. 2005. *Backed Artefact Use in Eastern Australia: A residue and use-wear analysis.* Unpublished PhD thesis, University of Queensland.

Rots, V. and B.S. Williamson 2004. Microwear and residue analyses in perspective: the contribution of ethnoarchaeological evidence. *Journal of Archaeological Science* 31:1287-1299.

Shafer, H. and R. Holloway 1979. Organic residue analysis in determining stone tool function. In B. Hayden (ed.) *Lithic Use-Wear Analysis*, pp. 385-399. New York: Academic Press.

Shanks, O.C., M. Kornfeld and D.D. Hawk 1999. Protein analysis of Bugas-Holding tools: new trends in immunological studies. *Journal of Archaeological Science* 26:1183-1191.

Sievert, A.K. 1990. Postclassic Maya ritual behaviour: microwear analysis of stone tools from ceremonial contexts. In B. Graslund, H. Knutsson, K. Knuttson and J. Taffinder (eds) *The Interpretative Possibilities of Microwear Studies*, pp. 147-157. Uppsala: Societas Archaologica Upsaliensis.

Sievert, A.K. 1992a. *Maya Ceremonial Specialisation: Lithic tools from the sacred cenote at Chichen Itza, Yucatan.* Madison: Prehistory Press.

Sievert, A.K. 1992b. Use-wear analysis of chipped stone tools. In C.C. Coggins (ed.) *Artifacts from the Cenote of Sacrifice, Chichen Itza, Yucatan*, pp. 182-187. Cambridge, Mass: Peabody Museum of Archaeology and Ethnology, Harvard University.

Slack, M., R. Fullagar, J. Field and A. Border 2004. New Pleistocene ages for backed artefact technology in Australia. *Archaeology in Oceania* 39(3):131-137.

Smith, M.A. 2004. The grindstone assemblage from Puritjarra rock shelter: investigating the history of seed-based economies in arid Australia. In T. Murray (ed.) *Archaeology From Australia*, pp. 168-186. Melbourne: Australian Scholarly Publishing.

Sobolik, K.D. 1996. Lithic organic residue analysis: an example from the Southwestern Archaic. *Journal of Field Archaeology* 23(4):461-469.

Turner, M., A. Anderson and R. Fullagar 2001. Stone artefacts from the Emily Bay Settlement Site, Norfolk Island. In A. Anderson and P. White (eds) *The Prehistoric Archaeology of Norfolk Island, Southwest Pacific*, pp. 53-66. Records of the Australian Museum, Supplement 27. Sydney: Australian Museum.

Tuross, N. and T.D. Dillehay 1995. The mechanism of organic preservation at Monte Verde, Chile, and one use of biomiolecules in archaeological interpretation. *Journal of Field Archaeology* 22(1):97-110.

Van Peer, P., R. Fullagar, S. Stokes, R.M. Bailey, J. Moeyersons, F. Steenhoudt, A. Geerts, T. Vanderbeken, M. De Dapper and F. Geus 2003. The Early to Middle Stone Age transition and the emergence of modern human behaviour at site 8-B-11, Sai Island, Sudan. *Journal of Human Evolution* 45:187-193.

Veth, P., R. Fullagar and R. Gould 1997. Residue and use-wear analysis of grinding implements from Puntutjarpa Rockshelter in the Western Desert: current and proposed research. *Australian Archaeology* 44:23-25.

Villa, P., A. Delagnes and L. Wadley 2005. A late Middle Stone Age artifact assemblage from Sibudu (KwaZulu-Natal): comparisons with the European Middle Paleolithic. *Journal of Archaeological Science* 32:399-422.

Wadley, L., B.S. Williamson and M. Lombard 2004. Ochre in hafting in Middle Stone Age southern Africa: a practical role. *Antiquity* 78(301):661-675.

Wallis, L. and S. O'Connor 1998. Residues on a sample of stone points from the west Kimberley. In R. Fullagar (ed.) *A Closer Look: Recent Australian studies of stone tools*, pp. 149-178. Sydney: Archaeological Computing Laboratory, University of Sydney.

Weisler, M.I. and M. Haslam 2005. Determining the function of Polynesian volcanic glass artifacts: results of a residue study. *Hawaiian Archaeology* 10:1-17.

Whittaker, J.C. and G. McCall 2001. Hand-axe-hurling hominids: an unlikely story. *Current Anthropology* 42(4): 566-572.

Williamson, B.S. 1996. Preliminary stone tool residue analysis from Rose Cottage Cave. *South African Field Archaeology* 5:36-44.

Williamson, B.S. 1997. Down the microscope and beyond: microscopy and molecular studies of stone tool residues and bone samples from Rose Cottage Cave. *South African Journal of Science* 93:458-464.

Williamson, B.S. 2004. Middle stone age tool function from residue analysis at Sibudu Cave. *South African Journal of Science* 100(3-4):174-178.

Wood, A.R. 1998. Revisited: blood residue investigations at Cayönü, Turkey. In G. Arsebuk, M.J. Mellink and W. Schirmer (eds) *Light on Top of the Black Hill: Studies presented to Halet Cambel*, pp. 763-764. Istanbul: Ege Yayinlari.

Zarillo, S. and B. Kooyman 2006. Evidence for berry and maize processing on the Canadian plains from starch grain analysis. *American Antiquity* 71(3):473-500.

APPENDIX: ARCHAEOLOGICAL MICROSCOPIC ANALYSES OF RESIDUES ON STONE ARTEFACTS.

Year	Author(s)	Site, age (approx.)	#[1]	Mag.[2]	U-w[3]	Journal[4]	Summary
2006	Mazza, Martini et al.	Campitello Quarry, Upper Valdarno Basin, Italy, Middle Pleistocene	3	0.74-70x	Y	JAS	Flakes examined via low-power analysis. Revealed traces of hafting resin on one artefact and abundant resin on another. Subsequent chemical analyses suggest *Betulaceae* trees as the origin of the resin/tar.
2006	Fullagar, Field et al.	Kuk Swamp, Papua New Guinea, <10,220BP	55	6-1000x	Y	JAS	Examined 55 artefacts to assess residue presence and preservation, then examined 12 artefacts for taro and yam starch specifically – 11 of these had starches.
2006	Haslam	Palau, Micronesia; Copan, Honduras	-	100-1000x	-	JAS	Few details on number or age of analysed artefacts. Identified fungal elements (conidia, hyphae) and noted the potential for confusion with starch granules.
2006	Perry, Sandweiss et al.	Waynuna, Peru, 3600-4000BP	3	200x	-	Nature	Starch removed from unwashed grindstones and separated with CsCl. Discusses plant use and migration from various microfossils. Identified maize, arrowroot and possible potato starches.
2006	Zarillo & Kooyman	Alberta, Canada, pre-European contact	3	400-630x	Y	Am Ant	Analysed two washed grinding tools and one flake tool. Starch from choke cherry, saskatoon and maize was identified, with tentative starch identification of prairie turnip and Graminae
2005	Lombard	Sibudu Cave, South Africa, 51-62ka	50	50-500x	Y	JHE	Identified various plant, animal and mineral residues, including hafting resins. Posited hunting tool-use.
2005	Perry	Pozo Azul Norte-1, Venezuela, 1200-1500BP	7	10-400x	Y	LAA	Identified starches on quartz microliths; testing interpretations of microlith use based on ethnohistoric records.
2005	Weisler & Haslam	Molokai, Hawaii, 300-400BP — Henderson Is. Pitcairn Group, 400-1000BP	14 — 15	100-1000x	Y	HA	Studied small volcanic glass flakes from two Pacific islands in an initial cataloguing of past uses. Found plant tissue, feather, starch. Uses identified are not assumed to be universal.
2004	Fullagar & Jones	Rocky Cape, Tasmania, Australia, 6.7ka	150	50-500x	Y	AO	Identified 43 used artefacts with residues present including general plant tissue, wood, starch, skin, and bone. Some discussion of activity areas.
2004	Gibson, Wadley et al.	Rose Cottage Cave, South Africa, 60-68ka	48	50-800x	Y	SAH	Identified plant tissues, blood, ochre and modern contaminants. Starch grains were present on all artefacts. Discussion of hafting.

Year	Author(s)	Site, age (approx.)	#[1]	Mag.[2]	U-w[3]	Journal[4]	Summary
2004	Hardy	La Quina, France, ca. 40-70ka	300	50-500x	Y	Antiquity	106 artefacts had residues, which included plant tissue, pollen, raphides, hair, bone feather and possible blood. Plant residues observed on 10 of 16 previously washed tools. Discussion of plant vs animal exploitation.
2004	Iriarte, Holst *et al.*	Los Ajos, Uruguay, 3500BP	3	100-400x	-	Nature	Maize starch granules (total n=17) recovered from 3 milling stones. Methods reference Piperno *et al.* (2000).
2004	Lombard	Sibudu Cave, South Africa, 51-61ka	24	50-500x	Y	SAAB	Uses floral and faunal residue distributions to infer hafting and subsistence activities (i.e. hunting)
2004	Pavlides	Yombon, West New Britain, late Pleistocene	3	-	Y	-	In *A Pacific Odyssey*. Few details provided, but plant processing (wood/starch) identified.
2004	Pawlik	Bornheim-Sechtem, Rhine Valley, Germany, 3500BP	1	-	Y	-	In *Lithics in Action*. Flint artefact from a grave site identified as a 'prehistoric pocket lighter' through light microscopy and subsequent SEM-EDAX of the tool and an associated haematite fragment.
2004	Pawlik	Ullafelsen, Tyrolian Alps, Austria, 10.5-11ka Henauhof-Nord II, Germany, Mesolithic	110 -	-	Y	-	In *Lithics in Action*. 25 siliceous artefacts from Ullafelsen had tar-like hafting residues identified by light microscopy. SEM subsequently used to characterise the tar. No magnification data. No data on number of artefacts from Henauhof-Nord – these artefacts also had hafting residues with embedded plant fragments visible using SEM.
2004	Pearsall, Chandler-Ezell *et al.*	Real Alto, Ecuador, 2400-2800BP	17	312-500x	-	JAS	Samples collected by brushing, washing and sonication from ground stone and cobbles. Maize starch recovered from all tools; most effective method was sonication. Some starch separated with CsCl. See also Pearsall (2003) and Chandler-Ezell *et al.* (2006).
2004	Perry	Orinoco valley, Venezuela, 500-1000BP	10	10-400x	N	JAS	Studied ground stone and flake tools, by microscopy and removal of residue by spot sampling and sonication. Identified arrowroot, maize, ginger and grass (Poaceae) starch. Results compared with macrobotanical remains.
2004	Piperno, Weiss *et al.*	Ohalo II, Israel, 22.5-23.5ka	1	400x	N	Nature	Starch grains identified in residue removed from a grindstone. Identified grass-seed starches, including *Hordeum* sp. Earliest direct evidence for human grass-seed processing.

Year	Author(s)	Site, age (approx.)	#[1]	Mag.[2]	U-w[3]	Journal[4]	Summary
2004	Rots & Williamson	Southern Ethiopia, modern	18	50-800x	Y	JAS	Compared ethnographic with archaeological scrapers. Found very few residues on the archaeological tools, including starch, blood and ochre.
2004	Slack, Fullagar *et al.*	Old Lilydale Homestead (OLH), QLD, Australia, 15ka; GRE8, QLD, Australia, 15ka	1 2	200x	Y	AO	Backed artefacts; few specific details of the residue study are provided, however. Found starch and 'dark smears' on the OLH tool, and greasy films and starch granules on one of the GRE8 tools.
2004	Smith	Puritjarra rock shelter, NT, Australia	56	10-625x	Y	-	In *Archaeology From Australia*. Examined grindstones for starch. Also tested for blood (with Hemastix) and plant alkaloids. Observed starch most closely matched acacia; starch was also observed in abundance on non-cultural gravels and in site sediments.
2004	Wadley, Williamson *et al.*	Sibudu Cave, South Africa, 26-60ka	531	50-800x	Y	Antiquity	Focused on ochre residues, also observed plant residues (e.g. tissue, fibres, starch, exudate).
2004	Williamson	Sibudu Cave, South Africa, Middle Stone Age	412	50-800x	Y	SAJS	Residues found include starch, collagen, blood, ochre, fungi, hair, plant tissue. Results discussed in terms of plant vs animal processing.
2003	Babot & Apella	Tucuman, Argentina, 1.5-2ka	1	400x	N	Archaeo-metry	Maize starch identified on a grindstone, in conjunction with ground bone residue.
2003	Denham, Haberle *et al.*	Kuk Swamp, Papua New Guinea, 6.5-10ka	3	-	Y	Science	Identified *Colocasia esculenta* (taro) starch. Very limited report – does not include microscope or magnification information, or provide photographs.
2003	Haslam	Copan, Honduras, 1900-2050BP	150	100-1000x	Y	-	In *Phytolith and Starch Research*. Maize residues identified on obsidian and other artefacts via starch grains and plant tissue.
2003	Lu	Zengpiyan, China, 7-12ka	80	-	Y	-	In *Zengpiyan-A Prehistoric Cave in South China*. 80 pebble tools examined for starch. Grass family and taro starch found on 11 tools.
2003	Van Peer, Fullagar, *et al.*	Sai Island, Sudan, 180-220ka	2	-	Y	JHE	Very little information provided. Phytolith and starch granules in residues removed from quartzite cobbles.
2002	Akerman, Fullagar *et al.*	Keep River region, NT, Australia, late Holocene	15	100-1000x	Y	AAS	No specific dates reported. Identified resin, ochre and plant tissue. Archaeological results are discussed in conjunction with residue study of ethnographic material.
2002	Risberg, Bengtsson *et al.*	Various sites, Sweden, 4.6-5.7ka	6	1000x	Y	JNAS	Grindstone residues from three sites examined for siliceous particles (e.g. diatoms, phytoliths) and starch.
2002	Robertson	Deep Creek Shelter, NSW, Australia, 1-3.5ka	41	6-1000x	Y	Tempus	Only feather residues reported for these artefacts, although other residues were present (see Robertson 2005).

Year	Author(s)	Site, age (approx.)	#[1]	Mag.[2]	U-w[3]	Journal[4]	Summary
2001	Balme, Garbin et al.	Puntutjarpa Rockshelter, WA, Australia, modern-10ka	49	50-800x	Y	AA	Grindstone study. Most tools had starch residue, some had blood, ochre, hair, plant fibres. Results discussed in terms of seed grinding.
2001	Dominguez-Rodrigo, Serralonga, et al.	Peninj, Tanzania, ca. 1.5Ma	5	400x+	N	JHE	Identified siliceous phytoliths, calcium oxalate crystals and plant tissues on the artefacts and surrounding soils. Suggest a woodworking function for handaxes.
2001	Hardy, Kay et al.	Buran Kaya III, Crimea, Ukraine, 32-37ka	19	100-500x	Y	PNAS	Evidence of hafting and use on plants and animals. Residues include hair, feathers, plant tissue, starch grains, raphides. Some discussion of plant vs animal use. This study also re-presents some of the residue results from Hardy (1998).
2001	Turner, Anderson et al.	Emily Bay, Norfolk Island	15	10-1000x	Y	RAM	No dates provided. Analysed 10 basalt, 5 obsidian artefacts. 4 artefacts had starch (<1 μm). Some plant tissue residues.
2000	Green & Anson	Kainapirina (SAC), Watom Island, Papua New Guinea, 1850-2400BP	-	-	Y	NZJA	Very little information provided. Highly weathered obsidian was studied, resulting in little residue (or use-wear) information. See Hanslip (1999).
2000	Piperno, Ranere et al.	Aguadulce shelter, Panama, 5-7ka	19	100-400x	N	Nature	Grindstone study, using CsCl extraction. Manioc, yam, arrowroot and maize starch identified. Earliest direct evidence for root crop cultivation in the Americas.
1999	Davis	Mleiha, United Arab Emirates, 0-2200BP	17	500-1000x	Y	-	In Mleiha. Water extraction from ground stone artefacts. Starch found on all artefacts. Possible Hordeum or Triticum starch recovered.
1999	Kealhofer, Torrence et al.	FAO, Garua Island, West New Britain, Papua New Guinea, modern-5900BP FRL, West New Britain, Papua New Guinea, modern->10ka	3 14	6-500x	Y	JAS	For both sites, independent use-wear and phytolith studies assessed the value of phytolith residue analysis. Identified collagen, starch, fibres, resin, phytoliths. Results used to discuss changes in tool use over time.
1999	Shanks, Kornfeld et al.	Bugas-Holding, Wyoming, USA, 200-500BP	46	50-100x	Y	JAS	17 tools displayed visible microscopic residues, recorded as presence/absence only. Immunological tests for blood then carried out.
1998	Atchison & Fullagar	Jinmium, NT, Australia	3	500x	-	-	In A Closer Look. Examined three grinding tools for starch by water extraction and removal to microscope slide. Dates uncertain.
1998	Attenbrow, Fullagar et al.	SE Australia, <900BP	28	10-1000x	Y	-	In A Closer Look. Examined stone 'fish-hook files' for evidence of shell-working. Found some shell residues, also plant.

Year	Author(s)	Site, age (approx.)	#[1]	Mag.[2]	U-w[3]	Journal[4]	Summary
1998	Barton, Torrence et al.	Garua Is., Papua New Guinea, 200-5000BP	14	200x	Y	JAS	Residues removed from obsidian artefacts by sonication, as a concurrent use-wear 'blind test' was being conducted. Starches quantified but not identified.
1998	Brass	Kafiavana, Papua New Guinea, <10.7ka	200	6-1000x	Y	-	In *A Closer Look*. Identified residues in four categories – plant, blood, plant/blood and undiagnostic. Blood tested by subsequent Hemastix and immunology. 'Plant' includes starch and plant fibres.
1998	Garling	Cuddie Springs, NSW, Australia	4	100-500x	N	-	In *A Closer Look*. Microscopy used to identify possible blood residues, then haemoglobin crystallisation used to attempt species identification. No dates provided.
1998	Hardy	Starosele, Crimea, Ukraine, 40-80ka	116	100-500x	Y	-	In *The Middle Paleolithic of the Western Crimea*. Residues classified initially as plant/animal. Identified starch, plant tissue, raphides, possible mastic, hair and feather. See Hardy and Kay (1998) for combined discussion of use-wear and residue results.
1998	Loy	Hauslabjoch, Italy, 5300BP	6	50-1000x	Y	NS	Identified feather, blood, hair, collagen, plant hairs on the stone component of the toolkit of Otzi the Iceman.
1998	Loy & Dixon	Various sites, Eastern Beringia	36	200-1000x	N	Am Ant	No dates provided. Observed blood film, red blood cells on Beringian fluted points. Immunological and DNA tests confirmed mammalian blood, including mammoth.
1998	Piperno & Holst	Central Panama, 1.3-7ka	13	100-1000x	N	JAS	Observed manioc, maize, arrowroot, and legume starches, as well as Zea sp. and palm phytoliths.
1998	Wallis & O'Connor	Widgingarri Shelters 1 and 2, WA, Australia, <4970BP	42	-	-	-	In *A Closer Look*. Examined stone points for residues. No magnification information provided. Worked material assessed as plant/animal.
1997	Fullagar & David	Ngarrabullgan Cave, Qld, Australia, 930-5400BP and 32.5->37.2ka	257	25-1000x	Y	CAJ	Found hair, blood, starch, feather, collagen; results are presented as plant/animal. Discusses implications for social perceptions of the site.
1997	Fullagar & Field	Cuddie Springs, NSW, Australia, modern-33ka	33	Up to 1000x	Y	Antiquity	Found distinctive phytoliths, generally small starches and blood films on grinding stones. Contact material presented as plant/animal.
1997	Veth, Fullagar et al.	Puntutjarpa Rockshelter, WA, Australia, 435-10,170BP	2	-	Y	AA	No information reported on methods or magnification. Starch was observed on two grindstone fragments.
1997	Williamson	Rose Cottage Cave, South Africa, 9.5-13.5ka	30	-	Y	SAJS	Follows from and combined with data reported in Williamson (1996); reports plant/animal processing and DNA extraction (not sequencing)

Year	Author(s)	Site, age (approx.)	#[1]	Mag.[2]	U-w[3]	Journal[4]	Summary
1996	Sobolik	Hinds Cave, Texas, USA, 2-5ka	55	10-400x	Y	JFA	Residues observed at 10-15x in situ, then removed with a wooden scraper and examined at up to 400x.
1996	Williamson	Rose Cottage Cave, South Africa, 6ka	167	50-500x	Y	SAFA	Reports plant/animal processing, with a predominance of plant. Fungal residues were found on one third of artefacts, and non-formal tools had residues.
1995	Tuross & Dillehay	Monte Verde, Chile, 12.5-13.2+ka	7	4-20x	N	JFA	Residues screened via microscopy, but analyses conducted using immunology – no correlation between observed 'stains' and immunological results
1994	Loy	Mustang Branch and Barton sites, Texas, USA	108	100-1000x	Y	-	In *Human Ecology in the Middle Onion Creek Valley*. Results reported as plant/animal, with additional chemical tests (also examined 40 burned rocks).
1993	Barton & White	Balof 2, New Ireland, Papua New Guinea, <14ka	21	>100x	Y	AP	Identified tuber processing on most artefacts over the 14,000 year period. Animal residues were not present.
1993	Boot	Graman, NSW, Australia, 2-5.5ka	2722	24-1000x	Y	-	In *Archaeometry: Current Australasian research*. Initial low-magnification screening showed 720 artefacts with evidence of use, 165 had use or hafting residues.
1993	Fullagar	West New Britain, Papua New Guinea, >3.5ka	43	-	Y	-	In *Traces et Fonction*. Identified starch grains, phytoliths and blood. Discusses plant vs animal processing, change over time, and phytoliths in particular. No magnifications provided.
1993	Fullagar	West New Britain, Papua New Guinea, >3.5-4ka	1	10-500x	Y	-	In *Archaeometry: Current Australasian research*. Recovered phytoliths from adhering soil, and observed in situ phytoliths, starch, raphides, fungal remains and suspected blood.
1993	Furby, Fullagar *et al.*	Cuddie Springs, NSW, Australia, 19-28ka	-	-	Y	-	In *Sahul in Review*. Few specific details of the study provided. Red blood cells, animal tissue, mammalian hair and woodworking/plant processing residues identified on flaked stone. Starch found on a grindstone.
1993	Loy	Toad River Canyon, Canada, 2200BP	11	-	N	WA	Identified blood/hair/animal tissue, as well as some plant remains. Immunological and DNA tests were then employed.

Year	Author(s)	Site, age (approx.)	#[1]	Mag.[2]	U-w[3]	Journal[4]	Summary
1992	Fullagar	FEA, West New Britain, Papua New Guinea, recent-3500BP	55	6-500x	Y	-	In *Poterie Lapita et Peuplement*. Studied obsidian flakes from six sites, divided into three phases based on Lapita occupation. Results presented as plant vs animal processing, with discussion of changes in resource exploitation and site specialisation.
		FRL, West New Britain, Papua New Guinea, recent->3500BP	420				
		FRI, West New Britain, Papua New Guinea, recent-3500BP	63				
		FOF, West New Britain, Papua New Guinea, 2000->3500BP	31				
		FHC, West New Britain, Papua New Guinea, recent->3500BP	302				
		FGT, West New Britain, Papua New Guinea, recent->3500BP	85				
1992	Fullagar, Meehan *et al.*	Various sites, Northern Australia, <2ka	72	<300x	Y	-	In *Prehistorie de l'Agriculture*. Examined a variety of ethnographic and recent archaeological artefact materials (72 of which were archaeological stone). Plant and animal identifications.
1992	Hurcombe	Ortu Comidu, Sardinia, 2-3.5ka	12	250-625x	Y	-	In *Use-wear Analysis and Obsidian*. Residue morphology is described, rather than assigned to particular origins; possible plant.
1992	Loy & Hardy	Tabun Cave, Israel, 90ka	10	12-1000x	N	Antiquity	Microscopic and immunological study of blood residues. Concludes tools used for (unspecified) animal processing.
1992	Loy, Spriggs *et al.*	Kilu Cave, Solomon Is., <9ka & 20-28.7ka	47	500-1000x	Y	Antiquity	Identified taro (*Colocasia esculenta* and *Alocasia macrorrhiza*) starch grains and raphides.
1992	Sievert	Chichen Itza, Mexico	166	50-200x	Y	-	In *Maya Ceremonial Specialization*. No specific dates available, although likely >500BP. Identified resins, gold, pigment, plant tissue, possible blood on artefacts that had previously been chemically cleaned. See also Sievert (1990, 1992b).
1991	Loy	Unspecified Moa butchery sites, New Zealand, 1000BP	20	-	-	-	In *Archaeometry '90*. Very little information – no data on techniques or sites. Identified moa feather fragments on 5 tools.

Year	Author(s)	Site, age (approx.)	#[1]	Mag.[2]	U-w[3]	Journal[4]	Summary
1990	Hyland, Tersak et al.	Shoop, Illinois, USA, 8ka	45	<10x	N	Am. Ant.	Very low magnification initial screening of artefacts for subsequent Hemastix and immunological testing for blood. One artefact possessed blood (cervid) residue.
1990	Jones	Kutikina Cave, Tasmania, Australia, 15-17ka	-	-	Y	-	In *The World at 18,000BP*. Few details provided (no magnifications, number of tools). Found evidence of meat, bone, plant and wood working.
1990	Loy	Lake Mungo, NSW, Australia, 20-35ka.	-	-	Y	ANH	No details of methods and number of tools examined are provided. Identified starch grains and plant tissue, blood, sinew and periosteum.
1989	Fullagar	Ritamauda rockshelter (QBB), Papua New Guinea, 0-3500BP	161	6-400x	Y	-	In *A Crack in the Spine*. Found organic smears and fragments, fibres, plant tissue, silica particles. Palm, reed and wood working inferred.
1989	Hall, Higgins et al.	SE Queensland, Australia	10	6-400x	Y	-	In *Plants in Australian Archaeology*. No dates provided. Found starches and plant tissue on pounding tools (and slides made from the tools).
1989	Loy & Wood	Cayonu Tepesi, Turkey, 6.8-7.4ka	398	-	N	JFA	No details of magnification. Artefacts were screened microscopically in the field for blood, then residues removed for lab testing. Identified human, auroch and *Ovis* sp. blood.
1989	Newman & Julig	Cummins site, Thunder Bay, Canada, 7.5-9ka	36	10-40x	N	CJA	Artefacts initially screened using low-power microscopy. Potential blood residues identified, then tested immunologically.
1989	Richards	Thorpe Common rockshelter, Yorkshire, UK, 5.7-6.4ka	50	10-625x	N	-	In *Breaking the Stony Silence*. Blood residues identified on 41 tools through microscopy and chemstrip testing. Observed blood also in the soil matrix immediately adhering to some artefacts.
1988	Fullagar	Aire Shelter II, Victoria, Australia, <600BP	1814	-	Y	-	In *Industries Lithique*. Study reported briefly, no magnifications provided. Residues on 13 tools, including collagen.
1987	Fullagar	Birrigai rockshelter, ACT, Australia, <1000-21000BP	69	-	Y	AO	Identified residues on 9 of the 69 artefacts. Found blood, collagen, plant and possible ochre. Blood may not be from tool-use. No magnifications provided.
1987	Jones	Kutikina Cave, Tasmania, Australia, 17-20ka	1	300x	Y	AG	No information on number of artefacts analysed (only the results from one are reported). Found collagen and blood identified as red-necked wallaby.

Year	Author(s)	Site, age (approx.)	#[1]	Mag.[2]	U-w[3]	Journal[4]	Summary
1987	Loy	Barda Balka, Iraq, 75-125ka; Jarmo, Iran, 8-10ka; Unspecified site, Belgium, 20ka	3 5 2	-	-	-	In *Archaeometry: Further Australasian Studies*. Three studies mentioned very briefly – little detail on methods or results. Found blood cells and mammalian blood deposits.
1985	Frederickson	Whakamoenga Cave, New Zealand; Twilight Beach, New Zealand	138 61	15-50x	N	NZAAN	No dates provided. Examined obsidian flakes from two sites for blood residues via light microscopy, test-strips and SEM. Light microscopy identified fibrous material, and deposits of various colours. Subsequent SEM identified erythrocytes.
1985	Loy	Hidden Cave, Nevada, USA	1	5-800x	N	-	In *The Archaeology of Hidden Cave*. No dates provided. Identified bat hair and presumed bat blood, as well as feather barbules.
1984	McBryde	Graman, NSW, Australia, 1-5ka	-	-	Y	-	In *Recent Advances in Indo-Pacific Prehistory*. No data on magnifications provided. It is unclear how many artefacts were examined (possibly 900). Identified hafting 'stains' on 26 backed blades from 3 sites.
1983	Loy	Various sites, Canada, 1-6ka	104	12-500x	-	Science	Blood residues observed on 90 artefacts, as identified by a variety of tests; species identified via haemoglobin crystallisation.
1982	Coughin & Claassen	Vail, Maine, USA, Palaeo-Indian	27	160x	N	-	In *The Vail Site*. Initial in situ observation of 'tiny biological fragments', followed by removal to slides. Observed well-preserved phytoliths, and plant tissue.
1979	Shafer & Holloway	Hinds Cave, Texas, USA, 2-5ka	25	10-400x	Y	-	In *Lithic Use-wear Analysis*. Initial observation of plant fibres and epidermis in situ, then removal to slides (for 11 tools). Found phytoliths, hairs, starch, and more, some identifiable to species
1976	Briuer	Chevelon Canyon and Cayote Creek Pueblo, Arizona, USA, 700-8700BP	2551	30x	Y	Am Ant	Used low-magnification to screen artefacts, found 37 with 'organic use residues'. These were classified (using chemical tests and morphological inspection) as plant or animal.

1: # = the number of artefacts examined. This number is often much larger than the number of artefacts with residues.
2: Mag. = magnifications used by the analyst.
3: U-w = whether or not use-wear was also examined in the study [Y = yes; N = no; - = insufficient information].
4: Journal = the journal in which the study was published. See below for abbreviations. If the study was not published in a journal, the source is listed in the 'Summary' column.

Journal abbreviations:

AA	Australian Archaeology	LAA	Latin American Antiquity
AAS	Australian Aboriginal Studies	NS	New Scientist
AG	Australian Geographic	NZAAN	New Zealand Archaeological Association Newsletter
Am Ant	American Antiquity		
ANH	Australian Natural History	NZJA	New Zealand Journal of Archaeology
AO	Archaeology in Oceania	PNAS	Proceedings of the National Academy of Sciences
AP	Asian Perspectives		
CAJ	Cambridge Archaeological Journal	RAM	Records of the Australian Museum
CJA	Canadian Journal of Archaeology	RPP	Review of Palaeobotany and Palynology
HA	Hawaiian Archaeology	SAAB	South African Archaeological Bulletin
JAS	Journal of Archaeological Science	SAFA	South African Field Archaeology
JFA	Journal of Field Archaeology	SAH	Southern African Humanities
JHE	Journal of Human Evolution	SAJS	South African Journal of Science
JNAS	Journal of Nordic Archaeological Science	VHA	Vegetation History and Archaeobotany
		WA	World Archaeology

7

BUILDING A COMPARATIVE STARCH REFERENCE COLLECTION FOR INDONESIA AND ITS APPLICATION TO PALAEOENVIRONMENTAL AND ARCHAEOLOGICAL RESEARCH

Carol J. Lentfer
School of Social Science
University of Queensland
St. Lucia, QLD 4072 Australia
Email: c.lentfer@uq.edu.au

ABSTRACT

Indonesia has a very long record of hominin occupation involving at least three human species. It also has a rich diversity of plants and a suite of economically important starch-rich staples that include taro (*Colocasia esculenta*), yams (*Dioscorea* spp.), bananas (*Musa* spp.) and sago palms. However very little is known about the prehistory of plant exploitation in the island archipelago. Key archaeological issues that are often discussed and debated, but remain poorly understood, include human adaptations and economic strategies used in different and changing environments, how local economies contributed to the development of plant management systems within the southeast Asian/Pacific regions, and how they might have been influenced by plant management systems initiated and developed elsewhere. Crucial to a burgeoning focus on archaeobotany in island Southeast Asia is the establishment of comparative modern reference collections for the flora of the region. This paper discusses the 'Indonesian Starch Project', the relevance of a comparative modern starch reference collection to Indonesian archaeology and the methods and procedures used in its development. The collection is broad-based and focuses on economically and ecologically importat plant groups. To date 121 families and 451genera are represented in the collection.

KEYWORDS

starch granules, Indonesia, southeast Asia, comparative reference collection

INTRODUCTION

Tom Loy's first forays into starch research began in the early 1980s when he and Richard Fullagar observed starch granules on stone, shell and glass artefacts recovered from a variety of archaeological contexts in North America and Oceania (Fullagar 2006). Tom continued his pioneering starch research into the 1990s. His discovery and identification of Araceae starch granules and raphides on stone tools from the Solomon Islands (Loy *et al.* 1992) was most influential, being the first strong evidence that people were using stone tools to process starchy foods in the Pacific region as long as 28,000 years ago. Since that time, it has been recognised that starch granules can be well-preserved in a variety of archaeological contexts, for example in residues on stone tools, pottery, other artefacts and ecofacts (Torrence and Barton 2006), within dental calculus and coprolites (Englyst *et al.* 1992; Juan-Tresserras 1998), in cereal seeds and other preserved foodstuffs (Samuel 1994; Ugent *et al.* 1987), and in sediments (Haslam 2004; Lentfer *et al.* 2002; Therin *et al.* 1999). Consequently there has been a burgeoning interest in starch research in archaeology with significant outcomes for understanding the prehistory of plant exploitation and manipulation in many regions of the world (e.g. Barton and Paz 2007; Crowther 2005; Fullagar *et al.* 2006; Helena *et al.* 2007; Horrocks *et al.* 2004; Parr and Carter 2003; Piperno *et al.* 2004; Zarillo *et al.* 2008).

The application of starch research to Indonesian archaeology has huge potential for resolving many key archaeobotanical issues in the region. The Indonesian archipelago has a very long record of human occupation extending well beyond a million years and involving three distinct human species (Brown *et al.* 2004; Hutterer 1985; Morwood *et al.* 2004; O'Sullivan *et al.* 2001; Sémah *et al.* 2003). It also has a rich diversity of plants and a suite of economically important species including a range of cereal grasses, fruit and nut bearing trees, and plants with edible roots, tubers, corms and rhizomes. From the beginnings of human colonisation of the archipelago, plant exploitation would most probably have been extensive and complex, specifically adapted to different environments and availability of naturally occurring resources. The nature of plant exploitation, diet and plant food processing would have been influenced through time by climatic and environmental change, as well as by the introduction of new technologies, people, cultures and changing economies. The archipelago lies between the southeast Asian mainland and New Guinea, two well-known centres for independent origins of early agriculture (Bellwood 1997; Denham *et al.* 2003, 2004; Harris 1996; Jiang 1995) and spans the Wallace line, a well-established boundary between the southeast Asian and Australasian biogeographic regions. As indicated by current distributions and phylogenetic relationships of taros, yams, *Musa* section bananas and *Metroxylon* and *Caryota* sago palms (Bellwood 1997; De Langhe and De Maret 1999; Denham *et al.* 2004; Fullagar *et al.* 2006; Lebot 1999; Lentfer and Green 2004; Simmonds 1962; Spriggs 1996; Yen 1990), the archipelago has been influenced from both directions across the Wallace Line. Undoubtedly, plant manipulation in the Indonesian archipelago itself would have contributed significantly to domestication, and similar to adjacent regions, played a substantial role in the development of early agriculture.

Four issues to be addressed by archaeobtanical research include the following:

- the question of plant exploitation and foraging strategies of *Homo erectus*, *H. floresiensis* and early *H. sapiens* - what are the similarities and differences between species and changes through time?;
- the nature of plant exploitation and food processing strategies adopted by humans living in tropical rainforests, sometimes referred to as 'green deserts' (Bailey *et al.* 1991, Dentan 1991);
- the origins and spread of vegeculture and arboriculture; and
- the timing and dispersal of various plant cultivars, particularly how the introduction and dispersal of starch-rich root and tuber crops, sago palms, bananas, sugar cane, breadfruit, certain nut trees and cereal crops, might have been influenced by Austronesian expansion in the region.

There has been a keen interest in these issues for several decades spurred on by certain key publications such as Sauer (1953) and Harris (1996). However, systematic and focused archaeobotanical investigation has not been prioritised in mainstream archaeological research, due partly to a perceived notion that plant remains are poorly preserved in the tropics, especially wet environments. Consequently, there is a dearth of information about the prehistory of plant exploitation, particularly for Indonesia. Direct archaeobotanical information is rare, confined mostly to evidence for cereal cropping in the archipelago during the mid-to-late Holocene (e.g.

Glover and Higham 1996). Besides this, much of our current understanding relies on speculation and inference primarily based on the presence of artefacts with inferred plant related function or association (Glover 1986) and/or plant phylogeny, phytogeography and linguistics used for determining plant origins and patterns of plant dispersal (Bellwood 1996; Matthews and Naing 2005; Spriggs 1996).

Recently, following the proven value of starch and phytolith analyses in wet tropical regions, there has been escalating interest in incorporating systematic archaeobotanical investigation in island southeast Asian archaeology. The Niah cave project, Sarawak, the Liang Bua Project in Flores, Indonesia, and various East Timor archaeological projects are key examples where starch and phytolith analyses are being used routinely for palaeoenvironmental reconstruction and residue analyses. Fundamental to this research is a sound knowledge of the range and variation of the various microfossils in plants that occur in the region. There are two major comparative modern phytolith reference collections relevant to the region, however starch reference material is limited. This paper discusses the 'Indonesian Starch Project' and outlines the methods and procedures being used to establish the first major comparative modern starch reference collection and database for the flora of Indonesia.

COMPARATIVE STARCH REFERENCE COLLECTIONS

The establishment of comparative modern starch reference material is of paramount importance to starch research and has been integral to studies seeking to determine taxonomic diagnostics and to identify starch granule provenance. Most collections have been focused on food crops and plant resources relevant to industries, individual studies and regions. For the majority of studies starch storage organs (i.e. tubers, corms, rhizomes, roots, fruits and seeds) have been targeted (e.g. Reichert 1913). Other plant parts with less economic importance and/or assumed to have mainly transitory starch, which occurs in the chloroplasts of plant tissue and is thought to be non-diagnostic (Gott *et al.* 2006), have not been thoroughly examined. Given the general lack of readily accessible, broad-scale collections, therefore, it is often the case that the establishment of new or additional comparative reference collections tailored to suit particular research questions is a mandatory component of research design (Dickau *et al.* 2007; Field and Gott 2006; Loy *et al.* 1992; Perry 2004).

A few collections have been established for the tropical southeast Asian/Pacific region, including those created by Loy, Barton, Therin, Crowther, Fullagar and Horrocks for their various research projects. However, all of these collections, with the exception of Therin's from West New Britain, Papua New Guinea (see Lentfer and Therin 2006) have primarily targeted food plants aroid corms (Araceae) and yam tubers (Dioscoreaceae) in particular. Given that Therin's collection is limited, consisting of a few hundred samples only, comparative modern reference material available for the determination of the range of starch granule variation within and between plant parts for any particular species, and within and between plant species is in short supply. With the possible exception of well-studied and targeted species or parts of particular species, therefore, accurate taxonomic identification is problematic unless starch granules are embedded in identifiable tissue or associated with other diagnostic plant residues. Furthermore, without a sound knowledge of starch production, range of morphologies and variation in any particular ecological setting, definitive taxonomic identification of what are considered to be well-analysed species should be treated with caution until there is more comparative information. Until this time, the approach taken by Paz (2001) and Barton and Paz (2007) should be adopted whereby levels of confidence are recorded according to how closely the archaeological material matches available comparative reference material.

THE INDONESIAN STARCH PROJECT

The aims of the 'Indonesian Starch Project', initiated in 2002 as a collaborative project between Indonesian and Australian researchers, are to facilitate the development of ancient starch analysis in the Indonesian region by systematic and comprehensive sampling of Indonesian flora and to establish a database that allows for more effective application of starch analysis to archaeological, palaeoenvironmental and palaeoethnobotanical studies. Economically important starch-rich plants, other useful plants and a wide variety of other flora in the region are included to enable systematic investigation of starch presence in plants, evaluation of the diagnostic properties of starch granules, and the identification of plants and plant parts with diagnostic starch.

Plant collection, identification and storage

Where possible all parts of plants including leaves, flowers, fruits, seeds, stems, wood, roots, corms, tubers, rhizomes are sampled. Not all plant parts are included in herbarium collections. Hence, although it can be very useful and expedient to sample from established herbarium collections, it is often necessary to obtain supplementary material from living plants. Furthermore, starch gelatinises at temperatures above 50° Celsius, causing it to swell, break and lose its birefringence properties. Therefore, heated starch is often not useful for diagnostic analysis and collection from herbariums where plant specimens have been heat-treated at temperatures above 50° Celsius for long-term safe storage can be problematic.

Currently, there are over 800 plant accessions in the collection. One hundred and twenty one families and 451 genera are represented (see summary list of families and species in Table 1). Field sampling during the initial stages of this project was undertaken in Flores, West Timor, Java and Bali giving priority to known useful plants and important ecological indicators (Figures 1 to 4). Subsequently, these collections have been expanded by sampling in the Bogor Botanic Gardens in west Java where the focus was on obtaining accessions of useful and non-useful species from plant families with economically important members and medicinal plants. Other accessions presently held in the reference collection include wild and domesticated bananas collected in Papua New Guinea and Flores, aroid and yam samples from the Philippines and Sarawak, and additionally, a range of useful comparative plant species from Fiji.

Table 1. List of plant families and genera included in the comparative starch reference collection. (P/I: pending identification; *Diploid and polyploid cultivars are included in *Musa*)

Family	Genus	No. Species
Acanthaceae	Acanthus	1
Acanthaceae	Andrographis	1
Acanthaceae	Asystasia	1
Acanthaceae	Barleria	1
Acanthaceae	Clinacanthus	1
Acanthaceae	Gendarussa	1
Acanthaceae	Graptophyllum	1
Acanthaceae	Pseuderanthemum	1
Actinidiaceae	Saurauia	3
Agavaceae	Pleomele	3
Aizoaceae	Sesuvium	1
Amaranthaceae	Deeringia	1
Amaryllidaceae	Crinum	1
Amaryllidaceae	Curculigo	3
Amaryllidaceae	Eucharis	1
Anacardiaceae	Anacardium	1
Anacardiaceae	Lannea	1
Anacardiaceae	Mangifera	2
Anacardiaceae	Pentaspadon	1

Family	Genus	No. Species
Anacardiaceae	P/I	1
Annonaceae	Annona	1
Annonaceae	Cananga	1
Annonaceae	Polyalthia	1
Annonaceae	Saccopetalum	1
Anthericaceae	Chlorophytum	1
Apiaceae	Centella	1
Apocynaceae	Alstonia	1
Apocynaceae	Cerbera	1
Apocynaceae	Pagiantha	1
Apocynaceae	Plumeria	1
Apocynaceae	Thevetia	1
Aquifoliaceae	Ilex	2
Araceae	Homalomena	1
Araceae	Acorus	1
Araceae	Aglaonema	5
Araceae	Alocasia	6
Araceae	Amorphophallus	3
Araceae	Anadendrum	2
Araceae	Anchormanes	1
Araceae	Anthurium	4

Araceae	*Colocasia*	3		Compositae	*Eupatorium*	1
Araceae	*Culcasia*	1		Compositae	*Galinsoga*	1
Araceae	*Cyrtosperma*	2		Compositae	*Pterocaulon*	1
Araceae	*Dieffenbachia*	8		Compositae	*Sida*	1
Araceae	*Dracontium*	1		Compositae	*Sonchus*	1
Araceae	*Homalomena*	3		Compositae	*Synedrella*	1
Araceae	*Monstera*	1		Compositae	*Vernonia*	3
Araceae	*Philodendron*	2		Compositae	*Wedelia*	1
Araceae	*Raphidaphora*	1		Compositae	P/I	-
Araceae	*Rhodospathia*	1		Compositae	P/I	-
Araceae	*Schismatoglottis*	4		Compositae	P/I	-
Araceae	*Scindapsus*	1		Compositae	P/I	-
Araceae	*Spathiphyllum*	2		Convallariaceae	*Ophigon*	2
Araceae	*Xanthosoma*	3		Convallariaceae	*Tupistra*	1
Araliaceae	*Arthrophyllum*	1		Convolvulaceae	*Ipomoea*	6
Araceae	P/I	-		Convolvulaceae	P/I	-
Araceae	P/I	-		Convolvulaceae	P/I	-
Araliaceae	*Harmsiopanax*	1		Crassulaceae	*Kalanchoe*	1
Araliaceae	*Macropanax*	1		Cucurbitaceae	*Benincasa*	1
Araliaceae	*Trevesia*	1		Cucurbitaceae	*Coccinia*	1
Asclepiadaceae	*Asclepias*	1		Cucurbitaceae	*Cucurbita*	1
Asclepiadaceae	*Calatropis*	1		Cucurbitaceae	*Lagenaria*	1
Asclepiadaceae	*Dischidia*	1		Cucurbitaceae	P/I	-
Asclepiadaceae	*Finlaysonia*	1		Cunioniceae	P/I	-
Aspidiaceae	*Dryopteris*	1		Cunioniceae	P/I	-
Aspidiaceae	*Stenosemia*	1		Cyclanthaceae	*Carludovica*	1
Aspleniaceae	*Asplenium*	1		Cyclanthaceae	*Cyathea*	1
Asteliaceae	*Cordyline*	1		Cyperaceae	*Carex*	1
Athyroiodeae	*Athyrium*	1		Cyperaceae	*Cyperus*	2
Averrhoaceae	*Averrhoa*	2		Cyperaceae	*Lipocarpha*	1
Begoniaceae	*Begonia*	1		Cyperaceae	*Mapania*	1
Bignoniaceae	*Crescentia*	1		Cyperaceae	*Rhynchospora*	1
Bignoniaceae	*Kigelia*	1		Cyperaceae	*Scirpus*	1
Bignoniaceae	*Stereospermum*	1		Davaliaceae	*Nephrolepis*	1
Bignoniaceae	*Tabebula*	1		Dilleniaceae	*Dillenia*	2
Blechnaceae	*Blechnum*	1		Dioscoreaceae	*Dioscorea*	10
Bombacaceae	*Bombax*	1		Dracaenaceae	*Sanservieria*	5
Bombacaceae	*Durion*	1		Ebenaceae	*Diospyros*	1
Boraginaceae	*Cordia*	2		Equisetaceae	*Equisetum*	1
Bromeliaceae	*Ananas*	1		Euphorbiaceae	*Acalypha*	1
Burseraceae	*Canarium*	3		Euphorbiaceae	*Aleurites*	1
Butomaceae	*Limnocharis*	1		Euphorbiaceae	*Antidesma*	1
Cactaceae	*Opuntia*	1		Euphorbiaceae	*Bischofia*	1
Cannaceae	*Canna*	2		Euphorbiaceae	*Codiaeum*	1
Capparaceae	*Capparis*	3		Euphorbiaceae	*Euphorbia*	3
Caryophylaceae	*Drymaria*	1		Euphorbiaceae	*Excoecaria*	1
Celastraceae	*Elaeodendron*	1		Euphorbiaceae	*Glochidion*	1
Chloranthaceae	*Sarcandra*	1		Euphorbiaceae	*Jatropha*	3
Combretaceae	*Lumnitzera*	1		Euphorbiaceae	*Macaranga*	1
Combretaceae	*Quisqualis*	1		Euphorbiaceae	*Mallotus*	1
Combretaceae	*Terminalia*	2		Euphorbiaceae	*Manihot*	1
Commelinaceae	*Aneilema*	1		Euphorbiaceae	*Melanolepis*	1
Commelinaceae	*Commelina*	1		Euphorbiaceae	*Omalanthus*	2
Commelinaceae	*Murdannia*	1		Euphorbiaceae	*Ricinis*	1
Commelinaceae	*Palisota*	1		Euphorbiaceae	*Sauropus*	1
Compositae	*Ageratum?*	1		Euphorbiaceae	P/I	1
Compositae	*Bidens*	2		Euphorbiaceae	P/I	-
Compositae	*Blumea*	2		Flacourtiaceae	*Scolopia*	1
Compositae	*Emilia*	1		Flagellariaceae	*Flagellaria*	1
Compositae	*Erechtites*	2		Gesneriaceae	*Cyrtandra*	1

Gleicheniaceae	*Gleichenia*	1		Leguminosae	*Archidendron*	1
Gramineae	*Andropogon*	1		Leguminosae	*Bauhinia*	1
Gramineae	*Bambusa*	3		Leguminosae	*Caesalpinia*	1
Gramineae	*Bambusa*	2		Leguminosae	*Cajanus*	1
Gramineae	*Bothriochloa*	1		Leguminosae	*Calliandra*	1
Gramineae	*Centotheca*	1		Leguminosae	*Canavalia*	1
Gramineae	*Chloris*	1		Leguminosae	*Cassia*	2
Gramineae	*Coix*	1		Leguminosae	*Centrosema*	1
Gramineae	*Cymbopogon*	1		Leguminosae	*Crotalaria*	2
Gramineae	*Dendrocalamus*	1		Leguminosae	*Cynometra*	1
Gramineae	*Echinochloa*	1		Leguminosae	*Derris*	2
Gramineae	*Eleusine*	1		Leguminosae	*Desmodium*	1
Gramineae	*Eragrostis*	1		Leguminosae	*Erythrina*	1
Gramineae	*Erianthus*	1		Leguminosae	*Glyricidia*	1
Gramineae	*Heteropogon*	1		Leguminosae	*Indigofera*	1
Gramineae	*Imperata*	1		Leguminosae	*Leucaena*	1
Gramineae	*Neololeba*	1		Leguminosae	*Mimosa*	1
Gramineae	*Oplismenus*	1		Leguminosae	*Moghania*	2
Gramineae	*Oryza*	1		Leguminosae	*Mucuna*	1
Gramineae	*Panicum*	2		Leguminosae	*Parkia*	1
Gramineae	*Paspalum*	3		Leguminosae	*Peltophorum*	1
Gramineae	*Pennisetum*	1		Leguminosae	*Phaseolus*	2
Gramineae	*Saccharum*	2		Leguminosae	*Pithecellobium*	1
Gramineae	*Setaria*	1		Leguminosae	*Pometia*	1
Gramineae	*Shizostachyum*	1		Leguminosae	*Pongamia*	1
Gramineae	*Sorghum*	3		Leguminosae	*Pongamia*	1
Gramineae	*Spinifex*	1		Leguminosae	*Pterocarpus*	1
Gramineae	*Sporobolus*	1		Leguminosae	*Sesbania*	1
Gramineae	*Themeda*	3		Leguminosae	*Tamarindus*	1
Gramineae	*Thuarea*	1		Leguminosae	*Tetrapleura*	1
Gramineae	*Thysanolaena*	2		Leguminosae	*Vigna*	1
Gramineae	*Tribulus*	1		Leguminosae	P/I	-
Gramineae	*Urochloa*	1		Leguminosae	P/I	-
Gramineae	*Zea*	1		Leguminosae	P/I	-
Guttiferae	*Garcinia*	3		Liliaceae	*Asparagus*	1
Gramineae	P/I	-		Liliaceae	*Aspidistra*	2
Gramineae	P/I	-		Liliaceae	*Cordyline*	1
Gramineae	P/I	-		Liliaceae	*Herreria*	1
Heliconiaceae	*Heliconia*	5		Liliaceae	*Pleomele*	1
Hypericaceae	*Hypricum*	1		Lomariopsidaceae	*Teratophyllum*	1
Icacinaceae	*Gonocaryum*	1		Lythraceae	*Lagerstroemia*	1
Icacinaceae	*Platea*	1		Lythraceae	*Pemphis*	1
Iridaceae	*Neomarica*	3		Magnoliaceae	*Manglietia*	1
Labiatae	*Coleus*	1		Magnoliaceae	*Michelia*	1
Labiatae	*Hyptis*	3		Malvaceae	*Abutilon*	1
Labiatae	*Ocimum*	1		Malvaceae	*Gossypium*	2
Labiatae	*Orthosiphon*	1		Malvaceae	*Hibiscus*	3
Labiatae	P/I	-		Malvaceae	*Sida*	1
Labiatae	P/I	-		Malvaceae	*Thespesia*	1
Labiatae?	P/I	-		Malvaceae	*Urena*	1
Lauraceae	*Cinnamomum*	1		Marantaceae	*Calathea*	4
Lauraceae	*Cryptocarya*	1		Marantaceae	*Maranta*	1
Lauraceae	*Litsea*	3		Marantaceae	*Monotagma*	1
Lauraceae	*Persea*	1		Marantaceae	*Phrynium*	1
Lecythidaceae	*Barringtonia*	3		Marattiaceae	*Angiopteris*	1
Leeaceae	*Leea*	1		Melastomataceae	*Medinilla*	1
Leguminosae	*Abrus*	1		Melastomataceae	*Melastoma*	2
Leguminosae	*Acacia*	2		Melastomataceae	*Osbeckia*	1
Leguminosae	*Adenanthera*	1		Meliaceae	*Dysoxylum*	1
Leguminosae	*Albizia*	2		Menispermaceae	*Anamirta*	1

Menispermaceae	Arcangelisis	1		Rubiaceae	Borreria	1
Menispermaceae	Tinospora	1		Rubiaceae	Lepisanthes	1
Menispermaceae	P/I	-		Rubiaceae	Morinda	2
Moraceae	Artocarpus	2		Rubiaceae	Mycetia	1
Moraceae	Ficus	16		Rubiaceae	Ophiorrhiza	1
Moringaceae	Moringa	1		Rubiaceae	P/I	-
Musaceae	Ensete	1		Rubiaceae	P/I	-
Musaceae	Musa	24*		Rubiaceae	P/I	-
Musaceae	Strelitzia	1		Rutaceae	Acronychia	1
Myristicaceae	Myristica	2		Rutaceae	Citrus	1
Myrsinaceae	Aegiceras	1		Rutaceae	Clausena	1
Myrtaceae	Eucalyptus	1		Rutaceae	Eudoia	2
Myrtaceae	Eugenia	1		Rutaceae	Feroniella	1
Myrtaceae	Psidium	1		Rutaceae	Glycosimis	1
Myrtaceae	Syzygium	4		Rutaceae	Pleiospermum	1
Nephrolepidaceae	Nephrolepis	1		Rutaceae	Swinglea	1
Orchidaceae	Calanthe	1		Rutaceae	P/I	-
Orchidaceae	Eria	2		Rutaceae ?	P/I	-
Orchidaceae	Goodyera	1		Sabiaceae	Meliosma	1
Orchidaceae	Malaxis	1		Salicaceae	Salix	1
Orchidaceae	Vanilla	1		Salvadoraceae	Azima	1
Orchidaceae	P/I	-		Santalaceae	Santalum	1
Oxalidaceae	Oxalis	1		Sapindaceae	Blighia	1
Palmae	Areca	2		Sapindaceae	Cardiospermum	1
Palmae	Arenga	6		Sapindaceae	Nephelium	1
Palmae	Borassus	1		Sapindaceae	Schleichera	1
Palmae	Calamus	2		Sapotaceae	Chrysophyllum	1
Palmae	Caryota	3		Sapotaceae	Planchonella	1
Palmae	Cocos	1		Saxifragaceae	Itea	1
Palmae	Corypha	1		Saxifragaceae	Polyosma	2
Palmae	Cyrtostachys	2		Selaginellaceae	Selaginella	1
Palmae	Elaeis	1		Solanaceae	Brugmansia	1
Palmae	Eugeissona	1		Solanaceae	Datura	1
Palmae	Licuala	4		Solanaceae	Solanum	2
Palmae	Livistonia	1		Sonneratiaceae	Sonneratia	1
Palmae	Metroxylon	1		Sterculiaceae	Heritiera	1
Palmae	Nypa	1		Sterculiaceae	Melochia	1
Palmae	Oncosperma	2		Sterculiaceae	Pterospermum	1
Palmae	Phoenix	1		Sterculiaceae	Theobroma	1
Palmae	Pinanga	1		Theaceae	Adinandra	1
Palmae	Ptychosperma	2		Thelypteridaceae	Sphaerostephanos	1
Palmae	Salacca	2		Thymelaeaceae	Phaleria	1
Palmae	Verschaffeltia	1		Tiliaceae	Grewia	1
Pandanaceae	Freycinetia	3		Tiliaceae	Microcos	2
Pandanaceae	Pandanus	10		Tiliaceae	Schoutenia	1
Phormiaceae	Dianella	2		Tiliaceae	P/I	-
Piperaceae	Piper	6		Ulmaceae	Celtis	1
Plantaginaceae	Plantago	1		Urticaceae	Cypholopus	1
Podocarpaceae	Podocarpus	1		Urticaceae	Debregeasia	1
Polygalaceae	Polygala	1		Urticaceae	Dendrocnide	1
Polygonaceae	Polygonum	2		Urticaceae	Leucosyke	1
Polypodaceae	Diplazium	1		Urticaceae	P/I	-
Polypodaceae	Drypoteris	1		Urticaceae	Pipturus	1
Polypodiaceae	P/I	-		Urticaceae	Villebrunea	1
Rhamnaceae	Ziziphus	1		Verbenaceae	Avicennia	1
Rhizophoraceae	Bruguiera	1		Verbenaceae	Clerodendron	3
Rhizophoraceae	Ceriops	1		Verbenaceae	Lantana	1
Rhizophoraceae	Rhizophora	1		Verbenaceae	Premna	1
Rosaceae	Prunus .	2		Verbenaceae	Stachytarpheta	2
Rosaceae	Rubus	3		Verbenaceae	Tectona	1

Verbenaceae	*Teijsmanniodendron*	1		Zingiberaceae	*Kaempferia*	1
Verbenaceae	*Vitex*	1		Zingiberaceae	*Languas*	1
Verbenaceae	P/I	-		Zingiberaceae	*Tapeinocheilos*	2
Verbenaceae	P/I	-		Zingiberaceae	*Zingiber*	4
Verbenaceae	P/I	-		Zingiberaceae	P/I	-
Verbenaceae	P/I	-		Zingiberaceae	P/I	-
Verbenaceae	P/I	-		Zingiberaceae	P/I	-
Zingiberaceae	*Alpinia*	4		Zygophyllaceae	*Tribulus*	1
Zingiberaceae	*Amomum*	5		** 27 further accessions	P/I	-
Zingiberaceae	*Catimbium*	2				
Zingiberaceae	*Costus*	12				
Zingiberaceae	*Cucurma*	7				
Zingiberaceae	*Etlingera*	2				
Zingiberaceae	*Globba*	1				
Zingiberaceae	*Hedychium*	4				
Zingiberaceae	*Hornstedtia*	1				

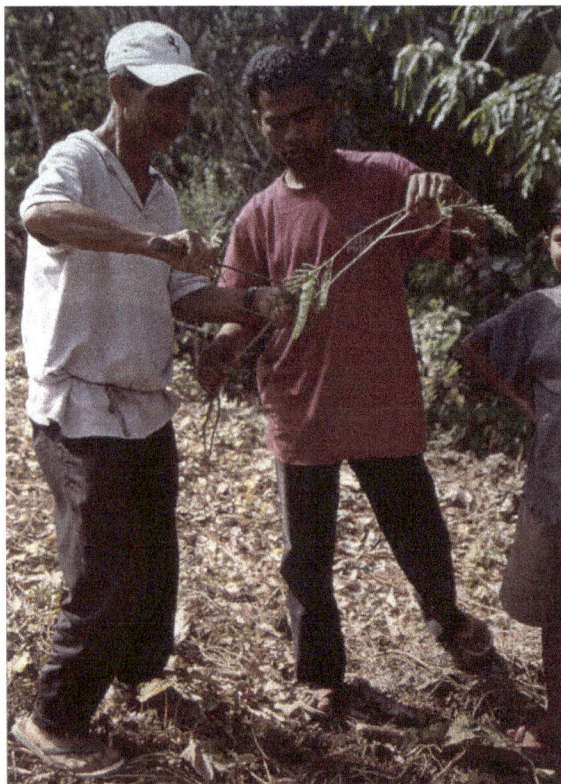

Figure 1. Collecting plants at Liang Bua, Flores Indonesia (photo: C. Lentfer).

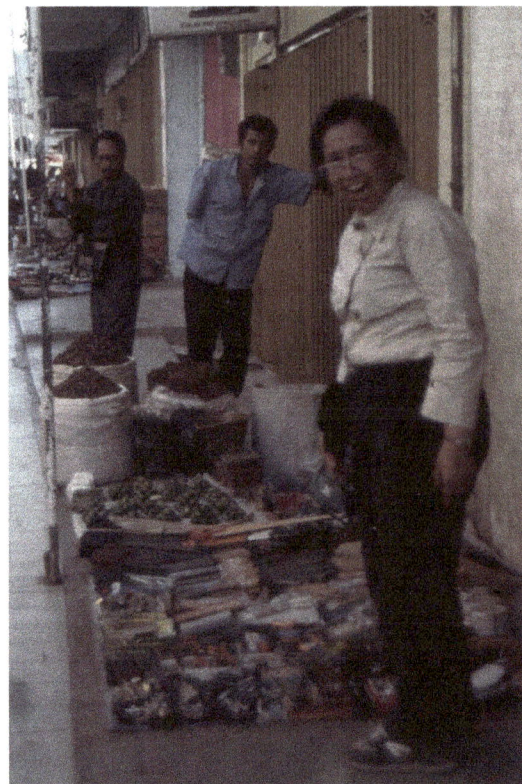

Figure 2. Dr Polhaupessy sampling plant products from markets in Kupang, West Timor (photo: C. Lentfer).

Plant specimens, including roots and other starch storage organs, are prepared following conventional herbarium procedures. The specimens are immersed in methylated spirits or ethyl alcohol at 70% dilution to avoid fungal growth during field collection and stored between paper in plant presses. This technique allows specimens to be stored wet for considerable periods in humid tropical conditions. For long-term storage they are slowly dried at temperatures not exceeding 35° Celsius. Taxonomic identification of specimens has been undertaken in the field and/or laboratory and cross-checked against herbarium specimens. Accessions are presently stored at the Bogor Herbarium, Indonesia and the University of Queensland, Australia.

Figure 3. Display of medicinal plant products beside garden house in West Flores (photo: C. Lentfer).

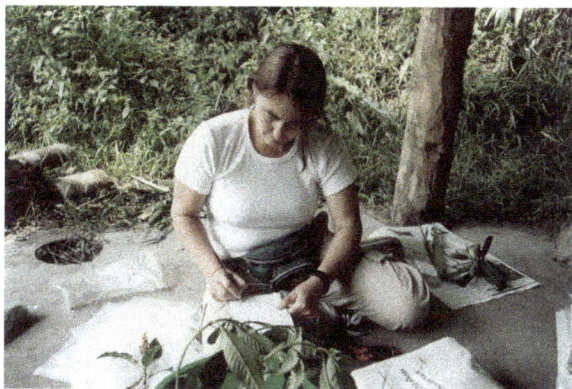

Figure 4. The author identifying plants collected from Rana Mese, West Flores and preparing herbarium specimens (photo: J. Collins).

Preparation of starch for microscopic analysis

Following this, sub-samples of all plant specimens including all parts of plants are stored in vials in 70% ethyl alcohol. Following this sub-samples of plant parts are prepared separately by grinding with a pestle and mortar in distilled water. Ground residue is stored in separate vials in 70% ethyl alcohol. For microscopic analysis 100 µl sub-sample of ground residue is mounted onto microscope slides and dried at temperatures at or below 35° Celsius. Corners of coverslips are secured onto glass slides with nail polish. Prior to analysis dried samples are re-hydrated with distilled water to allow for rotation of granules.

Starch granule analysis

All slide residues are analysed using light microscopy at x400, x600 and x1000 magnification. Qualitative analysis of starch granule assemblages is used initially. Presence or absence of starch is noted for every plant part for each plant specimen. Descriptors for all different types of granules in assemblages are recorded according to abundance, type (i.e. whether compound or simple), size range, and morphological characteristics of granules (see the 12 attributes listed in Table 2). During the initial recording stage attribute descriptions are separated by a slash (e.g. a/s/s/sr/hsph/0/cc/s/t/c/sv/stel). Photographs of all different starch granule morphotypes present in each plant part are taken with microscope dedicated digital cameras using non-polarized and polarized light. Figures 5 to 11 show examples from the collection of starch granules found in significant economic plants (photos by C. Lentfer and A. Crowther). Descriptors are transferred to an electronic database to facilitate searches and comparative analyses. Field collection records, plant uses, the presence or absence of starch granules in plant parts, and images of starch granules are also stored in an electronic database. The microscope slides and sub-samples in vials are presently stored at the University of Queensland and a sub-set is held at the Geological Research and Development Centre, Bandung, Indonesia.

The second stage of analysis is the determination of the diagnostic value of starch granules for species/plant tissue differentiation. This involves rigorous morphometric analysis of starch granule assemblages using quantitative statistical methods. It will be especially important for plants and tissues with starch granules that are similar and difficult to differentiate.

Table 2. List of attributes used for the description of starch granules

1. Abundance		2. Type		3. Size (Max. dim.)		4. 2D shape		5. 3D shape	
absent (0)	a	simple	s	< 5 µm	vs	subround	sr	hemispherical	hsph
rare									
(<20 granules)	r	compound	c	>5-10 µm	s	round	r	elongate hemisph.	el.hsph
common (>20-50)	c	semicompound	sc	>10-20 µm	m	ovate	ov	spherical	sph
abundant (>50-100)	ab			>20-50 µm	l	sub-ovate	sub.ov	subspherical	subsph
v. abundant (>100)	vab			>50 µm	vl	triangular ovate	tri.ov	ovoid	ov
						rectangular ovate	rect.ov	sub-ovoid	sub.ov
						elongate ovate	el.ov	triangular ovoid	tri.ov
						irregular ovate	irr.ov	rectangular ovoid	rect.ov
						irregular triangular ovate	irr.tri.ov	elongate ovate	el. ov
						elongate irregular ovate	el.irr.ov	irregular ovoid	irr.ov
						polygonal	plygl	irregular triangular ovoid	irr.tri.ov
6. Protrusion		**7. Facet**		**8. Texture**		irregular	irr	elongate irregular ovoid	el.irr.ov
present	+	none observed	0	wrinkle	wr	elongate irregular	el.irr	globose	gl
absent	0	flat	fl	smooth	s	crescent	cr	polyhedral	plyhdl
		concave	cc	rough	r	kidney	kid	quadrilateral	qu
		multifaceted flat	mff	ridged	rdg	elongate kidney	el.kid	irregular	irr
		multifaceted concave	mfc			triangular	tri	elongate irregular	el.irr
						square	sq	globose elongate	gl. el
						bell	bell	kidney	kid
						unilobate	unilob	elongate kidney	el. Kid
								cone	cone
								disc	di
								bell	bell
								unilobate	unilob
9. Lamellae		**10. Hilum - Position**		**11. Hilum - Type**		**12. Hilum - Fissure**		**Additional notes**	
present	+	eccentric	e	large vacuole	lv	none	0		
absent	0	centric	c	small vacuole	sv	stellate	stel		
		highly eccentric	he	crystal	cr	simple	s		
				slot	sl	open slot	os		
				irregular	irr	open irreg.	oi		
				papilla	pap	tri	tri		
						disc	d		

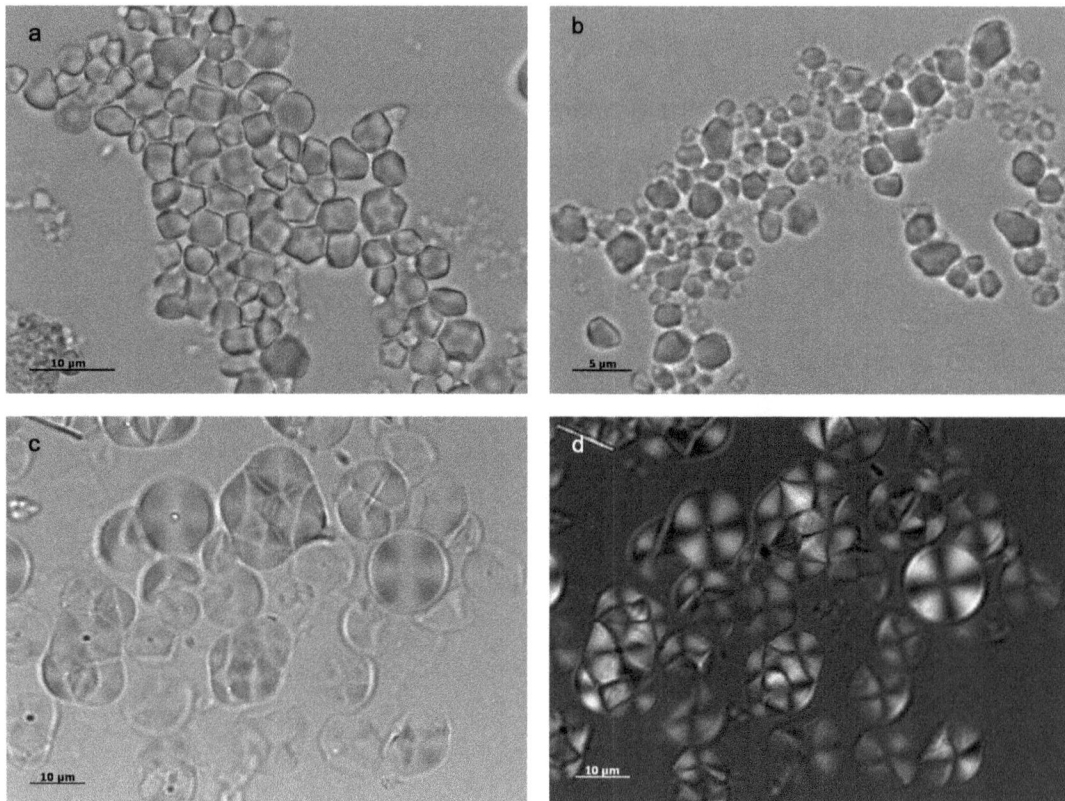

Figure 5. (a) *Alocasia* **sp. starch, showing typical multifacetted morphology (CL uncoded, corm). (b)** *Colocasia esculenta* **starch, showing typical multi-facetted morphology and small granule size (CL uncoded, corm). (c and d)** *Cyrtosperma chamissonis* **starch, showing simple and compound granule forms (c) and centric extinction cross, visible in cross-polarised light (d) (TL-IV69-4tuber) (scale bars in a, c and d = 10 µm; b = 5 µm).**

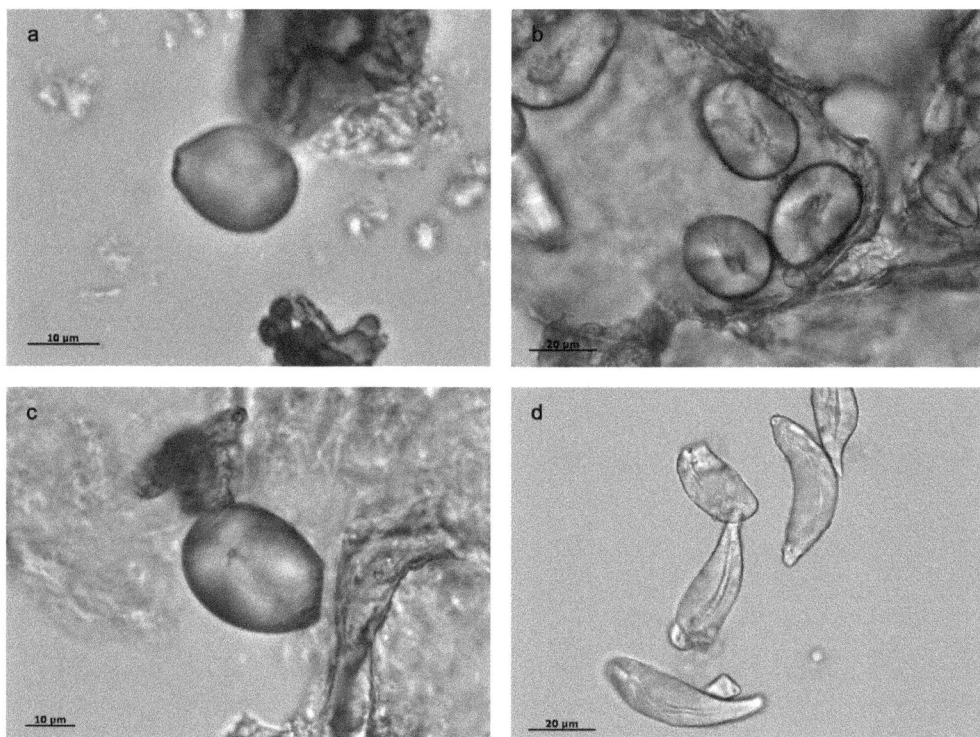

Figure 6. (a) *Caryota rumphiana* **(CL05160wood). (b)** *Eugeissonia utilis* **(CL05150wood). (c)** *Metroxylon sagu* **(CL05176trunk). (d)** *Salacca affinis* **(CL05182trunk) (scale bars in a and c = 10 μm; b and d = 20 μm).**

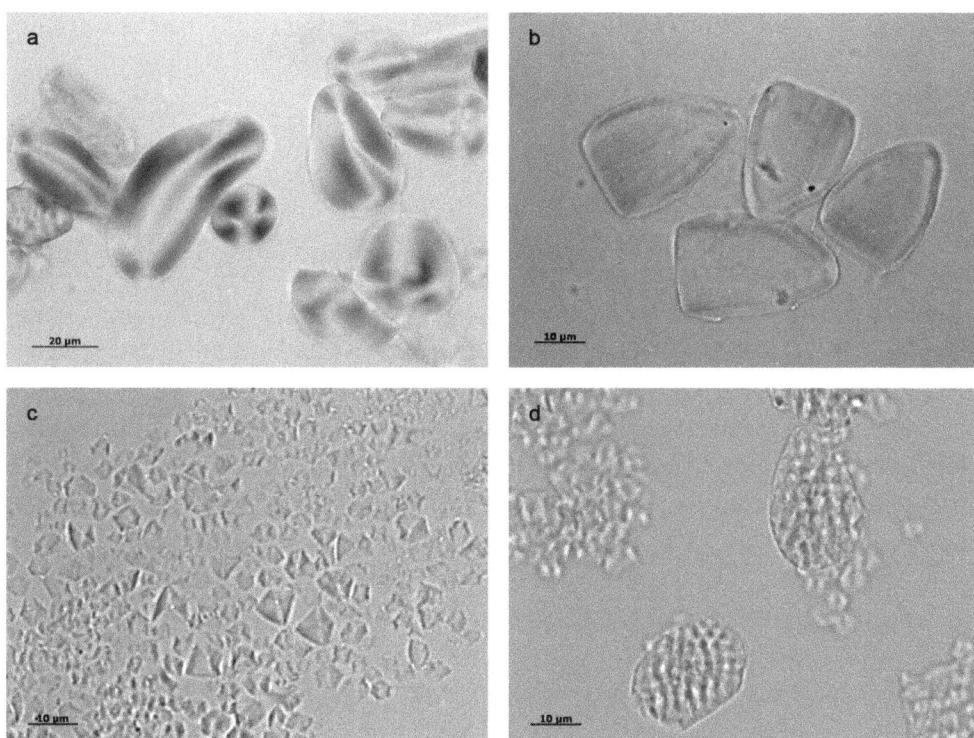

Figure 7. (a) Rounded ovoid to elongate-ovate starch forms in *Dioscorea alata* **(TL-IV69-8tuber). (b) Flat-based, rectilinear-triangular starch type produced by** *Dioscorea bulbifera* **(TL-IV69-14tuber). In addition to lamaellae, these granules have strong longitudinal striae that are very visible at the distal end. (c) Small, angular (multi-facetted) granules from** *Dioscorea esculenta* **(TL-IV69-7tuber). (d) Ovate-shaped compound grains in** *Dioscorea esculenta* **(TL-IV69-7tuber) (scale bar in a = 20 μm; b, c and d = 10 μm).**

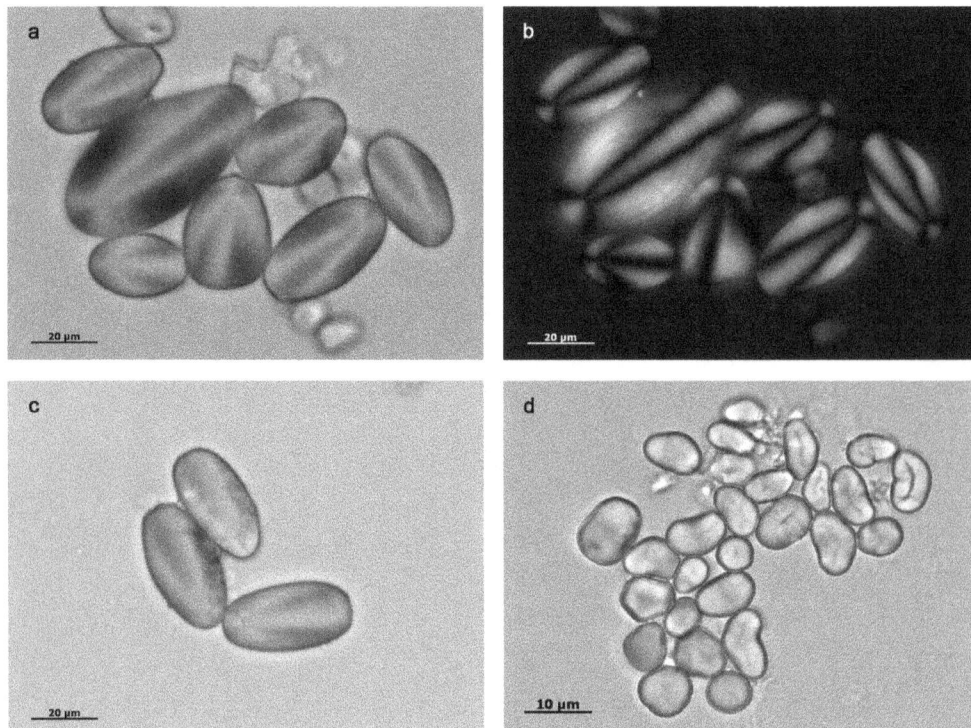

Figure 8. (a and b) Starch from *Heliconia caribea* (CL05197stem), showing typical rounded-rectilinear ovate granule form observed in all *Heliconia* sp. examined. (c and d) *Heliconia dasyantha* starch granules, showing the two distinctive starch types that occur within the seed (CL05288seed) (scale bars in a, b and c = 20 μm; d = 10 μm).

CURRENT STATUS OF THE REFERENCE COLLECTION

The comparative modern starch reference collection and database supplements other such collections of starch, phytolith, pollen, parenchyma and macrobotanical remains. It has enormous potential to contribute to our scientific knowledge in a wide variety of academic research disciplines including archaeology plant phylogenetic studies, palaeobotany, palynology and studies involving climatic and environmental change in the tropics throughout the Quaternary. Included in the accessions are plant families, genera and species common to tropical regions throughout the world, and as such, the reference collection and database have relevance to Indonesia, the wider Indo/Pacific region, and tropical regions elsewhere. By incorporating a broad range of economically and ecologically important plants and plant groups hitherto unexamined, and by treating plant parts as separate entities, it provides for rigorous comparative analyses of starch granules, thereby building onto our knowledge about the range and variation of morphologies and their diagnostic value for palaeobotanical analysis. The first stage of analysis involves recording descriptors and images of starch granules. This is nearing completion for a number of economically important plant families including the Araceae, Dioscoreaceae, Heliconiaceae, Musaceae, Zingiberaceae, Palmae and Pandanaceae (see Table 3). Analysis of other plant families is in progress. Following initial qualitative analyses, morphometric analysis using quantitative statistical methods will be undertaken. This will be on an ad hoc basis as the need arises, giving precedence to plant groups relevant to specific research problems.

The comparative modern starch reference collection is a growing archaeobotanical resource and will be expanded as new accessions are obtained. A record of accessions and work in progress is included in unpublished reports presented to the Pacific Biological Foundation (Lentfer 2008). I thank Tom Loy's enterprise, his long-standing and strong interest in ancient starch research and the groundwork he laid for the establishment of the Archaeological Microscopy Laboratory in the School of Social Science at the University of Queensland where the bulk of the analytical work for the 'Indonesian Starch Project' is being undertaken.

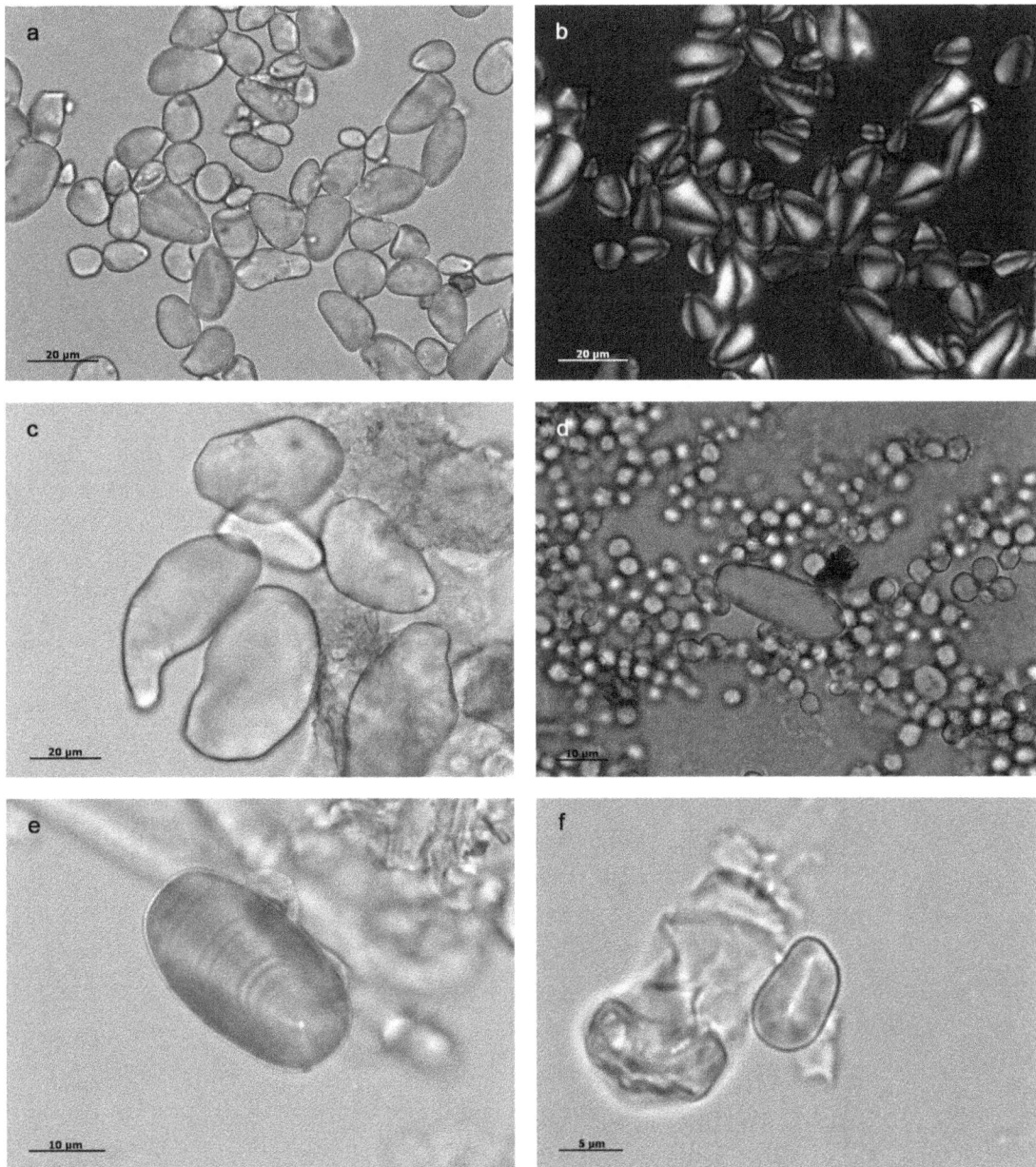

Figure 9. (a and b) *Musa banksii* **(WNB10fruit). (c)** *Musa acuminata x schizocarpa* **(ES10fruit). (d)** *Musa acuminata* **var.** *cerifera* **(CL05190fruit) produces two distinctive starch granule forms in the fruit; a larger, ovate to elongate ovate, eccentric type and a comparatively much smaller spherical to sub-spherical, centric type. (e and f) Two starch types present in** *Musa acuminata* **(CL_F/03/96leaf); cf.** *Heliconia dasyantha* **seed (Figure 8c-d). (Scale bars in a, b and c = 20 μm; d and e = 10 μm; f = 5 μm).**

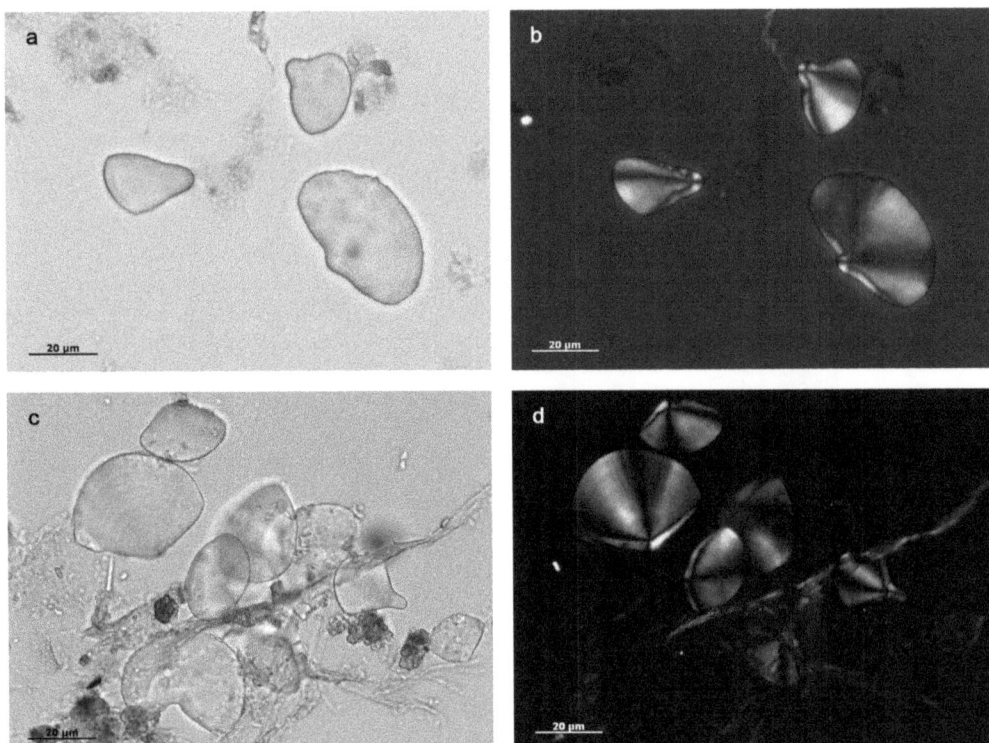

Figure 10. (a and b) *Musa halabanensis* (CL05194corm) and (c and d) *Musa sanguinea* both produce similar types of very distinctive, 'conical-bivalve' granules which may be diagnostic (scale bars = 20 μm).

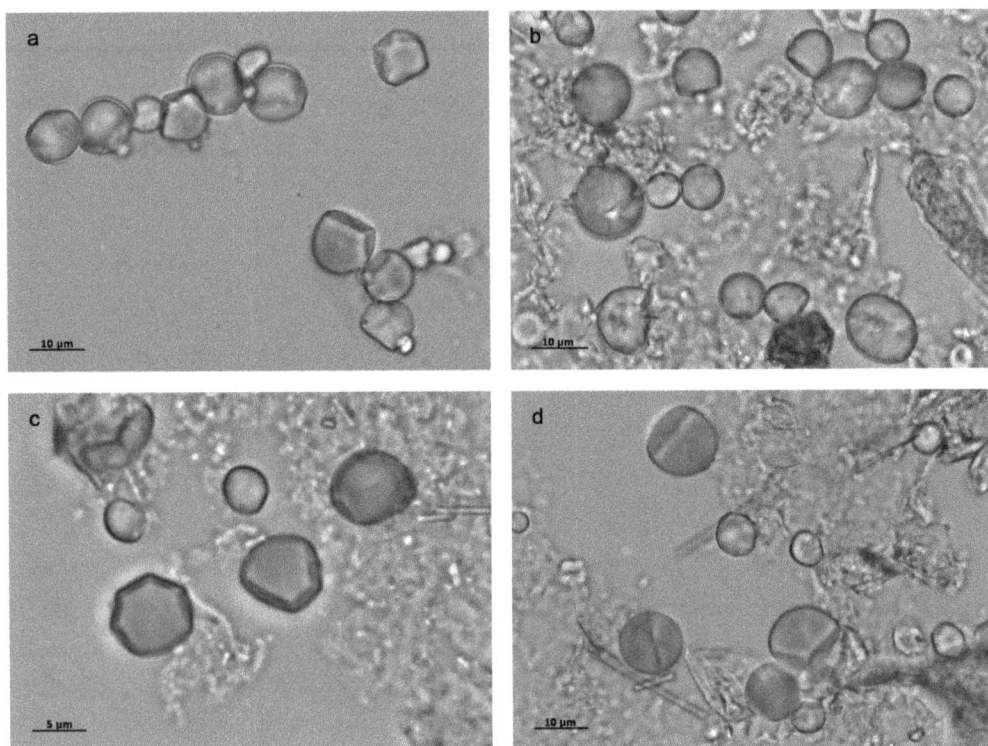

Figure 11. (a) *Pandanus dubius* (CL05341wood). (b) *Pandanus* sp. maluka (CL05140wood). (c and d). *Pandanus furcatus* (CL05131wood) (scale bars in a, b and d = 10 μm; c = 5 μm).

Table 3. Plants and plant parts that have been examined in the first stage of the analysis

Accession No.	Species	Plant Part
Araceae		
7075	*Acorus calamus*	lf,st,rz,rt
5016	*Aglaonema marantaifolia*	pet,lf,st,rt
5038	*Aglaonema novoguineense*	pet,lf,st,rt
5013	*Aglaonema oblanceotam*	pet,lf,st,rt
5014	*Aglaonema pictum*	pet,lf,st,rt
5015	*Aglaonema simplex*	pet,lf,st,rt
NC	*Aglaonema sp.*	lf,cm,rt
5022	*Alocasia alba*	lf,st,rt
48	*Alocasia culcullata*	st
18	*Alocasia denudata*	st,cm
5020	*Alocasia gigantea*	lf,cm,sd
5021	*Alocasia macrorrhizos* var. *diversifolia*	lf
5023	*Alocasia macrorrhizos* var. *diversifolia*	pet,st,fl
NC	*Alocasia macrorrhizos*	lf,st,cm,rt
58	*Alocasia princeps*	cm
NC	*Alocasia* sp. 'elephant ear'	st,rt
5019	*Amorphophallus blumei*	pet,lf,st,rt,cm,sd
6915	*Amorphophallus campunulatus*	cm
5018	*Amorphophallus variabilis*	lf,st,rt,cm,sd
5027	*Anchomanes hookeri*	pet,lf,cm,rt
NC	*Anthurium* sp.	lf,st,rt
91	Unknown	cm
5029	*Colocasia esculenta*	pet,lf,cm,rt
PM1	*Colocasia esculenta*	cm
53	*Colocasia esculenta*	cm
92	*Colocasia esculenta*	cm
6990	*Colocasia esculenta*	cm
NC	*Colocasia esculenta*	cm
5042	*Culcasia manii*	pet,lf,st,rt
694	*Cyrtosperma chamissonis*	cm
5089	*Cyrtosperma johnstonii*	pet,lf,rt,fl
5011	*Dieffenbachia amoena*	pet,lf,st,rt
5012	*Dieffenbachia arvida*	pet,lf,st,rt
NC	*Dieffenbachia exotica*	lf,st,cm,rt
5010	*Dieffenbachia howmanii*	pet,lf,st,rt
5009	*Dieffenbachia splendens*	pet,lf,st,rt
5028	*Dracontium gigas*	pet,cm,rt
5032	*Homalomena cordata*	pet,rt
5033	*Homalomena cordata*	pet,lf,st,rt
NC	*Homalomena* sp.	lf,st
5037	*Homalomona humilis*	pet,lf,st,rt,fl
5039	*Homalomona pendula*	pet,st,rt
NC	*Monstera deliciosa*	lf,rt
5030	*Philodendron gloriosum*	pet,lf,st,rt
NC	*Philodendron selloum*	lf,cm,rt
NC	*Raphidophora* sp.	lf,st,rt
5031	*Rhodospathia latifolia*	pet,lf,st
5034	*Schismatoglottis lancifolia*	pet,lf,st,rt,fr,
5035	*Schismatoglottis neoguineensis*	pet,st,rt
5036	*Schismatoglottis pumila*	pet,lf,st,rt
5040	*Schismatoglottis rupestrius*	pet,lf,st,rt,fl
5045	*Scindapsus pictus*	pet,lf,st,rt
5043	*Spathiphyllum commulatum*	pet,lf,st,rt,fl

NC	*Spathiphyllum* sp.	lf,cm,rt
NC	*Spathiphyllum* sp.	lf,tb,rt
NC	*Syngonium* sp.	lf,rt,st
5026	*Xanthosoma hastifolium*	pet,lf,st,cm,rt
5024	*Xanthosoma nigrum*	st,cm
5025	*Xanthosoma sagittifolium*	lf,st,pet
692	*Xanthosoma sagittifolium*	cm
Dioscoreaceae		
698	*Dioscorea alata*	tb
5088	*Dioscorea alata*	lf,st,pet,tb
7062	*Dioscorea alata*	lf
5086	*Dioscorea bulbifera*	pet,lf,st,blbl,tb,rt
6914	*Dioscorea bulbifera*	tb
697	*Dioscorea esculenta*	lf,st,pet,tb
6918	*Dioscorea esculenta*	tb
696	*Dioscorea esculenta*	tb
5087	*Dioscorea hispida*	pet,tb,rt
7060	*Dioscorea hispida*	pet,tb,sk
6913	*Dioscorea nummularia*	st
6912	*Dioscorea pentaphylla*	st
6910	*Dioscorea rotundata*	st
F/03/22	*Dioscorea* sp.	lf,st,tb
F/03/17	*Dioscorea* sp.	lf,st,tb
F/03/04	*Dioscorea* sp.	lf,tb
NC	*Dioscorea* sp.?	tb
6911	*Dioscorea trifida*	st
Heliconiaceae		
5197	*Heliconia caribea*	lf,psst,rt,fl
5196	*Heliconia collinsiana*	lf,psst,cm,rt
5288	*Heliconia dasyantha*	lf,psst,cm,fl,fr,sd
5287	*Heliconia indica*	lf,cm,rt,fl,fr
Musaceae		
MB1	*Ensete glaucum*	fr,sd
F/03/13	*Musa* AAA	psst,cm,rt,fr
F/03/96	*Musa acuminata*	lf,st
ES6	*Musa acuminata* ssp. *banksii*	fr
M7	*Musa acuminata* ssp. *banksii*	lf,psst,cm,fl,fr,sk,sd
ES11	*Musa acuminata* ssp. *banksii*	lf,psst,cm,fr,sk,sd
M8	*Musa acuminata* ssp. *banksii* x *schizocarpa*	lf,br,fl,fr,sk,sd
5190	*Musa acuminata* var. *cerifera*	lf,psst,cm,rt,fl,fr
ES10	*Musa acuminata* var. *schizocarpa*	fr,sd
ES10	*Musa acuminata* x *schizocarpa*	br,psst,fr,sk,sd
WNB10	*Musa banksii*	fr,sk,sd
5195	*Musa borneensis*	lf,psst,cm
NI14	*Musa* Fe'i	fr,sd
ES5	*Musa* Fe'i	fr
5194	*Musa halabanensis*	lf,psst,cm
WH1	*Musa ingens*	lf,rs,psst,rt,fl,fr
WH2	*Musa ingens*	psst,fl,fr,sk,sd
WH3	*Musa ingens*	sd
MB3	*Musa maclayi*	lf,psst,fr,sd
MB5	*Musa maclayi*	lf,br,psst,fl,fr,sk
5193	*Musa ornata*	lf,rt,cm
M5	*Musa peekeli*	lf,br,psst,sk,sd
M5	*Musa peekelii* ssp. *angustigemma*	fr
5192	*Musa salaccensis*	lf,psst,cm,rt
5191	*Musa sanguinea*	lf,psst,cm,rt,fl,fr
M4	*Musa schizocarpa*	lf,psst,fl,sd

MB2	*Musa schizocarpa*	lf,br,psst,fl,fr,sk
ES3	*Musa schizocarpa*	fr,sk
ES5	*Musa schizocarpa*	sd
ES4	*Musa schizocarpa*	fr,sk,sd
7106	*Musa* sp.	pet,lf,psst,rt
NC	*Musa* sp. 'halevudi'	lf,psst
5175	*Musa velutina*	lf,rt,fr,sk,sd
F/03/90	*Musa velutina?*	rt,fl,fr
NC	*Strelitzia reginae*	fr
Palmae		
7010	*Areca catechu*	fr
5162	*Arenga brevipes*	lf,wd,rt
5149	*Arenga obtusifolia*	rs
5156	*Arenga pinnata*	wd,rt
7013	*Arenga pinnata*	fr,sd
F/03/24	*Arenga pinnata?*	wd
5154	*Arenga tremula*	lf,rs,wd
5155	*Arenga undulatifolia*	lf,wd,rt
5187	*Borassus flabellifer*	wd,rt,sd
F/02/38	*Calamus javensis*	st
5158	*Caryota cummungli*	rs,wd,ped,fr,rt
5179	*Caryota intis*	rs,rt,sd
5160	*Caryota rumphiana*	wd
5177	*Cocos nucifera*	rs,wd
7097	*Corypha utan*	wd
5157	*Cyrtostachys microcarpa*	lf,rs,wd,rt
5184	*Cyrtostachys renda*	wd
5178	*Elaesis guineensis*	fr
5150	*Eugeissona utilis*	lf,rs,wd,rt,fr
5161	*Licuala rumphii*	rt
5183	*Licuala spinosa*	rs,wd,rt
5148	*Livistonia hasseltii*	sd,st
5176	*Metroxylon sagu*	pet,lf,wd,rt
5186	*Nypa fruticans*	lf,rs,fr
5159	*Oncosperma tigillarium*	rs,rt
5151	*Phoenix* sp.	rs,wd
5145	*Pinanga coronata*	wd,rt,fr
7102	*Pinanga coronata*	pet,lf,ped,wd,bk
5146	*Ptychosperma macarthii*	lf,wd,rt,fl
5153	*Ptychosperma propinquum*	wd
5182	*Salacca affinis*	lf,rs,wd,rt
5168	*Salacca zalacca*	sd
5181	*Salacca zalacca*	rs,wd,fr
5147	*Verschaffeltia splendida*	wd
Pandanaceae		
7084	*Freycinetia insignis*	lf,wd,fr
F/02/54	*Freycinetia* sp.	lf
5136	*Pandanus affinis*	lf,wd,rt
5341	*Pandanus dubius*	lf,wd
5131	*Pandanus furcatus*	lf,pet,wd,rt
5143	*Pandanus kurzii*	rt
5135	*Pandanus multifurcatus*	pet,lf,st,wd
5123	*Pandanus nitidus*	pet,lf,st,rt
5124	*Pandanus papilio*	pet,lf,st,rt
5142	*Pandanus polycephalus*	pet,lf,wd,fr,sd
5139	*Pandanus pygmaeus*	lf,wd,rt
5145	*Pandanus* sp.	lf,wd,rt,fr,sd
5140	*Pandanus* sp.	lf,wd,rt

5137	*Pandanus* sp.	pet,lf,wd,rz,fr
5120	*Pandanus* sp.	pet,lf,st
5134	*Pandanus* sp.	pet,lf,wd,rt,fr,sd
5121	*Pandanus* sp.	pet,lf,st,rt,rt
5144	*Pandanus* sp.	pet,lf,st,wd
F/03/92	*Pandanus* sp.	rt
F/03/067	*Pandanus* sp.	st
5333	*Pandanus* sp.	pet,lf,st,rt,fr,sd
7099	*Pandanus* sp.	lf,st,wd,rt
5122	*Pandanus* sp.	pet,lf,st,rt
5132	*Pandanus* sp.	pet,lf,wd,rt
5138	*Pandanus tectorius*	lf,wd,rz,rt
7004	*Pandanus tectorius*	lf
F/03/02	*Pandanus* sp.	rt
Zingiberaceae		
5063	*Alpinia katsumadai*	pet,st,rz,fr,sd
5063	*Alpinia katsumadai*	lf
5059	*Alpinia romburghiana*	pet,lf,st,rz,rt
5052	*Alpinia schumannia*	pet,lf,st,rz,fl
5052	*Alpinia schumannia*	ped
5060	*Alpinia speciosa/zerumbet*	st,rz
5057	*Amomum aculeatum* var. *sulianum*	pet,st,rz,rt,fl
5211	*Amomum compactum*	st
5066	*Amomum compactum*	pet,rz,rt
5053	*Amomum lappaceum*	pet,lf,rz
5050	*Amomum meglalochellos*	pet,lf,st,rz,fl
5065	*Amomum truncatum*	st,rz
5062	*Catimbium malaccensis*	st,rz
5064	*Catimbium speciosum*	lf,st,rz
5075	*Costus afer*	lf,st,rz,rt
5079	*Costus discolor*	st,rz
5051	*Costus lucanusianus*	pet,st,rz,rt,fl
5077	*Costus malortianus*	pet,lf,st,rz,rt,fl
5080	*Costus megalobrachtea*	rt
5048	*Costus mexicanus*	pet
5073	*Costus niveus*	rz
5073	*Costus niveus*	pet,st
5074	*Costus rumphianus*	st,rz
5078	*Costus speciosus*	rz,rt
5076	*Costus spiralis*	fl
5079	*Costus villosissimus*	pet,lf,st,rz,rt
F/03/96	*Cucurma* sp.	lf,st,rz,rt
7063	*Cucurma viridiflora*	pet,st,rz,rt
7073	*Cucurma zedoaria*	lf,rz,rt
5055	*Curcuma aeroginosa*	pet,lf,rz,rt
5061	*Curcuma longa*	pet,sr,rz,rt
5056	*Curcuma longa*	st,rz,rt
5083	*Curcuma roscoeana*	lf,st
5205	*Curcuma zedoaria*	pet,rt
5069	*Etlingera elattior*	st,rz
7074	*Globba marantina*	lf,st,rz,rt
5047	*Hedychium coronarium* var. *flavescens*	pet,rz,rt
5084	*Hedychium elatum*	pet,st,rz,fl
581	*Hedychium poccineum*	lf,st
5082	*Hedychium poccinium*	pet,rz,rt
5046	*Hedychium roxburghii*	pet,lf,st,rz,rt
5046	*Hedychium roxburghii*	pet,lf,st,rz,rt
5067	*Hornstedtia minor*	pet,lf,st,rz,rt
7059	*Kaempferia galanga*	lf,st,rz

5058	*Languas galanga*	pet,lf,st,rz,rt
7057	*Languas galanga*	lf,st,rz,rt
5071	*Tapeinocheilos ananassae*	lf,st,rz,fl,sd
5070	*Tapeinocheilos punglus*	pt,st,rz
5070	*Tapeinocheilos purgus*	pet,lf,rz,rt
5200	*Zingiber aromaticum*	pet,lf,st,rz,rt
5054	*Zingiber gramineum*	pet,rz
5054	*Zingiber gramineum* var. *robustum*	pet,lf,st,rz,rt
7049	*Zingiber oderatum*	pet,lf
5049	*Zingiber odoriferum*	pet,lf,st,rz,rt
NC	*Zingiber officinale*	rz
5201	*Zingiber ottensi*	rz,rt
7077	*Zingiber* sp.	lf,rz,rt
F/03/32	Unknown	rz
WNB	Unknown	fl,rz

KEY: br (bract), blbl (bulbil), cm (cm), fl (flower, fr (fruit), lf (leaf), ped (peduncle),pet (petiole), psst (pseudostem), rs (rachis), rt (root), rz (rhizome), sd (seed), sk (skin), st (stem), tb (tuber)

ACKNOWLEDGMENTS

This project has been in joint collaboration with Dr Netty Polhaupessy from the Geological Research and Development Centre, Bandung, Indonesia. Funding for the project has been provided by the Pacific Biological Foundation and supplemented with funding from the Australian Research Council. I thank the Geological Research Development Centre for their support, the Bogor Botanic Gardens, and a number of Australian Universities including the Southern Cross University (Lismore, NSW), the University of Queensland (Brisbane, Qld.) and the University of New England (Armidale, NSW). Other Institutions lending support for this project include Arkenas (Jakarta), the Provincial Government of the Eastern Lesser Sunda Islands, the local governments at Ruteng, Labuan Bajo, and Reo, and the Indonesian Conservation Department. I am indebted to all the people who helped us collect plants and who conveyed their knowledge of plants to us for our records. I greatly appreciate the assistance of Dr Netty Polhaupessy, Professor Soejono, Mr. Abraham G., Thomas Sutikna, Rokus Awe Due, Wahyu Saptomo, Jatmiko and Stefanis from Hotel Sindah, Ruteng. I greatly appreciate the assistance provided by Dr Huw Barton, especially pertaining to formulation of the diagnostic attributes for starch granules. I also thank Professor Mike Morwood, Associate Professor Bill Boyd and Jacqui Collins and Robert Neal for their kind assistance. Finally, I thank Alison Crowther, Cassandra Venn, Heath Anderson and Jasmine Murray for all their hard work in the laboratory.

REFERENCES

Bailey, R. C., M. Jenicke and R. Rechtman 1991. Reply to Colinvaux and Bush. *American Anthropologist* 91:59-82.

Barton, H. and V. Paz 2007. Subterranean diets in the tropical rainforests of Sarawak, Malaysia. In T. P. Denham, J. Iriarte and L. Vrydaghs (eds) *Rethinking Agriculture: archaeological and ethnoarchaeological perspectives*, p.50-77. New York: Left Coast Press.

Bellwood P. 1996. The origins and spread of agriculture in the Indo-Pacific region: gradualism and diffusion or revolution and colonisation. In D. R. Harris, (ed.) 1996. *The Origins and spread of Agriculture and Pastoralism in Eurasia*, pp. 465-498. London: UCL Press.

Bellwood, P. 1997. *Prehistory of the Indo-Malaysian Archipelago*. Honolulu: University of Hawaii Press.

Brown, P., T. Sutikna, M. J. Morwood, R. P. Soejono, Jatmiko, E. Wayhu Saptomo and Rokus Awe Due 2004. A new small-bodied hominin from the Late Pleistocene of Flores, Indonesia. *Nature* 431:1055-1061.

Crowther, A. 2005. Starch residues on undecorated Lapita pottery from Anir, New Ireland. *Archaeology in Oceania* 41:62-66.

De Langhe, E. and P. De Maret 1999. The banana: its significance in early agriculture. In C. Gosden and J. Hather (eds.), *The Prehistory of Food. Appetites for Change*, pp. 377-396. London: Routledge.

Denham, T., S. Haberle and C. Lentfer 2004. New evidence and revised interpretations of early agriculture in Highland New Guinea. *Antiquity* 78 (302):839-857.

Denham, T., S. G. Haberle, C. Lentfer, R. Fullagar, J. Field, N. Porch, M. Therin, B. Winsborough and J. Golson 2003. Multi-disciplinary evidence for the origins of agriculture from 6950-6440 cal. BP at Kuk Swamp in the Highlands of New Guinea. *Science* 301:189-193.

Dentan, R. K. 1991. Potential food sources for foragers in Malaysian rainforest: Sago, yams and lots of little things. *Journal of the Royal Institute of Linguistics and Athropology* 147:420-444.

Dickau, R., A. J. Ranere and R. G. Cooke 2007. Starch grain evidence for the preceramic dispersals of maize and root crops into tropical dry and humid forests of Panama. *Proceedings of the National Academy of Sciences* 104(9):3651-3656.

Englyst, H. N., S. M. Kingman and J. H. Cummings 1992. Classification and measurement of nutritionally important starch fractions. *European Journal of Clinical Nutrition* 46:S33-S50.

Field, J. and B. Gott 2006. Compiling a reference collection for studying Pleistocene grinding stones. In R. Torrence and H. Barton (eds) *Ancient Starch Research*, pp. 105-106. Walnut Creek, California: Left Coast Press.

Fullagar, R. 2006. History of starch research on stone tools. In R. Torrence and , H. Barton (eds) *Ancient Starch Research*, pp. 181-182. Walnut Creek, California: Left Coast Press.

Fullagar, R., J. Field, T. Denham and C. Lentfer 2006. Early and mid Holocene processing of taro (*Colocasia esculenta*), yam (*Dioscorea* sp.) and other plants at Kuk Swamp in the Highlands of Papua New Guinea. *Journal of Archaeological Science* 33:595-614.

Glover, I. 1986. *Archaeology in Eastern Timor, 1966-67*. Terra Australis 11. Canberra: Department of Prehistory, Australian National University.

Glover, I and C. F. W. Higham 1996. New evidence for early rice cultivation in south, southeast and east Asia. In D. R. Harris (ed.) 1996. *The Origins and spread of Agriculture and Pastoralism in Eurasia*, pp. 412-441 London: UCL Press.

Gott, B., H. Barton, D. Samuel and R. Torrence 2006. Biology of starch. In R. Torrence and H. Barton (eds), *Ancient Starch Research*, pp. 35-45. Walnut Creek, California: Left Coast Press.

Harris, D. R. (ed.) 1996. *The Origins and Spread of Agriculture and Pastoralism in Eurasia*. London: UCL Press.

Haslam, M. 2004. The decomposition of starch grains in soils: implications for archaeological residue analyses. *Journal of Archaeological Science* 31:1715-34.

Helena, C, C. Boyadjian, S. Eggers and K. Richard 2007. Dental wash a problematic method for extracting microfossils from teeth. Journal of Archaeological *Science* 34(10):1622-1628.

Horrocks, M., G. Irwin, M. Jones and D. Sutton 2004. Starch grains and xylem cells of sweet potato (Ipomoea batatas) and bracken (Pteridium esculentum) in archaeological deposits from northern New Zealand. *Journal of Archaeological Science* 31:251-258.

Hutterer, K.L., 1985. The Pleistocene archaeology of Southeast Asia in regional context. In G-J. Bartstra and W.A. Casparie (eds) *Modern Quaternary Research in Southeast Asia: Papers Read at Symposium 1 12th Congress of the Indo-Pacific Prehistory Association Philippines, 26th January-2nd February 1985*, pp. 1-23. Rotterdam: A.A. Balkema,.

Jiang, Q. 1995. Searching for evidence of early rice agriculture at prehistoric sites in China through phytolith analysis: an example from central China. *Review of Palaeobotany and Palynology* 89:481-485.

Juan-Tressaras 1998. La cerveza prehistorica: Investigaciones aqueobotanicas y y experimentales. In J.L. Maya, F. Cuesta and J. L. Cachero (eds) *Geno: Un Poblado del Bronce Final en el Bajo Segre (Lleida)*, pp. 239-252. Barcelona: Universitat de Barcelona.

Lebot,V. 1999. Biomolecular evidence for plant domestication in Sahul. *Genetic Resources and Crop Evolution* 46:619-628.

Lentfer, C. J. 2008. Building a Starch Reference Collection for Southeast Asia. Unpublished Report, Pacific Biological Foundation, Port Macquarie, N.S.W., Australia.

Lentfer, C. J. and R. C. Green 2004. Phytolith evidence for the terrestrial plant component at the Lapita Reber-Rakival site on Watom Island, Papua New Guinea. In V. Attenbrow and R. Fullagar (eds) *A Pacific Odyssey. Papers in Honour of Jim Specht. Records of the Australian Museum Museum. Supplement* pp. 75-88. Sydney: Australian Museum.

Lentfer, C. J., M. Therin and R. Torrence 2002. Starch grains and environmental reconstruction: A modern test case from West New Britain, Papua New Guinea. *Journal of Archaeological Science* 29:687-698.

Lentfer C. and M. Therin 2006. Collecting starch in Papua New Guinea. In R. Torrence and H. Barton (eds) *Ancient Starch Research*, pp. 99-101. Walnut Creek, California: Left Coast Press.

Loy, T. H., M. Spriggs and S. Wickler 1992. Direct evidence for human use of plants 28,000 years ago: starch residues on stone artifacts from the northern Solomon Islands. *Antiquity* 66:898-912.

Matthews, P. and K. W. Naing. 2005. Notes on the provenance of wildtype taros (Colocasia esculenta) in Myanmar. *Bulletin of National Museum of Ethnology* 29 (4):587-615.

Morwood, M.J., R. P. Soejono, R. G. Roberts, T. Sutikna, C. S. M. Turney, K. E. Westaway, W. J. Rink, J.- x. Zhao, G. D. van den Bergh, Rokus Awe Due, D. R. Hobbs, M. W. Moore, M. I. Bird & L. K. Fifield, 2004. Archaeology and age of a new hominin from Flores in eastern Indonesia. *Nature* 431:1087-1091.

O'Sullivan, P.B., M.J. Morwood, D. Hobbs, F. Aziz, Suminto, M. Situmorang, A. Raza and R. Maas, 2001. Archaeological implications of the geology and chronology of the Soa basin, Flores. *Indonesian Geology* 29(7):607–610.

Parr, J. F. and M. Carter 2003. Phytolith and starch analysis of sediment samples from two archaeological sites on Dauar Island, Torres Strait, Northeastern Australia. *Vegetation History and Archaeobotany* 12:131-141.

Paz. V. 2001. Archaeobotany and Cultural Transformation: Patterns of early plant utilisation in northern Wallacea. Unpublished PhD thesis. Cambridge: University of Cambridge.

Perry, l. 2004. Starch analyses reveal the relationship between tool type and function: An example from the Orinoco valley of Venezuela. *Journal of Archaeological Science* 31:1069-1081.

Piperno, D, R., Weiss, E., Holst, I. and Nadel, D. 2004. Processing of wild cereal grasses in the Upper Palaeolithic revealed by starch grain analysis. *Nature* 430:670-673.

Reichert, E. T. 1913. *The Differentiation and Specificity of Starches in Relation to Genera, Species, etc.* Washington DC: Carnegie Institution of Washington.

Samuel, D. 1996. Investigations of Ancient Egyption baking and brewing methods by correlative microscopy. *Science* 273:488-490.

Sauer, C.O. 1953. *Agricultural Origins and Their Dispersals*. New York: American Geographical Society.

Sémah, F., A. M. Sémah, and T. Simanjuntak 2003. More than a million years of human occupation in insular Southeast Asia: the early archaeology of eastern and central Java. In J. Mercader (ed.) *Under the Canopy: The Archaeology of Tropical Rain Forests*, pp. 161-190. New Brunswick: Rutgers University Press.

Simmonds, N. W. 1962. *The Evolution of Bananas*. London: Longmans.

Spriggs, M. 1996. What is Southeast Asian about Lapita? In T. Akazawa and Szathmary (eds) *Prehistoric Mongoloid Dispersals*, pp. 324-348. Oxford: Oxford University Press.

Torrence, R. and H. Barton (eds) 2006. *Ancient Starch Research*. Walnut Creek, California: Left Coast Press.

Therin, M., R. Fullagar and R. Torrence 1999. Starch in sediments: a new approach to the study of subsistence and land use in Papua New Guinea. In C. Gosden & J. Hather (eds) *Prehistory of Food*, pp. 438-462. London: Routledge.

Ugent, D., T. Dillehay, and C. Ramírez 1987. Potato remains from a Late Pleistocene settlement in southcentral Chile. *Economic Botany* 41:17-27.

Yen, D. E. 1990. Environment, agriculture and colonization of the Pacific. In D. E. Yen and J. M. J. Mummery (eds) *Pacific Production Systems: Approaches to economic prehistory, pp. 258-277. Occasional Papers in Prehistory* 18. Canberra: Department of Prehistory, Research School of Pacific Studies, Australian National University.

Zarrillo, S., D. M. Pearsall, J. S. Raymond, M. A. Tisdale and D. J. Quon 2008. Directly dated starch residues document early formative maize (Zea mays L.) in tropical Ecuador. *Proceedings of the National Academy of Sciences* 105:5006-5011.

8

MORPHOMETRIC ANALYSIS OF CALCIUM OXALATE RAPHIDES AND ASSESSMENT OF THEIR TAXONOMIC VALUE FOR ARCHAEOLOGICAL MICROFOSSIL STUDIES

Alison Crowther[1,2]

1. School of Social Science
The University of Queensland
St Lucia QLD 4072 Australia

2. Department of Archaeology
University of Sheffield
Northgate House, West Street
Sheffield S1 4ET United Kingdom
Email: alison.crowther@gmail.com

ABSTRACT

Plant microfossil analysis is an important tool for understanding prehistoric plant exploitation in the Pacific region, where macrobotanical remains rarely survive. Calcium oxalate raphides (needle-shaped plant crystals) have been identified in several tool-residue and sediment microfossil studies and in some cases classified to genera or species, yet their diagnostic potential has not been thoroughly assessed through quantitative reference collection analyses. This study presents a morphometric analysis of raphides from key Pacific economic plants including aroids, yams, palms, pandanus, banana (Musa), *Heliconia* and *Cordyline*, using both light and scanning electron microscopy. Variation in raphide size and shape was compared intra- and inter-taxa to assess whether significant differences exist. Results indicate that raphide size varies considerably within species, while similar shapes are found in a number of taxa, leaving little scope for taxonomic differentiation. The main exceptions are raphides produced by members of the Araceae (aroids), which may be identified at the genus level owing to their unique morphologies. These results suggest that raphides have some potential to be a useful taxonomic tool in archaeological micro- and macrobotanical studies, particularly if *Colocasia esculenta* (taro) and other aroids are present.

KEYWORDS

raphides, calcium oxalate crystals, microfossils, Pacific archaeobotany, microscopy, morphometric analysis

INTRODUCTION

Reconstructions of the plant exploitation and cultivation practices of prehistoric Pacific peoples have become increasingly reliant on studies of microbotanical remains recovered from site sediments and tool-use residues. Although analyses of starch granules, phytoliths and pollen are driving these developments (e.g. Fullagar *et al.* 2006; Lentfer and Green 2004; Therin *et al.* 1999), previously overlooked microfossil types such as calcium oxalate raphides are now receiving wider attention (e.g. Crowther 2005, in press; Horrocks and Barber 2005; Horrocks and Bedford 2005; Horrocks and Nunn 2007; Horrocks and Weisler 2006; Horrocks *et al.* 2007, 2008a, 2008b).

Raphides are needle-shaped calcium oxalate crystals that are produced by higher plants for defence, calcium storage and structural strength (Franceschi and Horner 1980; Franceschi and Nakata 2005; Nakata 2003). Owing to their inorganic properties, they have potential to be preserved in archaeological sites where other plant remains, particularly soft tissues, generally do not survive. Of significance to Pacific archaeobotany is the fact that raphides are highly abundant in aroids (Araceae) such as taro (*Colocasia esculenta*), swamp taro (*Cyrtosperma chamissonis*) and elephant ear taro (*Alocasia macrorrhiza*), which are major food staples across the region. Raphides have therefore become an important indicator of aroids in the archaeological record, and have been identified to this group in several microfossil and tool-residue studies (see references cited above). These studies have rarely accounted for the fact that raphides are also produced by many other economically important plants, including the yams, palms, bananas, and pandanus, to name just a few (see review by Crowther in press). If raphides are to continue playing a role in Pacific archaeobotanical studies, their diagnostic potential must be assessed and reliable classification keys must be established.

The diagnostic value of raphides was first demonstrated by Loy *et al.* (1992; see also Loy 1994), who investigated whether their presence in association with starch granules on stone tools could aid in residue identification. They determined that raphides are particularly abundant in aroids and either much rarer or absent in the other plants analysed (including various yams, sago, cycad and sweet potato) (Loy *et al.* 1992:910). Even though this observation was based on the analysis of a limited range of reference materials, the presence of large quantities of raphides in residue or microfossil assemblages has been interpreted in several studies since as indicating the probable presence of aroids (e.g. Crowther 2005; Horrocks and Bedford 2005; Horrocks and Nunn 2007; Horrocks and Weisler 2006; Horrocks *et al.* 2008b). Quantitative comparisons of raphide concentrations between economic plants are required to assess whether abundance has any diagnostic value in these cases.

Secondly, Loy *et al.* (1992) noted that the size, cross-section and termination shape of raphides produced by different species varied so as to be 'seemingly distinctive ... to genus' (Loy *et al.* 1992:900-910). The specific configuration of these morphological attributes were only briefly described (Loy *et al.* 1992:900, 905-906; see summary in Table 1), and have rarely been incorporated into more recent raphide identifications (cf. Fullagar *et al.* 2006). Finally, Loy *et al.* (1992:900) observed what they referred to as 'whisker' raphides in aroids, which were defined loosely as very thin, sometimes curved raphides that possibly represent an immature form. Bradbury and Nixon (1998) also described 'thick' (defined as ~ 3 μm wide) and 'thin' (~ 0.5 μm wide) raphides within some aroids. Whisker raphides are one of the most common raphide types observed in microfossil assemblages from the Pacific region and have been interpreted in all cases as indicating the presence of aroids (e.g. Crowther 2005; Horrocks and Bedford 2005; Horrocks and Barber 2005; Horrocks and Nunn 2007; Loy *et al.* 1992), even though it is presently unclear if they represent a distinctive aroid type. Although the findings of Loy *et al.* (1992) were essentially preliminary, in that they were based on light microscopic observations of raphides from only the storage organs (i.e. roots and tubers) of a limited range of taxa, they nevertheless highlighted the potential for differentiating raphides based on morphometric characteristics.

Subsequent studies have discussed limitations imposed on raphide analyses by the often fragmented and degraded state of archaeological crystals (e.g. Crowther 2005:64-65; Horrocks and Bedford 2005:70-71). For example, morphologically distinctive features such as terminations may not be present on archaeological raphides and size can be difficult to estimate from broken crystals. While cross-sectional shape has been observed in some studies using light microscopy (e.g. Fullagar *et al.* 2006:605; Loy *et al.* 1992), raphides are often too small for such features to be resolved unambiguously with a light microscope (see also discussion in Horrocks and Nunn

Table 1. Summary of Loy *et al.*'s (1992) description of raphides from various aroid and yam species

Species	Termination		Cross-section	Other descriptions
	End A	End B		
Aroids (general)	ns[a]	ns	Square	Small "whisker" raphides.
Alocasia macrorrhiza	Pointed	Pointed	Square	Elongate
Amorphophallus campanulatus	Pointed	Square	'X'-shaped at pointed end	Asymmetric
Colocasia affinis	ns	ns	ns	Small and simple raphides.
Colocasia esculenta	Sharply tapered	Short double-bevelled	Square	Short, small, simple lath-like.
Colocasia fallax	Blunt	Blunt	ns	Centrally constricted.
Colocasia gigantea	ns	ns	Square	Small, simple, lath-like, whisker raphides > 200 μm long.
Cyrtosperma chamissonis	ns	ns	ns	Large, well developed.
Xanthosoma sagitiffolium	ns	ns	ns	Large, well developed.
Dioscorea esculenta	Pointed	Pointed	Round	Long, smoothly tapered, larger than aroid raphides.
[a] ns – not specified by Loy *et al.* (1992)				

2007:742). One avenue that microfossil analysts have previously overlooked is to use scanning electron microscopy (SEM) for higher resolution imaging of morphology, as discussed below.

To address these issues, a modern reference collection of economically important plants from the Pacific Islands was analysed to assess whether raphides from different plant taxa can be differentiated on the basis of morphometric features. The study aimed not only to provide descriptions of raphides across a range of taxa using both light and scanning electron microscopy, but also to make quantitative statements about the degree of intra- and inter-species variability in size and shape. An additional objective was to clarify morphological descriptions of aroid raphides by Loy *et al.* (1992) and to provide further comment on the possible taxonomic significance of whisker raphides. Ideally, raphides from specimens growing in different environments would also be compared, however such analyses were beyond the scope of this study. The results of the study are briefly compared with other published raphide descriptions (e.g. Bradbury and Nixon 1998; Sakai and Hanson 1974) in order to assess the potential for environmentally-determined variation.

RAPHIDE MORPHOMETRIC VARIATION

Raphides are one of five types of calcium oxalate crystals produced by higher plants. Other crystal types include rosette-shaped druses, pencil-shaped styloids, block-shaped aggregates called crystal sand, and prisms (Horner and Wagner 1995). Raphides form in bundles of tens to thousands of crystals in specialised idioblast cells, from which they can be ejected as a defence mechanism (e.g. Arnott and Pautard 1970; Gardner 1994; Middendorf 1983; Sunell and Healey 1981). They can occur in any plant organ or tissue, including stems, leaves, roots, tubers and seeds (Horner and Wagner 1995).

Current evidence suggests that calcium oxalate crystal production is at least partly genetically controlled (Kausch and Horner 1982) and that the shape and location of crystals within specific tissues may have taxonomic significance (e.g. Bouropoulos *et al.* 2001; Cervantes-Martinez *et al.* 2005; Hartl *et al.* 2007; Horner and Wagner 1995; Lersten and Horner 2000, 2008; Webb 1999). Morphometric studies of individual crystals are limited and have mainly been undertaken on druses within the Cactaceae (e.g. Jones and Bryant 1992; Monje and Baran 2002). These studies have shown that druses produced by different cactus genera may be differentiated based on size and shape, with the latter being the more significant variable (Jones and Bryant 1992; Monje and Baran 2002). Monje and Baran (2002) demonstrated that druse morphologies are determined by chemical composition, so that some cactus groups produce exclusively calcium oxalate monohydrate druses, which have a more stellate appearance owing to their monoclinic crystal system, while others produce calcium oxalate dihydrate druses with typically tetragonal crystallites. Crystal size on the other hand tends to vary more within a single plant, depending on the function of the cell or tissue in which it formed as well as the amount of calcium available during crystal formation (Franceschi and Nakata 2005). Jones and Bryant (1992:234-236) found that in the stem of a single cactus species there was a consistent reduction in the size of druses from the central pith region towards the outer epidermis/hypodermis, and that size was therefore not as useful for taxonomic differentiation as morphology. They also found that druses from the stem tissues of species growing in different habitats were very consistent in size and shape, indicating that these characteristics are more genetically than environmentally determined (Jones and Bryant 1992:236).

Botanical studies report that higher plants produce at least four different morphological raphide types, referred to here as Types I-IV (e.g. Cody and Horner 1983:328; Horner and Tilton 1980; Horner and Wagner 1995; see also Crowther in press). These types are defined based on sectional and termination shape, as shown in Figure 1, as well as crystallographic characteristics such as twinning. Type I is the most common form, and comprises four-sided single crystals with two symmetrical pointed ends. Type II raphides, which are also four-sided, have one pointed and one bidentate or forked end (Prychid and Rudall 1999:726). This type of raphide has so far only been recorded in the Vitaceae (Cody and Horner 1983; Webb 1999). The bidentate end is formed by crystal twinning, which represents a dislocation or change in orientation of the crystal lattice along a plane within the raphide (Arnott 1981; Arnott and Webb 2000:133).

Mature Type III crystals have six to eight sides and symmetrical pointed ends (Tilton and Horner 1980). Owing to their multi-faceted surface, they are sometimes described as appearing elliptical or circular in cross-section when viewed under a light microscope (Prychid and Rudall 1999:726; Loy *et al.* 1992:900). Type III raphides have been reported in the Agavaceae (e.g. *Cordyline* spp.) (Wattendorff 1976), Typhaceae (Horner *et al.* 1981), Hyacinthaceae (Tilton and Horner 1980) and Dioscoreaceae (Crowther in press). Horner *et al.* (1981) demonstrated in a study of raphides from *Typha angustifolia* L. (Typhaceae) that this type of crystal develops in several morphological stages: immature four-sided crystals with chisel-shaped ends pass through a twelve-sided intermediate phase before further growth and extension along different margins produces a mature crystal that is hexagonal at its now pointed ends and octagonal in the central

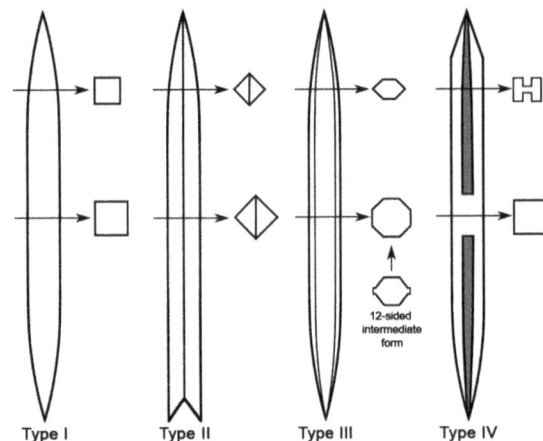

Figure 1. Diagram of the four known raphide morphological types, showing different terminations and cross-sections (after Horner and Wagner 1995:58).

region (see Figure 1). Immature crystals of this type, if present in a microfossil assemblage, may therefore be confused with Type I raphides. Wattendorff (1976:304) noted that Type III raphides have lateral 'wings' at the two smaller-angled edges, owing to unequal angles between the facets. This feature appears to correlate with the intermediate twelve-sided form described by Horner *et al.* (1981), as illustrated in cross-section in Figure 1.

Type IV raphides are twinned crystals with H-shaped cross-sections and asymmetrical ends (e.g. Bradbury and Nixon 1998; Kostman and Franceschi 2000). One end of this raphide type terminates at a sharp point while the other is blunt or wedge-shaped. A medial groove along two of the crystal's opposite faces gives this type its overall H-shaped cross-section, although the solid mid-point at which the groove meets is four-sided and sometimes visible under polarised light. Type IV raphides can also develop barbs or wing-like projections on their pointed ends as they mature (Bradbury and Nixon 1998:614; Prychid and Rudall 1999:726; Prychid *et al.* in press; Sakai *et al.* 1972). The groove in some species (e.g. *Pistia stratiotes*) is asymmetrical, being narrower on one side of the central bridge than the other (Kostman and Franceschi 2000:173). The H-shaped form is thought to be unique to raphides from the Araceae (Prychid and Rudall 1999; Prychid *et al.* in press), which may be of considerable value for identifying aroid crystals in archaeological assemblages. Barbs have however been noted in other taxa, albeit rarely (Cody and Horner 1983:319).

ASSESSING TAXONOMIC SIGNIFICANCE

The simplest method of assigning taxonomic significance to a microfossil is to determine whether any potentially diagnostic morphometric types occur within a plant taxon. For a raphide type to be diagnostic, it must occur in all members of a taxon at a certain level (e.g. family, genus, species) but be absent from all non-members (Madella *et al.* 2005). Madella *et al.* (2005) thereby distinguish *diagnostic* from *observed*, which refers to a type known to be present in, but not necessarily specific to, a particular taxon. If taxonomic significance cannot be assigned to a single morphometric type, the frequencies of different types within an assemblage may have significance (Madella *et al.* 2005; Piperno *et al.* 2000, 2004). Multivariate analyses of a combination of morphometric variables have also been shown to be useful for discriminating taxa, particularly when there appear to be few or only minor differences between species (Torrence 2006:135). Using these methods, for example, starch granules have been confidently differentiated and identified in several comparative and archaeological studies (e.g. Loy *et al.* 1992; Torrence *et al.* 2004). A key assumption of starch analysis, though, is that taxonomically distinctive granules are produced in storage or reserve organs (which are generally also the most economically useful), while other organs such as leaves and stems produce mainly small, non-diagnostic types (e.g. Fullagar *et al.* 2006; Therin *et al.* 1997; Torrence 2006:128). Species identification of starch or other microfossil types would be considerably more difficult if greater intra-species variation occurred, unless it can be determined from which part of the plant the microfossils are derived. The process of establishing a taxonomic key for raphides must therefore begin by assessing whether significant morphometric variation exists within a plant before comparing variation between species.

METHODS AND MATERIALS

Raphides from a total of 20 plant species representing seven families were examined in this study (Table 2). These species were selected because of their likely economic importance during Pacific prehistory (e.g. Barrau 1958; Hather 1992; Massal and Barrau 1965; Whistler 1991; Yen 1974, 1976, 1991), as well as to represent a variety of known raphide-bearing families. Some aroid species that are either recent introductions to the Pacific (e.g. *Xanthosoma sagittifolium*) or not presently regarded as food sources (e.g. *Amorphophallus blumei, Cyrtosperma johnstonii*) were included for comparative purposes. Samples were collected either in Indonesia and Papua New Guinea by Dr Carol Lentfer (The University of Queensland), from gardens at The University

Table 2. List of plants and corresponding organs from which raphides were extracted and examined in this study

Family	Species	Sample Number	Organ
Araceae	*Alocasia macrorrhiza* (L.) G. Don var. rubra (Hassk.) Furtado	B05021	Flower, leaf, petiole, root, stem
	Amorphophallus blumei (Schott.) Engl.	B05019	Bulbil, leaf, petiole, root, tuber
	Amorphophallus campanulatus Blume (syn. paeonifolius)	IV-69-15	Tuber
	Colocasia esculenta (L.) Schott	B05029	Leaf, petiole, root, tuber
	Cyrtosperma chamissonis Schott	IV-69-5	Tuber
	Cyrtosperma johnstonii (Bull) NE Br.	B05089	Flower, leaf, petiole, root, tuber
	Xanthosoma sagittifolium (L.) Schott	B05025	Leaf, petiole, root, stem
Arecaceae	*Caryota mitis* Lour.	B05179	Fruit, leaf, rachis, root, seed, stem
	Caryota rumphiana Blume ex Martelli	B05160	Leaf, rachis, stem, wood
	Cocos nucifera L. cv. Pinang	B05177	Leaf, rachis, root, wood
Dioscoreaceae	*Dioscorea bulbifera* L.	B05086	Bulbil, leaf, root, stem, tuber
	Dioscorea esculenta (Lour.) Burkill	B05085	Leaf, stem, tuber
	Dioscorea hispida Dennst.	B05087	Leaf, petiole, stem, tuber
Heliconiaceae	*Heliconia indica* Lam.	B05287	Flower, fruit, leaf, root, tuber
Laxmanniaceae	*Cordyline terminalis* (L.) Kunth	Local	Leaf, petiole, root, stem
Musaceae	*Musa acuminata* ssp. banksii (F. Muell.) Simmonds	ES6	Bract, fruit[a], leaf, petiole
	Musa ingens Simmonds	WH2	Bract, flower, leaf, petiole, pseudo-stem
	Musa maclayi F. Muell.	MB5	Bract, flower, fruit[a], leaf, petiole
Pandanaceae	*Pandanus dubius* Spreng	B05141	Leaf, root, wood
	Pandanus tectorius Soland. ex Park.	B05138	Leaf, root, wood

[a] Including seed and skin

Sample prefix provenance key:

B – Bogor Botanical Gardens, Indonesia

IV – Loy collection, provenance unknown

ES/WH/MB – Papua New Guinea

of Queensland, Australia by the author, or were from Dr Tom Loy's reference collection held at UQ (see Loy 1994). It should be noted that for the sake of simplicity all vegetative storage organs (e.g. tuber, corm, root and rhizome) are referred to throughout this study as tubers and were collectively analysed as such to facilitate their comparison. Likewise, pseudo-stems were included in the stem group.

Raphides were extracted by placing a small sample of tissue in a 1.5 ml vial, macerating the tissue in 70% ethanol for approximately 48 hours, then mechanically disrupting the tissue with a dental pick to free the raphides (after Webb *et al.* 1995). The ethanol-raphide suspension was agitated and a 50-100 µl aliquot removed to a microscope slide and allowed to air dry. A coverslip was applied and tacked down at each corner with clear nail polish. Slides were hydrated with water and examined in plane and cross-polarised light using an Olympus BX60 transmitted light microscope at magnifications between ×200 and ×1000 without the use of oil immersion.

Maximum length and width measurements for 50 randomly selected whole raphides were recorded per organ for each species. Length was measured to the nearest micron while width was measured to the nearest 0.5 µm. In cases where raphides were rare, multiple slides were prepared and an equal portion was randomly selected from each. If it was still not possible to record 50 using this method, all raphides on these slides were measured without replacement. Morphological features visible with a light microscope, such as termination shape and symmetry, grooves, bridges, facets, or wings were also described. Whisker raphides, which are defined here as raphides of any length with widths less than 0.5 µm, were excluded from the sample as it was generally difficult to determine if they were whole or fragmented owing to their fineness. Rather, each slide was scanned in transects to determine and separately record the maximum size of any whisker raphides present. The relative abundance of raphides in each sample was estimated on a qualitative scale from absent to very abundant (c. > 500 per slide) with intermediate levels of rare (c. < 20 per slide), common (c. < 100 per slide), and abundant (c. < 500 per slide). It is noted that these estimates of abundance are only indicative because sample sizes, including both the amount of tissue sampled from each organ and the amount of residue applied to each slide, were not consistent.

Classification of raphides to morphological type using SEM could only be undertaken on a small number of samples from each family owing to restricted access to instrumentation. All other species in each family were therefore tentatively classified to the same type based on the assumption that form is family-specific (see Horner and Wagner 1995 for discussion). Samples were prepared for SEM by applying a 50 µl aliquot of suspended raphide extract to a glass coverslip on an aluminum stub. The samples were air dried and coated with platinum using an EIKO IB-5 sputter coater for five minutes at 6 mA, which produced a 15 nm coating over the specimen surface. Raphides were examined at the Centre for Microscopy and Microanalysis, The University of Queensland, using a JEOL LV-6460 SEM in high vacuum at an accelerating voltage of 5 kV.

Statistical procedures

A multivariate analysis of variance (MANOVA) was undertaken to assess the degree of variation in raphide size (length and width) within and between plant species. MANOVA tests the null hypothesis (H_o) that all sample means are equal at a chosen confidence level (in this case 95% or where $p = 0.05$). The MANOVA was undertaken to test the experimental hypothesis that significant differences in raphide size occur *between species* but not *within species* in each morphological group. Within species tests were performed by comparing raphide sizes from each organ analysed for a particular plant. *Amorphophallus campanulatus* and *Cyrtosperma chamissonis* were excluded from this analysis because raphides from only a single organ were analysed (the tuber in both cases). Inter-species comparisons were made on an organ-to-organ basis so that, for example, all leaf samples are compared across species within each morphological group. This approach was considered more statistically and taxonomically meaningful than if each organ from each species was treated as an individual case and compared with all other organs of every species, even though fewer comparisons were thus able to be made. Although morphological variables could also have been included in this statistical model, it was decided to perform separate inter-species comparisons for each morphological group instead to facilitate interpretation of the results. Because Type I and immature Type III raphides are morphologically similar, these groups were combined as a single Type I/III group during this analysis in case they were misclassified. Further research is required to determine if there is a size threshold between immature and mature Type III raphides that can be used to differentiate the two groups.

All statistical analyses were performed using SPSS 13.0 for Macintosh. Before undertaking the multivariate analyses, descriptive statistics of the ranges, minimums, maximums, means and standard deviations were calculated for the length and width variables. The statistical soundness of each sample size was determined using the following equation, which calculates that the sample means are within 5% of the actual population means at a 90% confidence level (Madella *et al.* 2005:254; see also Drennan 1996):

$$n_{min} = \left(\frac{\sigma t}{ER} \right)^2$$

where n_{min} = the minimum adequate sample size;

σ = the standard deviation of the sample;

t = t-score at 90% confidence = 1.645; and

ER = the desired error range (in this case 0.05 × the sample mean).

Species were then grouped by morphological type and box-and-whisker plots comparing raphide lengths within each species were prepared to visualise the degree of variation.

The data were examined for conformity to the multivariate test assumptions and any necessary transformations were applied (see Field 2005:593; Pallant 2005:249). To equalize sample sizes, cases where less than 50 raphides were measured owing to rarity were excluded from the analysis. MANOVA is generally robust to violations of normality except in cases where such deviations are caused by outliers (Pallant 2005:249; Tabachnick and Fidell 1996). Multivariate outliers identified by calculating Mahalanobis distances (p > 0.001) were therefore excluded from the analysis (see Pallant 2005:250-251). Finally, Pillai's Trace statistic was used because of its robustness to modest violations of the test assumptions (Field 2005:594, 599). In addition to the multivariate analyses, follow-up univariate ANOVAs were performed for each dependent variable to determine if length and width contributed differently to any significant differences in size. Univariate outliers more than three standard deviations from the mean were excluded from these analyses (Field 2005:76). Post-hoc comparisons were not undertaken as it was unnecessary in light of the results to determine where significant differences occurred in each case (i.e. which organs within or between species had significantly different means and which formed homogenous sub-sets).

RESULTS

Morphological classifications

With the exception of the Pandanaceae and Araceae, all analysed families (Arecaceae, Dioscoreaceae, Heliconiaceae, Laxmanniaceae, and Musaceae) were found to produce Type III raphides. By light microscopy these raphides generally appeared simple and symmetrical, with smoothly-tapered terminations (Figure 2). In most cases sectional shape was not visible by light microscope, although 'wings' were sometimes detectable on some thicker, twelve-sided crystals (Figure 3). SEM analysis revealed that a sample population often included crystals at different stages of maturity (i.e. having immature four-sided forms, intermediate twelve-sided or 'winged' forms, and mature six-eight sided or faceted forms) (Figure 4). It was also noted that the terminations of some Type III raphides were finely serrated, as shown in Figure 5, but this feature was both rare and inconsistent within these samples. No other morphological features were observed on raphides from this group which could be used to further differentiate taxa.

Type I raphides were only observed in the Pandanaceae, and were present in both species analysed. These crystals had the same simple symmetrical morphology as the Type III forms when viewed by light microscopy. It was only possible to determine sectional shape with SEM (see examples in Figure 6).

All members of the Araceae produced Type IV raphides that differed considerably in terms of their morphologies and assemblage compositions (see Figure 7 to Figure 9 for examples from each species). Features differentiating aroid raphides from one another included whether or not the bridge and/or grooves were visible by light microscopy (which was generally directly related to overall raphide size), the location (centred/offset) and length (short/long compared to overall crystal length) of the bridge, and the shape of the short termination (continuous-wedge/protruding-wedge/tapered) (see Figure 10). The short wedge-shaped termination on asymmetrical crystals was either continuous (SWC) or protruding (SWP), while protruding forms had either square (SWP1) or inflected (SWP2) shoulders depending on crystal orientation. Barbs were very rarely observed and appear therefore to have little taxonomic value (cf. Figure 8b). The configuration of these morphological characteristics was generally consistent within each species regardless of the plant

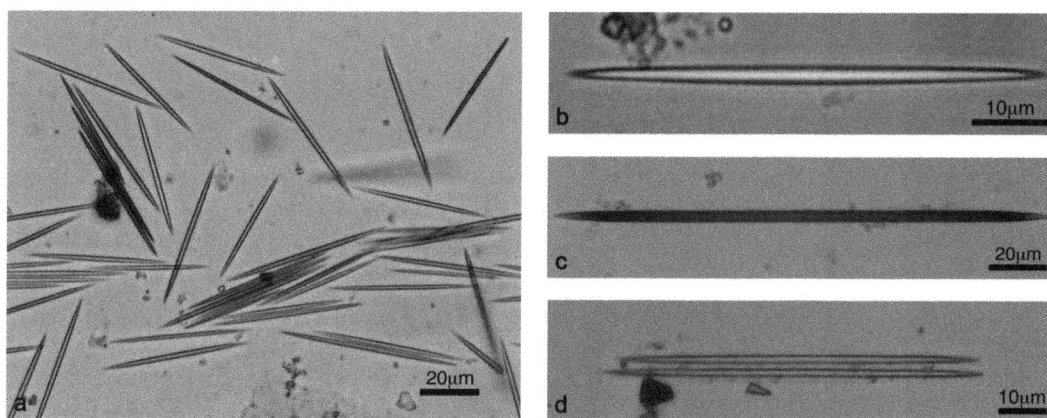

Figure 2. Light micrograph examples of Type III raphides from various species: (a) *Cordyline terminalis* **(leaf); (b)** *Dioscorea bulbifera* **(bulbil); (c)** *Heliconia indica* **(corm); and (d)** *Musa maclayi* **(bract) [all images transmitted light; (d) part-polarised].**

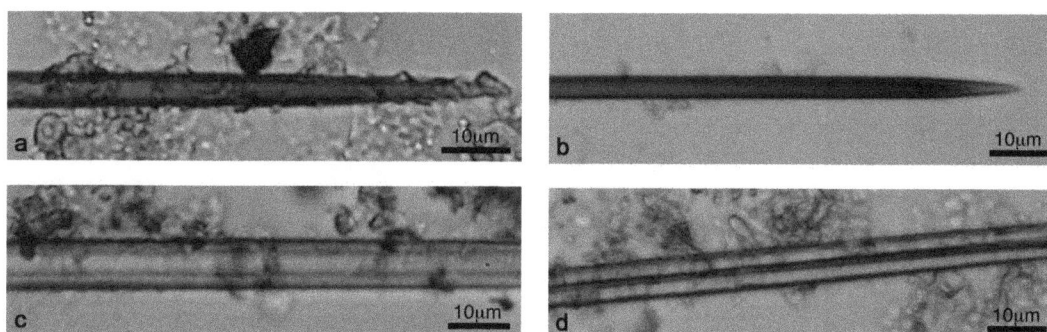

Figure 3. Examples of 'wings' visible on some thicker Type III raphides with light microscopy: (a) *Caryota rumphiana* **(leaf), (b)** *Heliconia indica* **(corm), (c)** *Caryota mitis* **(fruit); and (d)** *Musa acuminata* **(petiole) [all images transmitted light].**

part studied, although whether or not the bridge and grooves were visible by light microscopy tended to be more variable. The visibility of these features appeared to correspond with raphide size, as they were generally less distinctive on thinner (and usually shorter) raphides.

From these observations, a morphological key describing raphides typical of each aroid species was developed and is summarised Table 3. Using this key, for example, *Alocasia macrorrhiza* raphides may be differentiated from the other aroid species studied by their bridge, which is visible with light microscopy and relatively short compared to overall crystal length, as well as the shape of the short, wedge-shaped termination. By comparison, *Cyrtosperma* spp. (Figure 9a-c) and *Xanthosoma sagittifolium* (Figure 9d-e) have much longer bridges, while the bridge of *Colocasia esculenta* raphides (Figure 9f-g) is very rarely visible and not nearly as pronounced as that of *Alocasia*. *Amorphophallus* spp. typically had offset bridges and very distinctive protruding terminations, including some with inflected shoulders (Figure 8).

In some cases, the large size range within a single organ created the impression that more than one morphometric type existed within a single population. This phenomenon was observed in *Alocasia macrorrhiza* (flower), *Amorphophallus blumei* (leaf and petiole), *Colocasia esculenta* (petiole and root) and *Xanthosoma sagittifolum* (stem) (each morphometric type is described separately for these samples in Table 4). Scatterplots between the lengths and widths of these samples revealed that the populations were more or less continuously distributed along these variables and therefore are unlikely to represent discrete 'types'. The only exception was *Amorphophallus blumei*, in which bimodal populations with significantly different morphological features were present in both the leaf and petiole. The taxonomic implications of this variation are discussed later.

Figure 4. Examples of Type III raphides showing features observed during SEM: (a-b) mature facetted forms, (c) intermediate 'winged' forms, (d) transition to 6-sides at termination of a mature raphide, (e) transition from 12- to 6-sides at termination of an intermediate raphide, and (f-g) immature raphides with square sections [(a, g) *Cordyline terminalis***, (b)** *Musa maclayi***, (d, f)** *Caryota rumphiana***, (c, e)** *Dioscorea bulbifera***].**

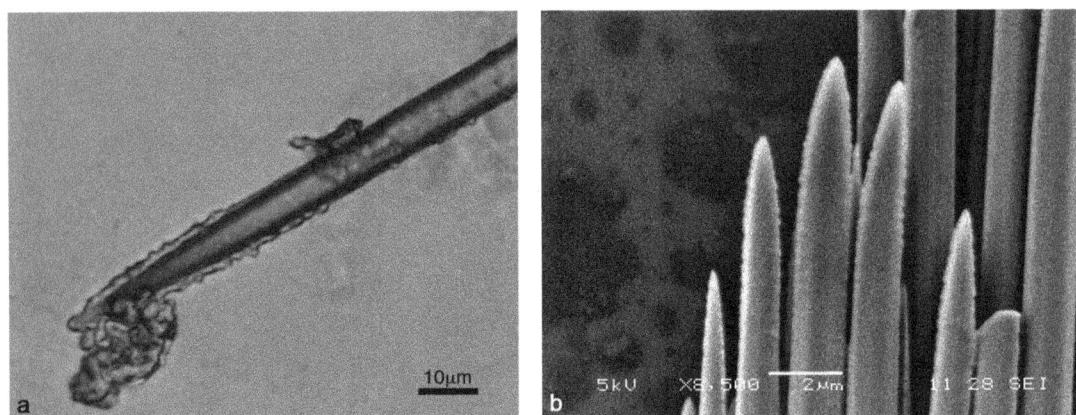

Figure 5. Examples of finely serrated terminations present on some Type III raphides [(a) *Caryota mitis***, transmitted light; (b)** *Cordyline terminalis***].**

Whisker raphides

Frequency and maximum length data for whisker raphides are summarised in Table 5. This raphide type was mainly observed in members of the Araceae (examples shown in Figure 11a, c) but was also present in several non-aroid taxa (Figure 11b, d). Curved whisker raphides (Figure 11c) were only observed in the Araceae, albeit rarely. Not all aroids contained whisker raphides—for example, they were completely absent from both species of *Cyrtosperma*, as well as *Xanthosoma sagittifolium*—but where present they were typically very abundant. Whisker

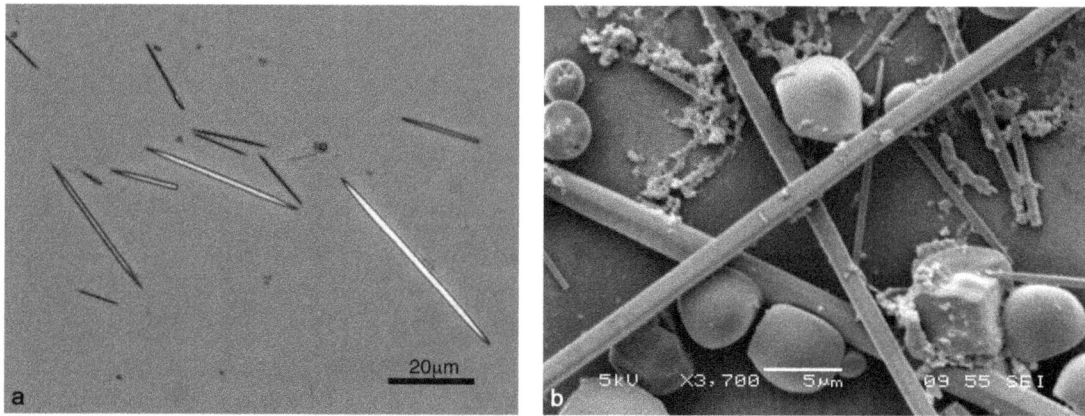

Figure 6. Type I raphides from *Pandanus dubius* [(a) part-polarised transmitted light; (b) SEM].

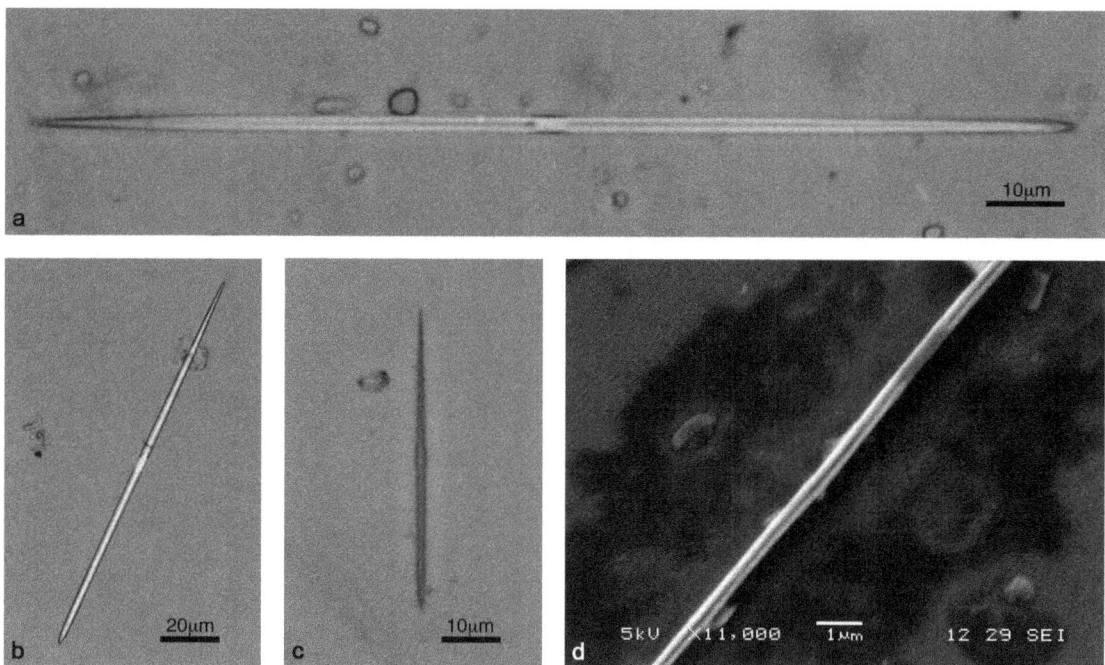

Figure 7. Examples of *Alocasia macrorrhiza* raphides showing: (a) grooves on larger types; (a-c) short, central bridge; and (d) whisker type with bridge and grooves which are visible only with SEM [(a-c) transmitted light; (a-b) part-polarised].

types were generally less frequent in non-aroid taxa and had shorter maximum lengths. They were the only raphide type present in the fruit of *Musa maclayi*, in which they were very abundant. It was found that maximum whisker length often exceeded that of non-whisker types, suggesting that they do not represent an immature crystal form as suggested by Loy *et al.* (1992:900). SEM analysis of whiskers from *Alocasia macrorrhiza* also revealed that they have mature Type IV morphologies, as shown in Figure 7d.

Results of statistical analyses

Box-and-whisker plots summarising the main descriptive statistics for raphide length are shown in Figure 12 and Figure 13 for the Type I/III and Type IV groups respectively. These plots indicate that although there is considerable overlap in the spread of raphide lengths, both within and between species, mean lengths are more varied. Some aroids (*Alocasia macrorrhiza* and *Amorphophallus* spp.) produce raphides with much greater size ranges, as does *Caryota mitis* (fruit, root and seed) in the Type I/III group. Detailed summaries of the descriptive statistics for both length and width, as well as estimates of raphide abundance within each plant, are presented in Table 6.

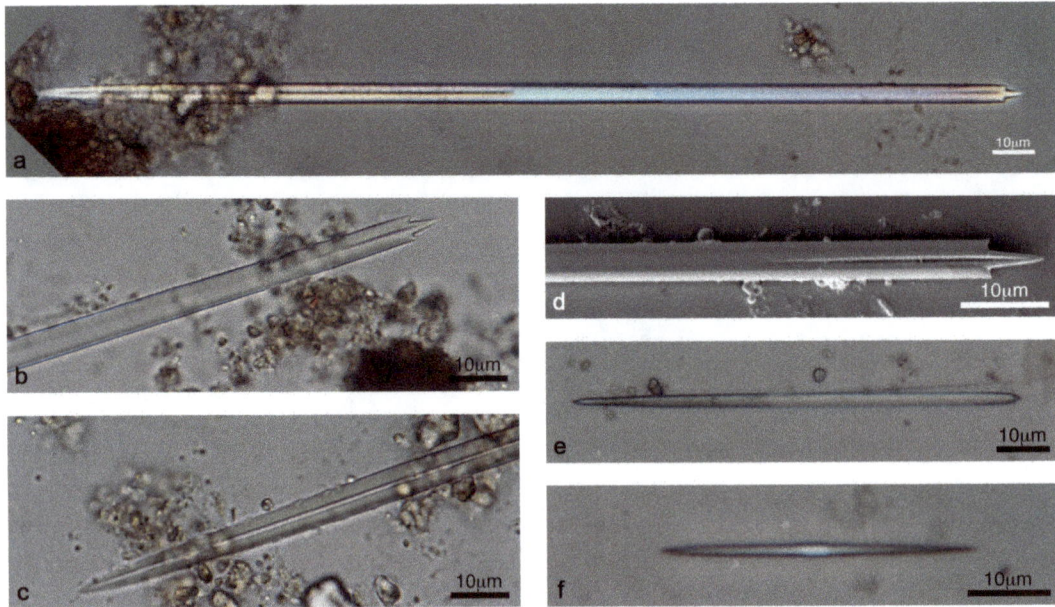

Figure 8. Examples of *Amorphophallus blumei* raphides showing morphological features visible on different sized crystals: (a-d) protruding wedge-shaped end, (a, e) long offset bridge and (a-d) grooves visible on larger types; (b-c) barbs visible on very large types; and (f) smaller, more symmetrical types with short, central bridges [all images except (c) transmitted light; (a) and (e) part-polarised].

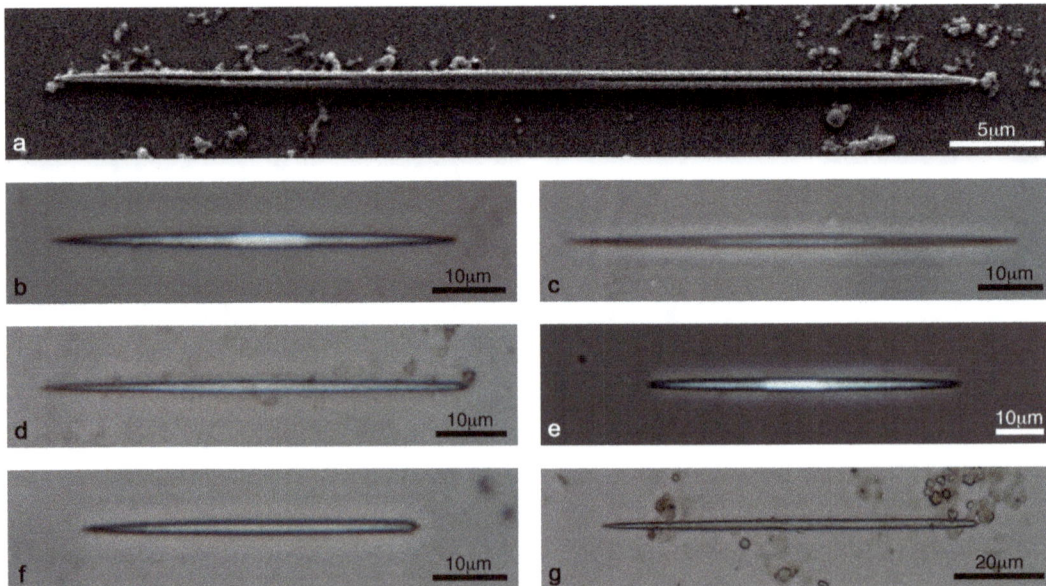

Figure 9. SEM and light microscopy examples of aroid raphides: (a-b) *Cyrtosperma johnstonii* and (c) *Cyrtosperma chamissonis* raphides showing similar symmetrical forms with a relatively long, centred bridge; (d-e) *Xanthosoma sagittifolium* raphides showing (d) asymmetrical type from the petiole with one short wedge-shaped end and lack of visible bridge, and (e) symmetrical type from the stem with visible long, centred bridge; and (f-g) examples of *Colocasia esculenta* raphides, showing asymmetrical form and lack of visible bridge or groove in light microscopy [all images except (a) transmitted light; (b, e) part-polarised].

The multivariate analyses revealed that there are highly significant differences ($p < 0.001$) between mean raphide sizes, both within individual species (Table 7) and between organs from different species within each morphological group (Table 8). The only exception was *Pandanus dubius*, $F_{(4, 294)} = 1.20$, $p = 0.312$, whose organ means cluster tightly around 40-47 µm (see

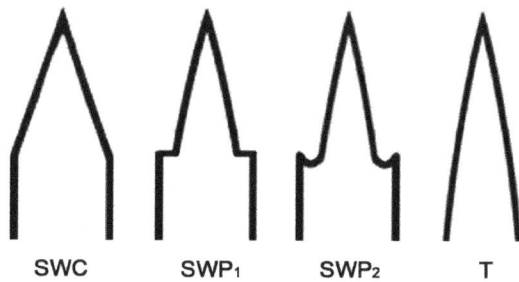

SWC   SWP₁   SWP₂   T

Figure 10. Diagram of termination types observed on Type IV raphides. The short wedge-shaped termination on asymmetrical crystals was either continuous (SWC) or protruding (SWP). Protruding forms have either square (SWP1) or inflected (SWP2) shoulders depending on crystal orientation. Terminations on symmetrical crystals are tapered (T).

Figure 12). The follow-up ANOVA tests indicated that variability in raphide size is generally significant for both length and width rather than just one of the two, although in some cases width was insignificant (see Table 7 and Table 8 for these results).

Minimum sample size calculations (Table 6) indicated that a sample of 50 raphides was adequate in about 50% of cases. The average minimum sample size required was 96, but this figure is inflated significantly by a few cases ($n = 15$) where much larger samples of between ~250 and 1000 were required. The particularly high n_{min} values calculated for *Amorphophallus blumei* leaf ($n_{min} = 797$) and petiole ($n_{min} = 986$) are most likely caused by the wide size range and bimodal distributions detected within these samples. If n_{min} values greater than 250 are excluded from the calculation, the average minimum sample size required is 56. The sample of 50 raphides analysed for each organ is therefore considered generally adequate for the purposes of this study.

DISCUSSION

The results of this study indicate that raphide morphology varies little across all taxa analysed aside from the Araceae, which produce a diagnostic form not currently observed in any other economically important Pacific family. Raphide form also differs significantly between aroid genera, which offers some scope for determining species of origin in contexts where only a limited range of aroid taxa are likely to occur. Size on the other hand varies considerably both within and between most species for all groups analysed. The high degree of intra-species heterogeneity in raphide size is a major limitation on using this variable for taxonomic studies.

Morphological variation

The most taxonomically meaningful raphide forms were observed within the Araceae which, as predicted from the literature review, was the only taxon found to produce Type IV raphides.

Table 3. General summary of key morphological features of raphides from the Araceae visible by light microscopy

Species	Bridge (if visible)	Grooves	Shape of the short termination
Alocasia macrorrhiza	Short, centred	Visible on thicker types	Continuous wedge
Amorphophallus blumei	Short, centred (thinner types); long, offset (thicker types)	Visible on thicker types	Protruding wedge (thicker types); continuous wedge or tapered (thinner types)
Amorphophallus campanulatus	Long, offset; rarely centred	Visible on thicker types	Protruding wedge
Colocasia esculenta	Not/rarely visible	Not visible	Continuous wedge
Cyrtosperma chamissonis	Rarely visible; long, centred	Not visible	Tapered
Cyrtosperma johnstonii	Long, centred	Not visible	Tapered
Xanthosoma sagittifollium	Long, centred	Not visible	Continuous wedge or tapered

Table 4. Summary of morphometric features of raphides from the Araceae observed during light microscopy

Species	Organ	Whisker raphides		Non-Whisker raphides					
		Freq[a]	Max L[b] (µm)	Freq[a]	L[c] (µm)	W[d] (µm)	Bridge[e]	Grv[f]	Term[g]
Alocasia macrorrhiza	Flower	A	–	C	35-95	1.0	NV	NV	Wk Assym
				R	60-115	1.5-2.5	RV, S, C	NV	SWC + T
	Leaf	Ab	100	R	45-200	1.5-3.5	V, S, C	V	SWC + T
	Petiole	Vab	100	Vr	150-240	3.0-5.0	V, S, C	V	SWC + T
	Root	Vab	105	R	170-280	3.0-5.5	V, S, C	V	SWC + T
	Stem	Vab	95	Vr	255-270	3.5-4.5	V, S, C	V	SWC + T
Amorphophallus blumei	Bulbil	Vab	70	C	85-225	1.5-5.0	V, L, O	V	SWP + T
	Leaf	Vab	75	C	20-90	1.0-1.5	Vx, S, C	NV	SWC + T
				R	150-290	2.5-6.0	V, L, O	V	SWP + T
	Petiole	Ab	75	C	20-100	1.0	Vx, S, C	NV	T
				C	150-250	3.0-4.5	V, L, O	V	SWP + T
	Root	Ab	65	R	130-260	2.0-5.5	V, L, O	V	SWP + T
	Tuber	Ab	85	C	140-265	2.0-6.0	V, L, O	V	SWP + T
Amorphophallus campanulatus	Tuber	Vab	80	C	110-170	2.0-3.5	V, L, O/C	V	SWP + T
Colocasia esculenta	Leaf	C	120	R	30-55	1.0-2.0	NV	NV	SWC + T
	Petiole	Ab	100	R	20-55	1.0	NV	NV	Wk Assym
				R	60-80	1.5-2.0	RV, S, C	NV	SWC/P + T
	Root	A	–	C	15-40	1.0	NV	NV	Wk Assym
				R	60-90	1.5-2.0	NV	NV	SWC + T
	Tuber	R	85	C	50-95	1.0-1.5	NV	NV	SWC + T
Cyrtosperma chamissonis	Tuber	A	–	C	45-70	1.0-1.5	RV, L. C	NV	T
Cyrtosperma johnstonii	Flower	A	–	Ab	40-80	1.0-1.5	RV, L. C	NV	T
	Leaf	A	–	Ab	25-60	1.0-1.5	Vx, L, C	NV	T
	Petiole	A	–	C	35-65	1.0-1.5	Vx, L, C	NV	T
	Root	A	–	C	20-50	1.0	Vx, L, C	NV	T
	Tuber	A	–	C	35-65	1.0-2.0	Vx, L, C	NV	T
Xanthosoma sagitifolium	Leaf	A	–	Vr	30-70	1.0	NV	NV	SWC + T
	Petiole	A	–	C	30-65	1.0-1.5	NV	NV	SWC + T
	Root	A	–	C	30-65	1.0-2.0	Vx, L, C	NV	T
	Stem	A	–	C	25-40	1.0	NV	NV	T
				Vr	50-70	1.5-2.5	Vx, L, C	NV	T

[a] Frequency (Freq) key: A – absent, Vr – very rare, R – rare, C – common, Ab – abundant, Vab – very abundant

[b] Max. L – maximum length

[c] L – length (range)

[d] W – width (range)

[e] Bridge key: visibility with light microscopy: NV – not visible, RV – rarely visible, Vx – mainly visible in cross-polarised light, V – visible; length: L – long, S – short; position: C – centered, O – offset

[f] Groove (Grv) visibility key: V – visible, NV – not visible

[g] Termination (Term) type key: Wk Assym – weakly asymmetrical; SWC – continuous short wedge; SWP – protruding short wedge; T – tapered

Table 5. Summary of frequency (Freq) and maximum length (Max. length) of non-aroid whisker raphides for each species where present

Family	Species	Organ	Freq[a]	Max. length (µm)
Dioscoreaceae	*Dioscorea hispida*	Leaf	Ab	10
Heliconiaceae	*Heliconia indica*	Root	R	45
Musaceae	*Musa maclayi*	Bract	C-Ab	60
		Leaf	C	65
		Fruit	Vab	70
Pandanaceae	*Pandanus dubius*	Root	Ab-Vab	50
		Wood	C-Ab	40
	Pandanus tectorius	Root	C	65
		Wood	Ab	50

[a] – Frequency key: R – rare; C – common; Ab – abundant; Vab – very abundant

Table 6. Summary of descriptive statistics for raphide length (L) and width (W), showing sample sizes (n), minimum adequate sample sizes (n_{min}), ranges, minimum (Min) and maximum (Max) values, means, standard deviations (St dev), and estimated frequency (Freq) of non-whisker types

Family: *Species*	Organ	L/W (µm)	n	n_{min}[a]	Range	Min (µm)	Max (µm)	Mean[b] (µm)	St dev[b]	Freq[c]
Araceae: *Alocasia macrorrhiza*	Flower	L	50	51	76	39	115	69.5	15.0	R-C
		W	50	73	1.5	1	2.5	1.0	0.5	
	Leaf	L	50	78	149	47	196	132.0	35.5	R
		W	50	59	2	1.5	3.5	2.5	0.5	
	Petiole	L	5	27	78	154	232	186.0	29.0	Vr
		W	5	75	2	3	5	3.5	1.0	
	Root	L	50	13	110	170	280	242.5	27.0	R
		W	50	23	2.5	3	5.5	4.5	0.5	
	Stem	L	4	1	15	255	270	264.0	6.5	Vr
		W	4	15	1	3.5	4.5	4.0	0.5	
Amorphophallus blumei	Bulbil	L	50	63	139	85	224	141.5	34.0	R-C
		W	50	68	3.5	1.5	5	3.0	0.5	
	Leaf	L	50	986	272	18	290	93.0	88.5	C
		W	50	636	5	1	6	2.0	1.5	
	Petiole	L	50	797	219	21	240	92.5	79.0	C
		W	50	473	3.5	1	4.5	2.0	1.5	
	Root	L	10	43	104	150	254	188.0	37.5	R
		W	10	106	3.5	2	5.5	3.5	1.0	
	Tuber	L	50	22	123	142	265	186.0	27.0	C
		W	50	43	4	2	6	4.0	1.0	
Amorphophallus campanulatus	Tuber	L	50	8	60	108	168	149.0	12.5	C
		W	50	30	1.5	2	3.5	3.0	0.5	

Family: *Species*	Organ	L/W (µm)	n	n_{min}[a]	Range	Min (µm)	Max (µm)	Mean[b] (µm)	St dev[b]	Freq[c]
Colocasia esculenta	Leaf	L	50	11	25	30	55	43.5	4.5	R
		W	50	47	1	1	2	1.0	0.5	
	Petiole	L	50	166	58	20	78	53.5	202.0	R
		W	50	95	1	1	2	1.5	0.5	
	Root	L	50	281	74	16	90	32.0	16.0	R-C
		W	50	36	1	1	2	1.0	0.5	
	Tuber	L	50	9	42	52	94	78.5	7.0	C
		W	50	38	1	1	2	1.5	0.5	
Cyrtosperma chamissonis	Tuber	L	50	13	25	45	70	59.0	6.5	C
		W	50	52	1	0.5	1.5	1.0	0.5	
Cyrtosperma johnstonii	Flower	L	50	20	40	40	80	55.5	7.5	Ab
		W	50	34	0.5	1	1.5	1.0	0.0	
	Leaf	L	50	44	32	25	57	43.0	8.5	Ab
		W	50	10	0.5	1	1.5	1.0	0.0	
	Petiole	L	50	10	30	36	66	50.0	5.0	C
		W	50	38	0.5	1	1.5	1.0	0.0	
	Root	L	50	31	28	20	48	33.0	5.5	C
		W	50	1	0	1	1	1.0	0.0	
	Tuber	L	50	16	30	35	65	44.0	5.5	C
		W	50	46	1	1	2	1.0	0.0	
Xanthosoma sagittifolium	Leaf	L	50	69	42	28	70	49.0	12.5	Vr
		W	50	1	0	1	1	1.0	0.0	
	Petiole	L	50	13	31	32	63	55.0	6.0	C
		W	50	29	0.5	1	1.5	1.0	0.0	
	Root	L	50	10	34	32	66	56.0	5.5	C
		W	50	26	1	1	2	1.5	0.0	
	Stem	L	50	72	47	25	72	33.0	8.5	R-C
		W	50	65	1.5	1	2.5	1.0	0.5	
Arecaceae: *Caryota mitis*	Fruit	L	50	37	254	196	450	352.0	65.0	Ab
		W	50	41	8	4.5	12.5	9.0	2.0	
	Leaf	L	50	125	99	31	130	69.0	23.5	Ab-Vab
		W	50	118	3	1	4	2.5	1.0	
	Rachis	L	4	32	38	72	110	95.0	16.0	R
		W	4	12	0.5	2.5	3	3.0	0.5	
	Root	L	50	295	312	88	400	172.5	90.0	R-C
		W	50	289	9	2	11	4.5	2.5	
	Seed	L	50	62	270	180	450	308.0	74.0	C-Ab
		W	50	49	7	5	12	8.5	2.0	
	Stem	L	50	115	129	41	170	90.0	29.5	C-Ab
		W	50	63	2.5	1	3.5	2.5	0.5	
Caryota rumphiana	Leaf	L	50	482	169	24	193	63.5	42.5	C
		W	50	453	4	1	5	1.5	1.0	
	Rachis	L	50	80	120	42	162	93.0	25.5	C
		W	50	51	2.5	1.5	4	2.5	0.5	
	Stem	L	50	93	104	36	140	95.0	28.0	C
		W	50	80	3	1	4	2.5	0.5	
	Wood	L	50	13	25	23	48	39.5	4.5	C
		W	50	50	1	1	2	1.0	0.0	
Cocos nucifera	Leaf	L	50	32	32	23	55	37.0	6.5	C
		W	50	26	0.5	1	1.5	1.0	0.0	
	Rachis	L	50	28	33	30	63	43.0	7.0	C
		W	50	82	1	1	2	1.5	0.5	
	Root	L	50	31	21	25	46	34.0	5.5	R-C
		W	50	23	0.5	1	1.5	1.0	0.0	
	Wood	L	50	27	30	25	55	38.0	6.0	C
		W	50	78	1	1	2	1.0	0.5	

Family: Species	Organ	L/W (µm)	n	n_{min} [a]	Range	Min (µm)	Max (µm)	Mean[b] (µm)	St dev[b]	Freq[c]
Dioscoreaceae: Dioscorea bulbifera	Bulbil	L	50	77	68	10	78	52.5	14.0	R-C
		W	50	79	1.5	0.5	2	1.5	0.5	
	Leaf	L	50	22	36	36	72	57.0	8.0	C
		W	50	85	1	1	2	1.0	0.5	
	Root	L	50	149	120	30	150	66.0	24.5	Ab
		W	50	129	2.5	1	3.5	1.5	0.5	
	Stem	L	50	34	38	35	73	49.5	9.0	C-Ab
		W	50	44	1	1	2	1.5	0.5	
	Tuber	L	50	84	91	22	113	73.0	20.5	Ab
		W	50	81	2	1	3	2.0	0.5	
Dioscorea esculenta	Leaf	L	50	84	55	30	85	51.5	14.5	C
		W	50	88	1	1	2	1.0	0.5	
	Stem	L	50	31	39	21	60	45.0	7.5	C
		W	50	49	1	0.5	1.5	1.0	0.0	
	Tuber	L	50	81	69	25	94	64.0	17.5	Ab
		W	50	92	1.5	1	2.5	1.5	0.5	
Dioscorea hispida	Leaf	L	50	24	50	40	90	57.5	8.5	C-Ab
		W	50	34	1.5	1	2.5	1.5	0.5	
	Petiole	L	50	23	31	32	63	45.5	6.5	C
		W	50	43	1	1	2	1.5	0.5	
	Stem	L	50	17	32	36	68	50.0	6.5	Ab
		W	50	25	1	1	2	1.5	0.0	
	Tuber	L	50	21	57	55	112	86.5	12.0	Ab
		W	50	21	1.5	1	2.5	2.0	0.5	
Heliconiaceae: Heliconia indica	Flower	L	50	48	69	43	112	78.0	16.5	C
		W	50	86	1.5	0.5	2	1.0	0.5	
	Fruit	L	50	112	180	55	235	125.0	40.0	C
		W	50	50	2.5	1.5	4	2.5	0.5	
	Leaf	L	50	145	174	10	184	96.5	35.5	C
		W	50	142	2.5	1	3.5	1.5	0.5	
	Root	L	50	690	175	25	200	43.5	35.0	Ab
		W	50	216	2.5	1	3.5	1.0	0.5	
	Tuber	L	50	108	130	46	176	96.0	30.5	C
		W	50	113	2.5	1	3.5	2.0	0.5	
Laxmanniaceae: Cordyline terminalis	Leaf	L	50	27	34	36	70	54.0	8.5	Vab
		W	50	62	1	1	2	1.5	0.5	
	Petiole	L	50	26	39	41	80	57.5	7.0	Vab
		W	50	62	1	1	2	1.5	0.5	
	Root	L	50	16	15	25	40	34.0	4.0	Vab
		W	50	1	0	1	1	1.0	0.0	
	Stem	L	50	38	63	40	103	68.0	12.5	Vab
		W	50	70	2	1	3	1.5	0.5	
Musaceae: Musa acuminata	Bract	L	50	61	98	47	145	89.5	21.5	C
		W	50	61	2	1	3	1.5	0.5	
	Fruit	L	2	66	19	45	64	54.5	13.5	R
		W	2	1	0	1	1	1.0	0.0	
	Leaf	L	50	35	60	30	90	57.0	10.0	R
		W	50	6	0.5	0.5	1	1.0	0.0	
	Petiole	L	50	263	177	29	206	110.5	54.5	R
		W	50	284	3	1	4	2.0	1.0	

Family: *Species*	Organ	L/W (µm)	n	n_{min}[a]	Range	Min (µm)	Max (µm)	Mean[b] (µm)	St dev[b]	Freq[c]
Musa ingens	Bract	L	50	27	43	25	68	46.5	7.5	Vab
		W	50	81	1	1	2	1.5	0.5	
	Leaf	–	0	–	–	–	–	–	–	A
	Flower	L	50	48	39	31	70	43.5	9.0	Vab
		W	50	84	1	1	2	1.5	0.5	
	Petiole	–	0	–	–	–	–	–	–	A
	Pseudo-stem	L	50	84	72	33	105	62.5	17.5	R
		W	50	109	2	1	3	1.5	0.5	
Musa maclayi	Bract	L	50	38	53	40	93	58.0	11.0	C-Ab
		W	50	23	0.5	1	1.5	1.0	0.0	
	Flower	–	0	–	–	–	–	–	–	A
	Leaf	L	50	36	57	25	82	62.5	11.5	C
		W	50	34	0.5	1	1.5	1.0	0.0	
	Fruit	–	0	–	–	–	–	–	–	A
	Petiole	L	50	23	49	75	124	97.0	14.0	Vab
		W	50	121	2	1	3	2.0	0.5	
Pandanaceae: *Pandanus dubius*	Leaf	L	50	259	74	10	84	40.5	19.5	C
		W	50	60	1	1	2	1.0	0.5	
	Root	L	50	85	59	21	80	47.0	13.0	Ab
		W	50	95	1.5	1	2.5	1.0	0.5	
	Wood	L	50	145	60	20	80	41.5	15.0	Vab
		W	50	86	1	1	2	1.0	0.5	
Pandanus tectorius	Leaf	L	50	571	105	15	120	43.5	31.5	C
		W	50	86	1	1	2	1.0	0.5	
	Root	L	50	77	60	25	85	5.0	14.0	Ab
		W	50	58	1	0.5	1.5	1.0	0.5	
	Wood	L	50	274	90	22	112	58.5	29.5	C-Ab
		W	50	98	1.5	1	2.5	1.0	0.5	

[a] Average n_{min} = 96.
[b] Rounded to the nearest 0.5 µm.
[c] Frequency (Freq) key: A – absent, Vr – very rare, R – rare, C – common, Ab – abundant, Vab – very abundant.

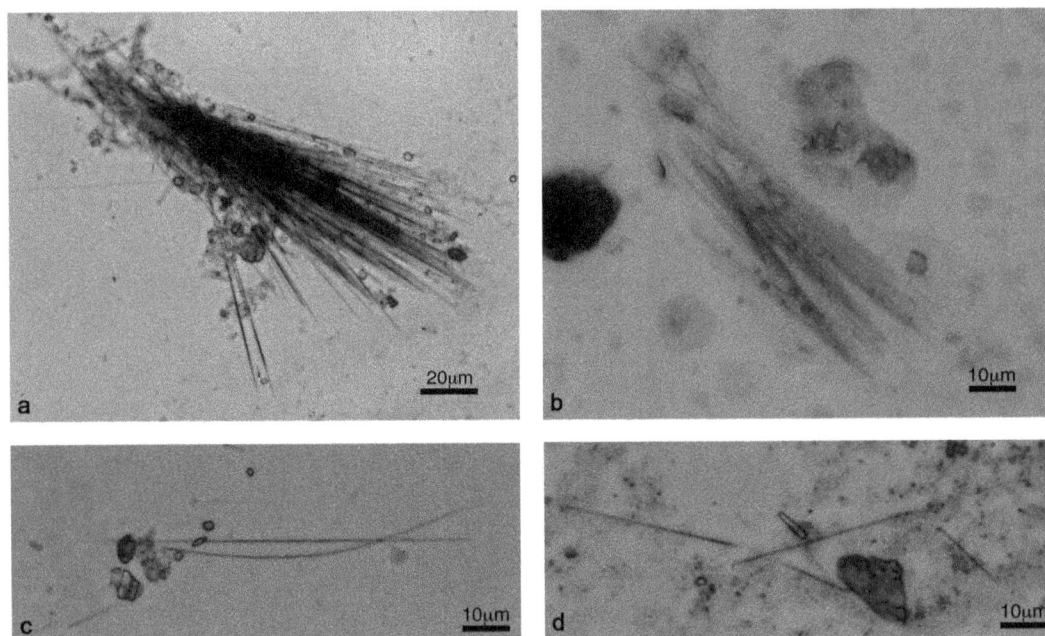

Figure 11. Examples of whisker raphides: (a) *Colocasia esculenta* (leaf); (b) *Musa maclayi* (fruit); (c) curved aroid whiskers (*Alocasia macrorrhiza*, stem); and (d) *Pandanus dubius* (root) [all images transmitted light].

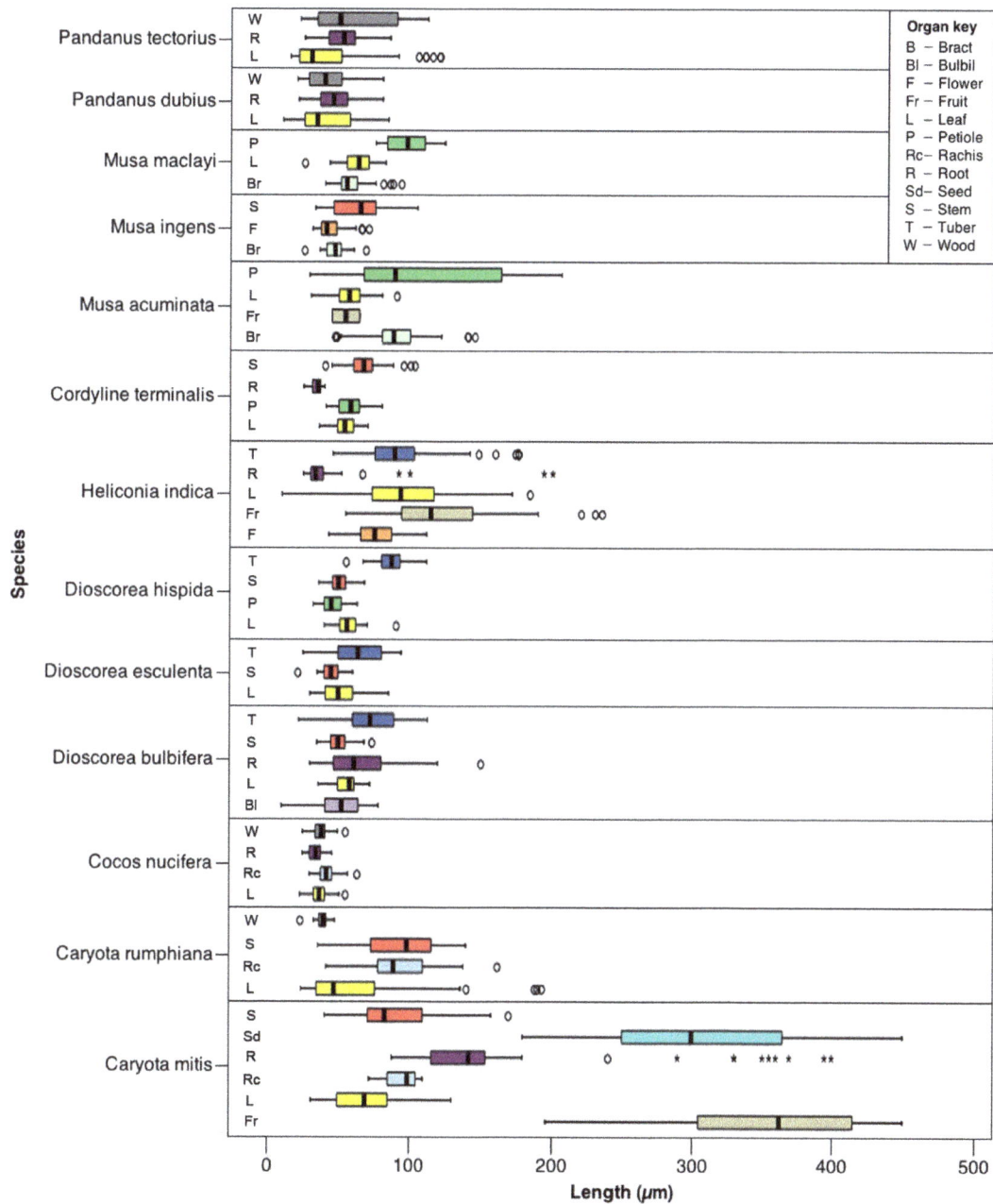

Figure 12. Box-and-whisker plot showing the spread of raphide lengths about the mean for species classified in the Type I/III morphological group.

All remaining taxa produce either Type I or Type III morphologies, which overlap depending on crystal maturity. The most diagnostic attribute of aroid raphides is the H-shaped cross-section or, as seen in plane view, the presence of longitudinal grooves on opposing faces of the crystal. These features are sometimes visible by light microscopy depending largely on raphide size, but can generally be seen on all aroid raphides, regardless of overall size or degree of fragmentation, using SEM. Although asymmetrical terminations are also unique to Araceae raphides, this feature is not common to all species (e.g. *Cyrtosperma* spp. have symmetrical tapered ends). Terminations may also be missing from fragmented archaeological crystals owing to their fragility, although the short wedge-shaped end, if present, may further support the identification of aroid raphides. Based on these observations, this study has demonstrated that it is possible to at least differentiate aroid from non-aroid taxa based on raphide morphology alone. Despite its broadness, this level of classification is significant given the importance of the edible aroids to past and present subsistence regimes in the Pacific Islands.

Table 7. Results of the intra-species MANOVA (Pillai's Trace) and ANOVA tests. Non-significant results are shown in bold

Species	MANOVA			ANOVA (length)			ANOVA (width)		
	df	F	p^a	df	F	p^b	df	F	p^b
Alocasia macrorrhiza	4	292	66.29	<0.001	2	146	529.16	<0.001	2	145	529.64	<0.001
Amorphophallus blumei	6	390	12.11	<0.001	3	196	25.27	<0.001	3	196	29.05	<0.001
Colocasia esculenta	6	386	41.78	<0.001	3	193	118.85	<0.001	3	194	18.59	<0.001
Cyrtosperma johnstonii	8	480	28.25	<0.001	4	242	89.44	<0.001	4	241	7.27	<0.001
Xanthosoma sagittifolium	6	384	111.20	<0.001	3	190	131.84	<0.001	3	194	139.31	<0.001
Caryota rumphiana	6	388	19.42	<0.001	4	245	210.46	<0.001	4	245	203.40	<0.001
Cocos nucifera	6	392	10.35	<0.001	3	194	48.18	<0.001	3	194	39.66	<0.001
Caryota mitis	8	488	43.32	<0.001	3	196	16.58	<0.001	3	196	16.99	<0.001
Dioscorea bulbifera	8	490	22.99	<0.001	4	243	18.54	<0.001	4	244	23.00	<0.001
Dioscorea esculenta	4	294	11.01	<0.001	2	146	24.07	<0.001	2	147	18.28	<0.001
Dioscorea hispida	6	390	55.86	<0.001	3	195	244.68	<0.001	3	195	47.57	<0.001
Heliconia indica	8	484	49.05	<0.001	4	243	59.47	<0.001	4	242	86.49	<0.001
Cordyline terminalis	6	392	32.29	<0.001	3	196	122.77	<0.001	3	195	39.68	<0.001
Musa acuminata	4	292	14.16	<0.001	2	146	31.14	<0.001	2	145	34.41	<0.001
Musa ingens	4	292	23.51	<0.001	2	147	35.25	<0.001	2	146	2.463	**0.089**
Musa maclayi	4	294	38.31	<0.001	2	145	176.50	<0.001	2	143	58.91	<0.001
Pandanus dubius	4	294	1.20	**0.312**	2	147	2.43	**0.091**	2	144	1.735	**0.180**
Pandanus tectorius	4	290	6.91	<0.001	2	147	4.34	0.015	2	146	1.807	**0.168**

[a] A significant difference occurs when $p < 0.05$.
[b] A significant difference occurs when $p < 0.025$ (using a Bonferonni corrected α-value for two dependent variables).

Table 8. Results of the inter-species MANOVA (Pillai's Trace) and ANOVA tests within the Type I/III and Type IV (Araceae) morphological group. Non-significant differences are shown in bold

Group	Organ	MANOVA			ANOVA (length)			ANOVA (width)		
		df	F	p^a	df	F	p^b	df	F	p^b
Type I/III	Bract	4	294	76.03	<0.001	2	146	124.73	<0.001	2	146	48.20	<0.001
	Flower	2	96	180.48	<0.001	1	98	166.38	<0.001	1	98	8.90	0.004
	Fruit	2	97	312.34	<0.001	1	98	441.01	<0.001	1	98	630.82	<0.001
	Leaf	22	1170	38.75	<0.001	11	584	27.364	<0.001	11	584	32.47	<0.001
	Petiole	6	392	28.27	<0.001	3	196	58.46	<0.001	3	196	12.77	<0.001
	Rachis	2	97	91.41	<0.001	1	98	182.05	<0.001	1	97	110.86	<0.001
	Root	12	678	37.13	<0.001	6	340	92.52	<0.001	6	339	97.48	<0.001
	Stem	12	686	33.99	<0.001	6	342	61.67	<0.001	6	341	77.04	<0.001
	Tuber	6	390	25.18	<0.001	3	196	22.65	<0.001	3	195	7.31	<0.001
	Wood	6	388	9.16	<0.001	3	195	15.75	<0.001	3	194	0.66	**0.581**
Type IV	Flower	2	96	24.37	<0.001	1	96	38.36	<0.001	1	96	1.140	**0.288**
	Leaf	8	484	19.60	<0.001	4	244	41.40	<0.001	4	242	42.30	<0.001
	Petiole	6	390	17.16	<0.001	3	193	11.54	<0.001	3	196	18.18	<0.001
	Root	6	388	68.14	<0.001	3	193	2128.40	<0.001	3	195	1080.84	<0.001
	Tuber	8	480	78.34	<0.001	4	241	1036.50	<0.001	4	243	390.02	<0.001

[a] A significant difference occurs when $p < 0.05$.
[b] A significant difference occurs when $p < 0.025$ (using a Bonferonni corrected α-value for two dependent variables).

In addition to being distinct from all other analysed taxa, aroid raphides were found to differ significantly from one another in terms of the size and location of the central bridge, visibility of grooves by light microscopy and termination shape (see the taxonomic key presented in Table 3 above). The configuration of these features was reasonably consistent within a single species, which is important for taxonomic purposes, and appears to differ between taxa at the genus rather than species level. Although species identifications are highly desirable for residue and microfossil studies, genus-level identifications of aroid raphides are still very useful in the Pacific context. Once again, the main limitation for taxonomic identification is that the presence and configuration of these features can only be observed on whole crystals, which are rarely found in archaeological contexts owing to their fragility.

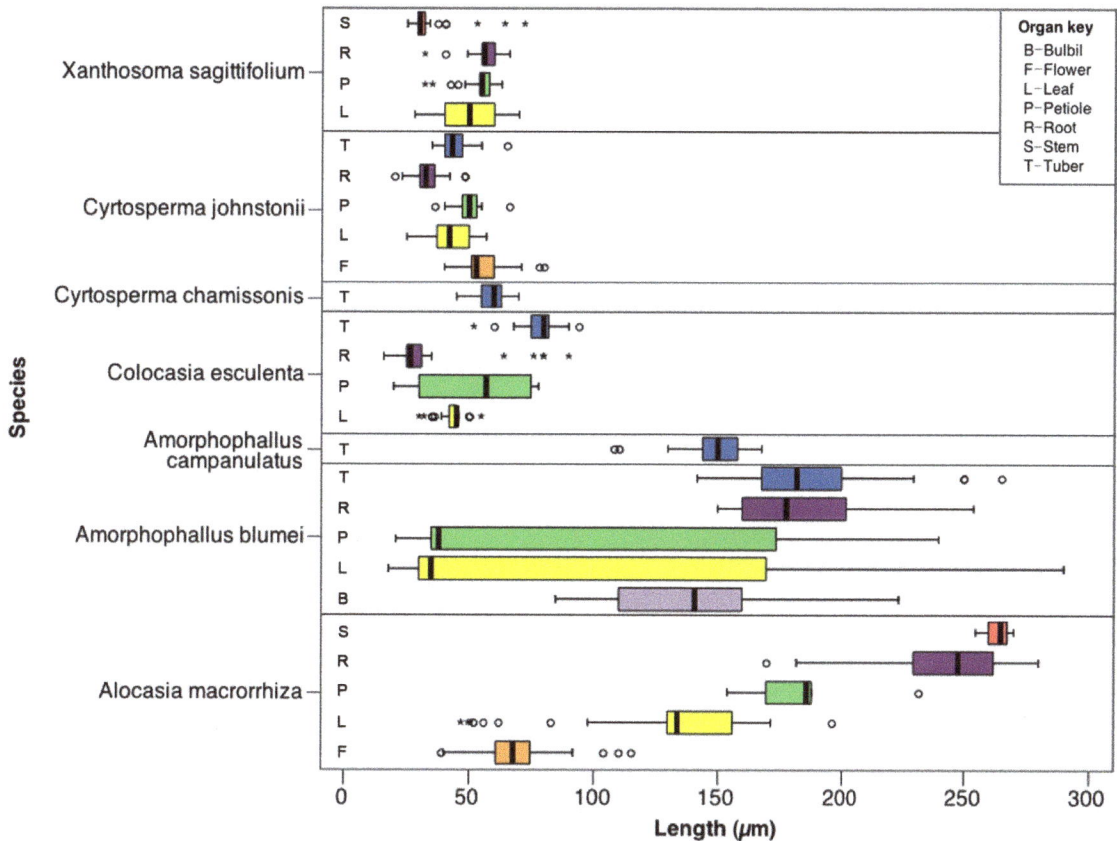

Figure 13. Box-and-whisker plot showing the spread of raphide lengths about the mean for species classified in the Type IV morphological group.

Even though similar raphide forms were generally observed within each species regardless of the plant part studied, minor exceptions did occur that warrant further discussion. For one, the short central bridge typical of *Alocasia macrorrhiza* raphides was not visible by light microscopy on crystals from the flower, possibly owing to their relative thinness. Because the flower of this species is not likely to contribute significant quantities of raphides to the archaeological record, however, this deviation is not considered problematic for taxonomic determinations. *Amorphophallus blumei* was the only other species within which significantly different raphide forms were observed. Its typical form has a long bridge that is offset to one end, but the leaf and petiole also produce a shorter, thinner raphide type with a short, central bridge, similar to that of *Alocasia macrorrhiza*. In cases where large enough raphide assemblages are recovered from archaeological contexts, it may be possible to differentiate these taxa based on the presence of the larger, more diagnostic types, but where only the thin type is found, taxonomic differentiation may be more difficult. Further SEM studies to determine if the smaller *A. blumei* raphides have the same distinctive protruding termination (SWP) observed on the larger crystals may also prove useful. A consideration in this instance is that *Amorphophallus blumei* is not presently an economically important species in the Pacific and is less likely to occur in archaeological assemblages from this region if current use reflects past exploitation. The closely related species *A. campanulatus* is on the other hand both widespread and of minor importance as a food source, and must therefore be investigated more thoroughly to establish if it also produces these different raphide forms.

Additional research is also required to test the hypothesis that aroid raphide morphology is determined at the genus rather than species level as only a limited number of intra-genus comparisons were made in this study. For example, although similar forms were observed within *Amorphophallus blumei* and *A. campanulatus*, as well as within *Cyrtosperma chamissionis* and *C. johnstonii*, only a single comparative sample within these genera was analysed. Interestingly, the distinctive protruding terminations observed on both *Amorphophallus* species were also reported on raphides from *Amorphophallus rivieri* by Sakai *et al.* (1984) and in several *Amorphophallus* species by Prychid *et al.* (in press), lending support to the notion that raphide morphology is genera-specific. It may be that calcium oxalate crystal morphology in general is determined at the genus level, considering the similar observation by Jones and Bryant (1992) that druse forms produced by cacti are shared at this level.

The final factor to consider with regard to morphological variation is the possible influence of environmental factors on form. Although this issue was not investigated directly by this study, the observed features of aroid raphides are generally consistent with other published accounts. For example, Bradbury and Nixon (1998) also described the short bridge of *Alocasia macrorrhiza* raphides, the symmetry and comparatively long bridge of *Cyrtosperma chamissonis* raphides, as well as the absence of 'thin' (whisker) types from this species. Likewise, Sakai and Hanson (1974:742) noted that while the bridge of *Alocasia macrorrhiza* raphides is visible by light microscopy, it is not visible on either *Colocasia esculenta* or *Xanthosoma sagittifolium* raphides. Even though the bridge was observed on both root and stem raphides from *X. sagittifolium* in this study (see Table 4), Sakai and Hanson did not examine root samples (only corm, stem, petiole and leaf), and raphides with visible bridges were only rarely present in the stem (Table 4). In addition to indicating that morphology is more genetically than environmentally determined, these published observations further validate the applicability of the taxonomic key developed in this study.

By comparison, no potentially diagnostic traits were observed on any Type I and III raphides that could further differentiate taxa from these groups. Although only the Pandanaceae produced exclusively Type I crystals, this form overlaps with immature Type III crystals produced by other non-Pandanaceae taxa. In cases where more mature Type III crystals are present in archaeological assemblages, *Pandanus* spp. may therefore be discounted as possible taxa of origin, but where only Type I crystals are present (particularly in small assemblages), it will be more difficult to rule out the possibility that they are derived from plants with Type III. Further research is required to determine whether any significant differences in size or shape occur between mature Type I and immature Type III raphides, which might enable confident separation of these groups.

Size variation

The results of the multivariate tests indicate that raphide size is significantly affected by the organ in which it is produced, which is highly problematic for taxonomic purposes. Not only were there significant differences in the mean sizes of raphides within each species, but size ranges were often highly variable at the species level and also overlapped considerably between species. It is therefore very difficult to demonstrate any meaningful relationship between raphide size and taxa of origin. Owing to the large degree of intra-species size variation, the significant differences detected between species (when like organs were compared) have little taxonomic value except in situations where raphides are derived from a *single known organ*. This is rarely the case for microfossils extracted from bulk sediment samples, which by-and-large represent whole-plant assemblages, but is feasible in contexts where raphides are directly associated with storage starch granules, such as in use-residues on artefacts. In these cases it may be possible to discriminate plant taxa more effectively if measures of raphide size (and shape) are included in the starch taxonomic key, rather than mere presence/absence measures as used in previous studies (e.g. Loy *et al.* 1992; Torrence *et al.* 2004). This could be further tested through discriminant function analysis of combined starch and raphide data for reference materials.

Clearly apparent from the size data is the fact that large raphides (i.e. longer than ~ 250 µm) are altogether very rare, while raphides longer than 150-200 µm are less common among the Araceae (these were only observed in *Alocasia macrorrhiza* and *Amorphophallus* spp.). Threshold values may be useful for identifying key taxa if they are found to produce raphides in exceptionally large size ranges. Larger and more comprehensive reference collections must be analysed if such identifications are to be supported.

Implications for previous archaeological raphide analyses

Many of the criteria used previously to identify raphides from *Colocasia esculenta* and other Pacific aroids in microfossil assemblages, including abundance and the presence of the thin, whisker-type have been shown here to be unreliable. Rather, it has been demonstrated that even though raphides are often very abundant in the aroids, they also occur in comparably high frequencies in other taxa. In fact, the highest density across all samples was observed in the wood of *Pandanus dubius* rather than in the Araceae, although inconsistent sample sizes may have biased this finding. Additionally, no appreciable trend was detected between raphide frequency and either organ or species of origin, suggesting that raphide abundance is not strictly taxonomically or biologically related. It was also found that whisker raphides occur in a number of non-aroid taxa, including *Dioscorea hispida*, *Heliconia indica*, *Musa maclayi*, *Pandanus dubius* and *P. tectorius*, indicating that this type is not specific to the Araceae. Analysis of a wider range of reference materials may expand this list further. Curved whisker raphides, on the other hand, have only been observed in the Araceae thus far and may have some diagnostic value.

The results of archaeological analyses that have relied on criteria of raphide abundance and whisker raphide presence (e.g. Crowther 2005; Horrocks and Barber 2005; Horrocks and Bedford 2005; Horrocks and Nunn 2007; Horrocks and Weisler 2006; Horrocks *et al.* 2007; Loy *et al.* 1992) should be treated with caution until more detailed morphological analyses using SEM have been made. This is an important issue, as many of these studies have contributed critical evidence for the use and spread of *Colocasia esculenta* and other aroid genera as well as associated horticultural practices across the Pacific region during prehistory. While most of these reports have used the co-presence of other diagnostic microfossil types such as starch granules to identify these plants archaeologically, raphides have still been drawn on as key corroborating evidence to support these identifications. It is critical, therefore, that the identification of raphides in these studies be as independently secure as possible, for which reference to explicit morphometric criteria such as the presence or absence of a longitudinal groove (to differentiate aroid from non-aroid taxa) must be presented. This is particularly important in cases where an indirect relationship between the raphides and other microfossil types must otherwise be assumed (such as in bulk sediment samples that are likely to contain microfossils from a wide range of taxa), or where plant genus or species have been identified based on the presence of raphides alone. For example, the species identification of *C. esculenta* raphides in the absence of other diagnostic taro-type microfossils at an early wetland ditch in Motutangi, New Zealand (Horrocks and Barber 2005:Figure 6, 110) requires at least some discussion of morphometric attributes in order to be upheld.

CONCLUSION

This study has demonstrated that there is only limited potential for raphides from different economically-important Pacific plant taxa to be differentiated based on morphometric variables. Specifically, it was found that raphide size varies considerably within species, while similar shapes are shared among a number of taxa, leaving little scope for taxonomic differentiation. The key exceptions were members of the Araceae, which produce raphides that can be separated from all other taxa by the presence of a longitudinal groove, and from each other by the configuration and visibility of morphological features such as terminations and the central bridge. This study has firmly established that routine identifications of aroid raphides from Pacific samples are possible based on the presence of these features alone. Furthermore, in cases where crystals are fragmented or too small to resolve morphological features with light microscopy, it has been shown that such identifications can still be made using SEM. This is a very significant outcome for residue and microfossil studies, considering that aroids are traditionally a major staple in many Pacific Islands societies and that these plants are therefore likely to be represented in archaeological residue assemblages from the region. Previous identifications of aroid raphides from archaeological contexts may also be strengthened if reviewed in light of these criteria, until which time they should be regarded as tentative.

Although the scope for discriminating between non-aroid taxa based on morphometric variables appears limited at this stage, this potential may increase in contexts where only a restricted range of taxa are likely to be present. In these cases, size may have greater significance, particularly if larger and much rarer types are present. Stronger identifications may also be achieved in cases where raphides co-occur in *direct* association with other distinctive microfossils, such as starch granules or phytoliths, as a much wider range of attributes can then be used to differentiate taxa. This possibility could be further tested by incorporating the data from this study into multivariate statistical models developed for the taxonomic analysis of starch granules (e.g. Loy *et al.* 1992; Torrence *et al.* 2004).

ACKNOWLEDGEMENTS

I thank Carol Lentfer for generously providing plant reference materials, Kim Sewell and other staff at The University of Queensland's Centre for Microscopy and Microanalysis for assistance during SEM, and Michael Haslam and an anonymous reviewer for their helpful comments on an earlier draft of this paper. Funding for this study was provided by the School of Social Science, The University of Queensland.

REFERENCES

Arnott, H.J. 1981 An SEM study of twinning in calcium oxalate crystals of plants. *Scanning Electron Microscopy* 1981(III):225-234.

Arnott, H.J. and F.G.E. Pautard 1970 Calcification in plants. In H. Schraer (ed.) *Biological Calcification: Cellular and molecular aspects*, pp. 375-446. New York: Appleton-Century-Crofts.

Arnott, H.J. and M.A. Webb 2000 Twinned raphides of calcium oxalate in grape (vitis): implications for crystal stability and function. *International Journal of Plant Science* 161(1):133-142.

Ayensu, E.S. 1972 *Anatomy of the Monocotyledons VI*: Dioscoreales. Oxford: Clarendon Press.

Barrau, J. 1958 *Subsistence Agriculture in Melanesia*. Bernice P. Bishop Museum Bulletin 219. Honolulu: Bernice P. Bishop Museum.

Bouropoulos, N., S. Weiner and L. Addadi 2001 Calcium oxalate crystals in tomato and tobacco plants: morphology and in vitro interactions of crystal-associated macromolecules. *Chemistry - A European Journal* 7(9):1881-1888.

Bradbury, J.H. and R.W. Nixon 1998 The acridity of raphides from the edible aroids. *Journal of the Science of Food and Agriculture* 76:608-616.

Cervantes-Martinez, T., H.T. Horner, Jr, R.G. Palmer, T. Hymowitz and A.H.D. Brown 2005 Calcium oxalate crystal macropatterns in leaves of species from groups Glycine and Shuteria (Glycininae; Phaseoleae; Papilionoideae; Fabaceae). *Canadian Journal of Botany* 83(11):1410-1421.

Cody, A.M. and H.T. Horner 1983 Twin raphides in the Vitaceae and Araceae and a model for their growth. *Botanical Gazette* 144(3):318-330.

Crowther, A. 2005 Starch residues on undecorated Lapita pottery from Anir, New Ireland. *Archaeology in Oceania* 40:62-66.

Crowther, A. In Press Reviewing raphides: issues with the identification and interpretation of calcium oxalate crystals in microfossil assemblages. In A. Fairbairn and S. O'Connor (eds) *Proceedings of the 2005 Australasian Archaeometry Conference*. Canberra: ANU E Press.

Drennan, R.D. 1996 *Statistics for Archaeologists: A commonsense approach*. New York: Plenum.

Field, A. 2005 *Discovering Statistics Using SPSS*. 2nd edn. London: Sage Publications.

Fullagar, R.L.K., J. Field, T.P. Denham and C.J. Lentfer 2006 Early and mid Holocene tool-use and processing of taro (*Colocasia esculenta*), yam (*Dioscorea* sp.) and other plants at Kuk Swamp in the highlands of Papua New Guinea. *Journal of Archaeological Science* 33:595-614.

Franceschi, V.R. and H.T. Horner, Jr 1980 Calcium oxalate crystals in plants. *The Botanical Review* 46(4):361-427.

Franceschi, V.R. and P.A. Nakata 2005 Calcium oxalate in plants: formation and function. *Annual Review of Plant Biology* 56:41-71.

Hartl, W.P., H. Klapper, B. Barbier, H.J. Ensikat, R. Dronskowski, P. Müller, G. Ostendorp, A. Tye, R. Bauer and W. Barthlott 2007 Diversity of calcium oxalate crystals in Cactaceae. *Canadian Journal of Botany* 85(5):501-517.

Hather, J.G. 1992 The archaeobotany of subsistence in the Pacific world. *Archaeology in Oceania* 24(1):70-81.

Horner, H.T., A.P. Kausch and B.L. Wagner 1981 Growth and change in shape of raphide and druse calcium oxalate crystals as a function of intracellular development in *Typha angustifolia* L. (Typhaceae) and *Capsicum annuum* L. (Solanaceae). *Scanning Electron Microscopy* 1981(III):251-262.

Horner, H.T., and B.L. Wagner 1995 Calcium oxalate formation in higher plants. In S.R. Khan (ed.) *Calcium Oxalate in Biological Systems*, pp. 53-72. Boca Raton, Florida: CRC Press.

Horrocks, M. and I. Barber 2005 Microfossils of introduced starch cultigens from an early wetland ditch in New Zealand. *Archaeology in Oceania* 40:106-114.

Horrocks, M. and S. Bedford 2005 Microfossil analysis of Lapita deposits in Vanuatu reveal introduced Araceae. *Archaeology in Oceania* 40:67-74.

Horrocks, M., S. Bulmer and R.O. Gardner 2008a Plant microfossils in prehistoric archaeological deposits from Yuku rock shelter, Western Highlands, Papua New Guinea. *Journal of Archaeological Science* 35:290-301.

Horrocks, M., J.A. Grant-Mackie and E.A. Matisoo-Smith 2008b Introduced taro (*Colocasia esculenta*) and yams (*Dioscorea* spp.) in Podtanean (2700-1800 years BP) deposits from Mé Auré Cave (WMD007), Moindou, New Caledonia. *Journal of Archaeological Science* 35(1):169-180.

Horrocks, M., S.I. Nichol, P.C. Augustinus and I.G. Barber 2007 Late Quaternary environments, vegetation and agriculture in northern New Zealand. *Journal of Quaternary Science* 22(3):267-279.

Horrocks, M. and P.D. Nunn 2007 Evidence for introduced taro (*Colocasia esculenta*) and lesser yam (*Dioscorea esculenta*) in Lapita-era (c. 3050-2500 cal. yr BP) deposits from Bourewa, southwest Viti Levu Island, Fiji. *Journal of Archaeological Science* 34:739-748.

Horrocks, M. and M.I. Weisler 2006 Analysis of plant microfossils in archaeological deposits from two remote archipelagos: the Marshall Islands, eastern Micronesia, and the Pitcairn Group, southeast Polynesia. *Pacific Science* 60(2):261-280.

Jones, J.G. and V.M. Bryant 1992 Phytolith taxonomy in selected species of Texas cacti. In G. Rapp, Jr. and S.C. Mulholland (eds) *Phytolith Systematics: Emerging issues*, pp. 215-238. New York: Plenum Press.

Kausch, A.P. and H.T. Horner 1982 A comparison of calcium oxalate crystals isolated from callus cultures and their explant sources. *Scanning Electron Microscopy* 1982(I):199-211.

Kostman, T.A. and V.R. Franceschi 2000 Cell and calcium oxalate crystal growth is coordinated to achieve high-capacity calcium regulation in plants. *Protoplasma* 214:166-179.

Lentfer, C.J. and R.C. Green 2004 Phytoliths and the evidence for banana cultivation at the Lapita Reber-Rakival Site on Watom Island, Papua New Guinea. In V. Attenbrow and R.L.K. Fullagar (eds) A Pacific Odyssey: Archaeology and anthropology in the western Pacific. Papers in honour of Jim Specht, pp. 75-88. *Records of the Australian Museum, Supplement* 29. Sydney: Australian Museum.

Lersten, N.R. and H.T. Horner, Jr 2000 Calcium oxalate crystal types and trends in their distribution pattern in leaves of *Prunus* (Rosaceae: Prunoideae). *Plant Systematics and Evolution* 224:83-96.

Lersten, N.R. and H.T. Horner, Jr 2008 Crystal macropatterns in leaves of Fagaceae and Nothofagaceae: a comparative study. *Plant Systematics and Evolution* 271:239-253.

Loy, T.H. 1994 Methods in the analysis of starch residues on prehistoric stone tools. In J.G. Hather (ed.) *Tropical Archaeobotany: Applications and new developments*, pp. 86-114. London: Routledge.

Loy, T.H., M. Spriggs and S. Wickler 1992 Direct evidence for human use of plants 28,000 years ago: starch residues on stone artefacts from the northern Solomon Islands. *Antiquity* 66(253):898-912.

Madella, M., A. Alexandre and T. Ball 2005 International code for phytolith nomenclature. *Annals of Botany* 96:253-260.

Massal, E. and J. Barrau 1965 *Food Plants of the South Sea Islands*. New Caledonia: South Pacific Commission.

Middendorf, E.A. 1983 The remarkable shooting idioblasts. *Aroideana* 6(1):9-11.

Monje, P.V. and E.J. Baran 2002 Characterization of calcium oxalates generated as biominerals in cacti. *Plant Physiology* 128:707-713.

Nakata, P.A. 2003 Advances in our understanding of calcium oxalate crystal formation and function in plants. *Plant Science* 164:901-909.

Pallant, J. 2005 *SPSS survival manual: a step by step guide to data analysis using SPSS for Windows (Version 12)*. Crows Nest: Allen and Unwin.

Piperno, D.R., A.J. Ranere, I. Holst and P. Hansell 2000 Starch grains reveal early crop horticulture in the Panamanian tropical forest. *Nature* 407(6808):894-897.

Piperno, D.R., E. Weiss, I. Holst and D. Nadel 2004 Processing of wild cereal grains in the Upper Palaeolithic revealed by starch grain analysis. *Nature* 430:670-673.

Prychid, C.J., R.S. Jabaily and P.A. Rudall In Press Cellular ultrastructure and crystal development in *Amorphophallus* (Araceae). *Annals of Botany*.

Prychid, C.J. and P.A. Rudall 1999 Calcium oxalate crystals in monocotyledons: a review of their structure and systematics. *Annals of Botany* 84:725-739.

Sakai, W.S. and M. Hanson 1974 Mature raphid and raphid idioblast structure in plants of the edible aroid genera *Colocasia, Alocasia*, and *Xanthosoma. Annals of Botany* 38:739-748.

Sakai, W.S., M. Hanson and R.C. Jones 1972 Raphides with barbs and grooves in *Xanthosoma sagittifolium* (Araceae). *Science* 178:314-315.

Sakai, W.S., S.S. Shiroma and M.A. Nagao 1984 A study of raphide microstructure in relation to irritation. *Scanning Electron Microscopy* 1984(II):979-986.

Sunell, L.A. and P.L. Healey 1981 Scanning electron microscopy and energy dispersive x-ray analysis of raphide crystal idioblasts in taro. *Scanning Electron Microscopy* 1981(III):235-244.

Tabachnick, B.G. and L.S. Fidell 1996 *Using Multivariate Statistics*. 3rd edn. New York: Harper Collins.

Therin, M., R. Torrence and R.L.K. Fullagar 1997 Australian Museum starch reference collection. *Australian Archaeology* 44:52-53.

Therin, M., R.L.K. Fullagar and R. Torrence 1999 Starch in sediments: a new approach to the study of subsistence and land use in Papua New Guinea. In C. Gosden and J.G. Hather (eds) *The Prehistory of Food: Appetites for change*, pp. 438-462. One World Archaeology Vol. 32. London: Routledge.

Tilton, V.R. and H.T. Horner, Jr 1980 Calcium oxalate raphide crystals and crystalliferous idioblasts in the carpels of *Ornithogalum caudatum. Annals of Botany* 46:533-539.

Torrence, R. 2006 Description, classification, and identification. In R. Torrence and H. Barton (eds) *Ancient Starch Research*, pp. 115-143. Walnut Creek, California: Left Coast Press.

Torrence, R., R. Wright and R. Conway 2004 Identification of starch granules using image analysis and multivariate techniques. *Journal of Archaeological Science* 31:519-532.

Wattendorff, J. 1976 A third type of raphide crystal in the plant kingdom: six-sided raphides with laminated sheaths in *Agave americana* L. *Planta* 130:303-311.

Webb, M.A. 1999 Cell-mediated crystallization of calcium oxalate in plants. *Plant Cell* 11:751-761.

Webb, M.A., J.M. Cavaletto, N.C. Carpita, L.E. Lopez and H.J. Arnott 1995 The intravacuolar organic matrix associated with calcium oxalate crystals in leaves of Vitis. *The Plant Journal* 7(4):633-648.

Whistler, W.A. 1991 Polynesian plant introductions. In P.A. Cox and S.A. Banack (eds) *Islands, Plants and Polynesians: An introduction to Polynesian Ethnobotany*, pp. 41-66. Portland: Dioscorides Press.

Yen, D.E. 1974 Arboriculture in the subsistence of Santa Cruz, Solomon Islands. *Economic Botany* 28:237-243.

Yen, D.E. 1976 Agricultural systems and prehistory in the Solomon Islands. In R.C. Green and M. Cresswell (eds) *Southeast Solomon Islands Cultural History*, pp. 61-74. Bulletin 11. Wellington: The Royal Society of New Zealand.

Yen, D.E. 1991 Polynesian cultigens and cultivars: the question of origin. In P.A. Cox and S.A. Banak (eds) *Islands, Plants and Polynesians: An introduction to Polynesian Ethnobotany*, pp. 67-98. Portland: Dioscorides Press.

9

STARCH GRANULE TAPHONOMY: THE RESULTS OF A TWO YEAR FIELD EXPERIMENT

Huw Barton

School of Archaeology and Ancient History
University of Leicester
Leicester LE1 7RH United Kingdom
Email: hjb15@le.ac.uk

ABSTRACT

This paper reports the results of an experiment designed to investigate the preservation of starch granules on stone tools left exposed and buried in an open field for a total of two years. The study was undertaken to learn in more detail the conditions under which starch granules may be preserved or degraded in the archaeological record, as one of a set of long-term experiments of organic residue preservation. A total of eight stone flakes made from silcrete and silicified tuff were used to process a starchy tuber (sweet potato). Half of the sample was buried and half left on the surface, with samples collected after four months and the remainder after two years. The results indicate that starch granules persist as a residue on stone flakes left on the surface and buried for up to two years and that the physical condition of the starch granules is similar to that encountered in archaeological contexts. An unexpected result of this study was that starch granules appear to have survived in greater numbers on artefacts left on the ground surface, suggesting that rapid burial of tools in the past does not necessarily increase the likelihood of recovering ancient starch granules.

KEYWORDS

starch, taphonomy, sweet potato

INTRODUCTION

This paper reports the results of an experiment to investigate the preservation of starch granules on stone tools that were exposed or buried in an open field for up to two years. The study was undertaken to simulate the conditions under which starch granules may be preserved or degraded in the archaeological record. While starch granules have now been encountered in a variety of deposits, the mechanism of starch preservation over long timescales is still not well understood.

At the time that this study was initiated in 1993, few taphonomic experiments on the persistence of organic residues had been undertaken and these were solely concerned with the preservation of blood (Brown 1988; Gurfinkel and Franklin 1988), though there is now some published work on organic residue taphonomy (Barton *et al.* 1998; Jahren *et al.* 1997; Lu 2003; Therin 1998; Williamson 2006). Starch granules had largely been recovered from tools buried in caves situated in relatively dry environments and from open site middens and caves (Bruier 1976; Hall *et al.* 1989; Shafer and Holloway 1979; Ugent *et al.* 1981, 1982). The recovery of starch on stone tools from the Kilu cave, Solomon Islands (Loy *et al.* 1992), and Balof2, New Ireland (Barton and White 1993) in Melanesia, were early exceptions. Since then the number of studies investigating starch granules has increased dramatically, particularly over the last ten years. Starch has been recovered from a wide variety of burial conditions and does not appear to be limited in its preservation by climatic factors (see Barton and Matthews 2006 for a review).

While ancient starch analysis is now accepted by many researchers providing records of tool function, plant use and vegetation history, there is still much to learn about the mechanism(s) of starch preservation. In particular it would be desirable to have a clearer understanding of the rate at which starch granules degrade initially after deposition and to have better knowledge of the characteristics of organic residue deposits that may facilitate long-term preservation of starch granules.

Study area

The study area was located in the Hunter Valley, approximately 10 kilometres west of the town of Singleton, within the Warkworth Mining Lease, an open cut mine for the extraction of coal (Figure 1). The experimental site lies within a broad valley floor surrounded by low undulating terrain of moderate slope nearby a meander of the Hunter River. Vegetation cover of the study area consisted primarily of tall grasses with some small saplings; a typical Hunter Valley landscape cleared for grazing. The soil consisted of wind and water borne sandy soil from decayed sandstone bedrock, was poor and slightly acidic. A small creek nearby ran intermittently, consisting of a series of contiguous stagnant ponds, some of which were saline.

The location for the experiment was established as part of an ongoing archaeological salvage of areas thought to be adversely affected by construction of a temporary dam across Sandy Hollow Creek, which could lead to local flooding of the area. This study was originally planned as part of a broader taphonomic study undertaken by Laila Haglund (2002) to monitor the post depositional movement of artefacts in Hunter Valley soils.

EXPERIMENTAL AIMS AND METHOD

This study was originally designed to be one part of a more comprehensive investigation of organic residue taphonomy including stone tools used to work wood, meat and sweet potato. Only the results of the sweet potato are discussed here. This experiment was designed to examine what happens to starch over two time intervals (16 weeks and 108 weeks) and whether there is a difference in the rate of decay/survival of starches on tools that were either buried or left

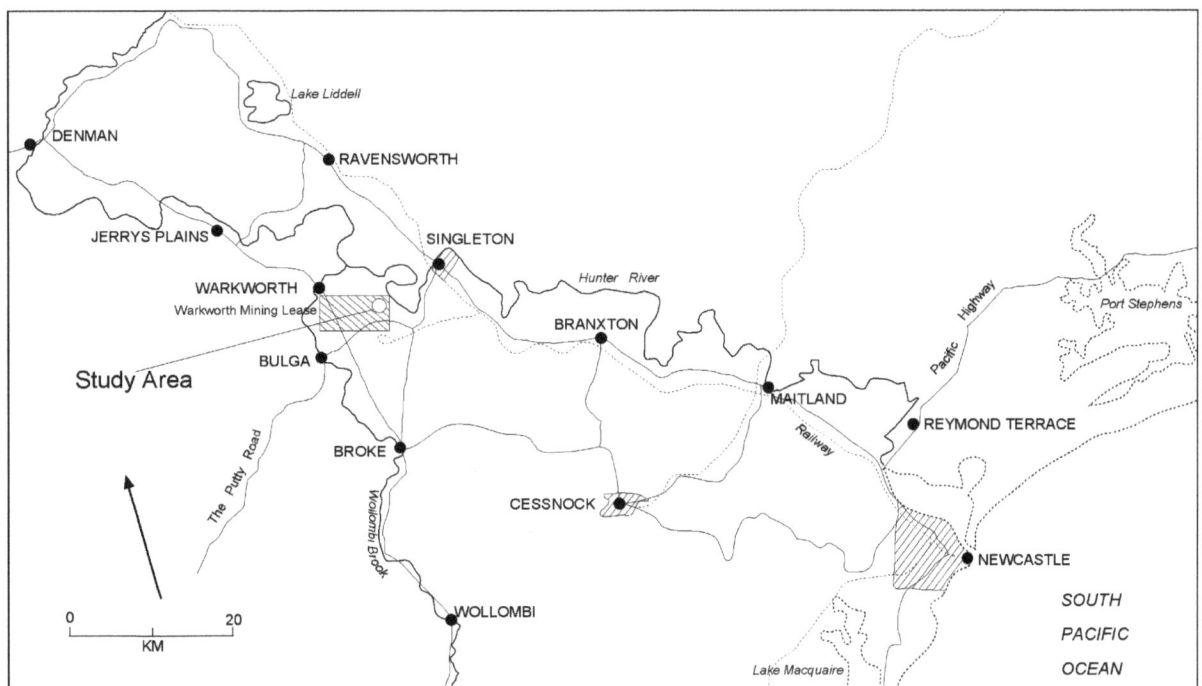

Figure 1. Map of study location.

on the ground surface. The experiment was initiated in the field on the 20/5/1993 and the final artefacts were collected on the 21/8/1995. The experimental sample was then boxed and remained unanalysed until 2007.

Experimental sample

The experimental sample consisted of eight artefacts made by the author from locally derived yellow silcrete and silicified tuff, two raw materials commonly used for the manufacture of stone flakes in the region. The experiment was duplicated on each type of raw material, and the sample was then divided into two groups (Group 1 and Group 2), where half of each Group was exposed on the ground surface and half buried. The Group 1 sample was left in the field for a total of 16 weeks (surface and subsurface) and the Group 2 tools for the maximum of 108 weeks (surface and subsurface) before collection. In this way, every combination of raw material, exposure time and exposure conditions were represented.

Each artefact was given a unique number inscribed onto a circular metal tag that was tied to the flake with metal wire (Figure 2). This system proved sufficient for the field study as no tags were detached over the two year period that the experiment was run. Two shallow trenches (one square was for the 16 week sample and the other for the 108 week sample) were excavated for the buried sample. These experimental squares were separated by an intervening distance of two metres. The squares were excavated into the upper A-unit Hunter Valley soils to a maximum depth of between 10 to 20 centimetres.

The artefacts to be buried were laid out on the trench floor with their metal tags raised above ground level and the trench backfilled. The surface sample was then laid out adjacent to the trench on the undisturbed surface (Figure 2). To ensure that these artefacts were not moved by sheet erosion or animals, a plastic mesh (approx 10 cm square) was laid out over the artefacts and securely pegged around its perimeter. This mesh was not disturbed during the experiment. Soon after being deposited, most of the surface artefacts were attracting the attention of a variety of insects, particularly ants, visibly seen removing surface residues.

At the completion of the experiment, each artefact was placed in a separate clip lock plastic bag and care was taken not to remove adhering sediment from the tool surface. A previous study (Barton *et al*. 1998) showed that adhering sediment was important in retaining use deposited

Figure 2. View of experimental area with tagged artefacts on the surface pegged down beneath plastic mesh and buried in test square.

organic residues. All tools had some sedimentary coating, including those left exposed on the ground surface. The study sample was then stored in plastic bags until 2007.

Microscopy

All artefacts in the study sample were initially scanned for likely patches of preserved organic matter under low magnification and then in greater detail at high magnification using a Zeiss Axioscop MAT reflected light microscope with X10, X20, X50 and X100 objectives. Identified starch granules were counted and measured. The analysis under transmitted light was undertaken using a Zeiss Axioscop MAT transmitted light microscope with an X63 oil immersion objective. This microscope is also set up for cross polarised and differential interference contrast illumination.

For this experiment, a single aqueous extraction, using ultra pure water and a micropipettor, was applied to the tool surface to remove a sample of residue (see Fullagar 2006 for details and variations of this approach). In this experiment, a 20 microlitre droplet of water was placed on the tool surface and left for approximately 30 seconds then agitated with the nylon pipette tip before retrieving the sediment. The sample was then placed on a glass microscope slide and allowed to air dry under a cover. The mountant used was Naphrax which has a refractive index of 1.73. To ensure that the results between all tools would be comparable, only a single extract per tool was undertaken. However, when dealing with archaeological material, more than one attempt at extraction might be necessary, or even more than one method might be employed, to increase yield or even to achieve a single positive result. A previous study (Barton *et al.* 1998) successfully undertook whole tool extractions using an ultrasonic bath. However, as further experiments are anticipated with this dataset it was not desirable to remove all residues present at this time.

Sweet potato starch

The worked material used to simulate tuber use in the past was commercially available sweet potato, *Ipomoea batatas*, bought in the local supermarket. The starch granules from sweet potato measured for this experiment ranged from three to 23 microns (Table 1). Granule shape is round to sub-round depending on orientation, with single or multiple facets (see Figure 3B). Some granules appear hemispherical when viewed from the side with a clear hilum that is eccentric to highly eccentric. The hilum usually has a distinct vacuole that may be roughly circular or with multiple fissures (see also Horrocks *et al.* 2004 for a description of archaeological specimens of *Ipomoea batatas*).

Table 1. Starch granule measurements

	No. of granules measured	Mean (µm)	Range (µm)
Ipomoea batatas	105	9	3.5-23.8
Tool #			
19 (J)	29	10.9	5.2-15.7
5 (K)	39	8.1	3.2-19.11
20 (V)	130	6.3	3.3-15.5
6 (Z)	2	5.9	5.6-6.3
7 (M)	15	8.5	4-14.4
21 (X)	1	9.8	-

RESULTS

All experimental artefacts were recovered at 16 weeks and there was no evidence of disturbance to the experimental area. On inspection after 108 weeks there was some disturbance to the central portion of the square containing the buried sample, though areas at the margins had not been affected. One of the buried artefacts, Tool #22, was not recovered during final collection. It is most likely that the missing artefact was physically removed either by animal activity or by an overly inquisitive visitor. Several artefacts from the surface sample had become partially buried by sediment from sheet erosion or by local flooding from the nearby creek.

Within the adhering sediment of all artefacts were high quantities of small rounded, transparent or opaque grains – probably quartz. These grains were typically one to two microns in size and appeared similar in form and had optical properties similar to very small starch granules (Figure 3A). The microscope was not rigged for cross-polarised illumination in reflected light mode so the presence or absence of the extinction cross could not be used for granule identification. Some organic bodies, such as fungal spores which produce a weak cross and otherwise appear starch-like (Haslam 2004; Loy 2006) and the identification of other features, such as a hilum is considered necessary for reliable starch recognition under reflected light.

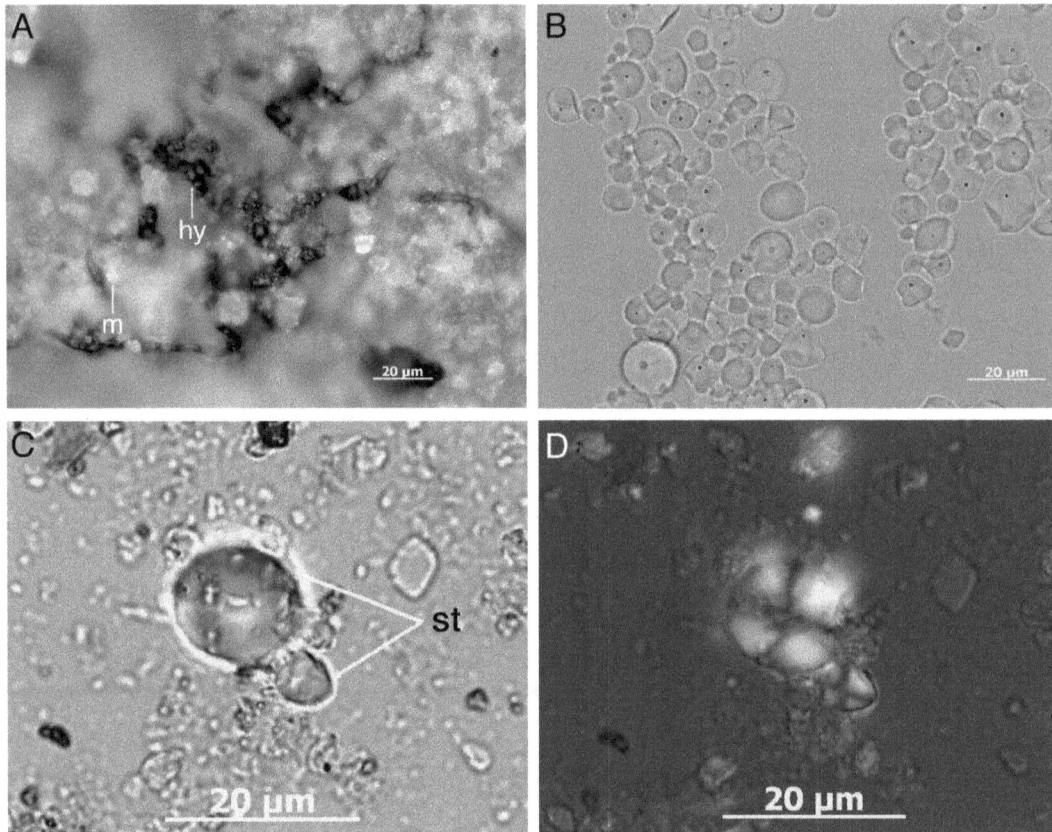

Figure 3. (A) Tool #7 (108 weeks, surface sample), typical view of tool surface under reflected light observation. Note distribution of fungus (hy) and large numbers of well rounded, bright, mineral grains (m). (B) *Ipomoea batatas* from modern reference collection. (C) Starch granules (st) extracted from surface of Tool #7 and also viewed in cross-polaried light (D).

Results from surface tools (Groups 1 and 2)

The total number of starch granules recovered from the surface sample was 196 compared with 132 granules from the buried sample (Table 2). Contrary to expectations, the results of this experiment reflect a general trend favouring preservation of starch residue on the surface tools; a pattern that is repeated in both Group 1 and Group 2 samples. If we accept these results as reflecting a general trend in residue preservation, the results seem to support early observations made by Loy (1987, 1990; and see also Gurfinkel and Franklin 1988), that rapid drying of residues following their deposition on a tool and the binding of residue with particulate matter in soils may be important factors facilitating long-term preservation (cf. Barton and Matthews 2006). It is not rapid burial itself that is necessarily responsible for residue preservation. Once a residue dries, it forms a hard plaque that may be relatively hydrophobic (Loy 1990:650) and resistant to a wide range of microbial attacks (Barton 2007; Barton and Matthews 2006). Rapid drying and desiccation have also been singled out as common factors in the preservation of other organic compounds over geological as well as archaeological timescales (Eglinton and Logan 1991). The physical scale of the residue might also be an important factor here; where the quantity of residue becomes too small to support a population of microbes.

Table 2. Starch granule counts from Warkworth experimental tools

Tool #	Raw material	Time (weeks)	Location	Reflected count§	Extract count	Total granule count
19	silcrete	16	surface	7	22	29
5	silicifed tuff	16	surface	0	151	151
20	silcrete	16	subsurface	0	130	130
6	silicified tuff	16	subsurface	0	2	2
7	silicified tuff	108	surface	0	15	15
21	silcrete	108	surface	0	1	1
8	silicified tuff	108	subsurface	0	0	0
22	Silcrete	108	subsurface	not recovered	-	-
			Totals	7	321	328

§Reflected refers to counts made directly on the tool surface before extraction to a slide

It might also be concluded that intermittent rewetting following rainfall is not in itself a major mechanism facilitating decay. Rainfall records from the town of Singleton (Figure 1) for the year of 1993 indicate that for the months of May to September, the region received a total of 279 mm rainfall, an average of 56 mm/month (Australian Bureau of Meteorology). Between May 1993 and August 1995 the region received a total of 1,214 mm rainfall. It is probable that water becomes an important factor in organic decay at the microscopic scale only when water acts as a medium for microbes - including fungus and bacteria (Eglinton and Logan 1991:320). Perhaps, when these tools lay on the surface, predominantly dry, in direct sunlight and bathed in ultraviolet radiation, this did actually aid in the long-term preservation of organic matter at the microscopic scale.

Results from buried tools (Groups 1 and 2)

A total of 132 starch granules were recovered from the buried sample. All of these were from Group 1 tools. Survival of starch in a buried context appears highly variable, and in this context at least, likely to result in one tool out of three having very low counts or possibly no starch residues at all. No starch granules were recovered from Group 2 sub-surface sample that were left for 108 weeks in the field. Starch granules appear to have been removed from tool surfaces at a far greater rate in the buried experimental sample than in the surface sample; where in fact the reverse had been anticipated. It was thought that constant exposure to the elements would be far more likely to remove starch from these tools via various weathering processes and in particular the constant re-wetting from rain.

The buried tools lay at a relatively shallow depth of 10-20 cm in what was essentially a loosely packed and recently disturbed soil horizon. It is possible that this soil was biologically very active throughout the experiment and at this depth would have contained primarily aerobic microorganisms, such as fungi. In fact, nearly all tools analysed (buried and surface samples) had traces of fungal activity on the tool surface in the form of hyphal filaments (Figure 3A). Some of these filaments had infested masses of starch granules, e.g. Tool #5 (Group 1: surface, Figure 4C) and cellulose tissue, Tool #7 (Group 2: surface). The high starch counts from Tool #20 indicate that in the right conditions, starch will be well preserved, even after 16 weeks burial in biologically active soils. The results from the sample left for 108 weeks is within the expected range for starch granules encountered in archaeological contexts (see above).

For reasons stated above, it was thought that preservation would be more likely on buried tools enclosed in soil than on tools exposed to the elements at the ground surface. We should be cautious not to over analyse this trend as there is considerable variation between individual tools. For example, Tool #19 from the surface had 29 granules whereas one of the buried tools, Tool #20 had a count of 130 granules and Tool #6 had a total count of only two granules. These results might reflect a wider range of variation between individual tools than is catered for in this

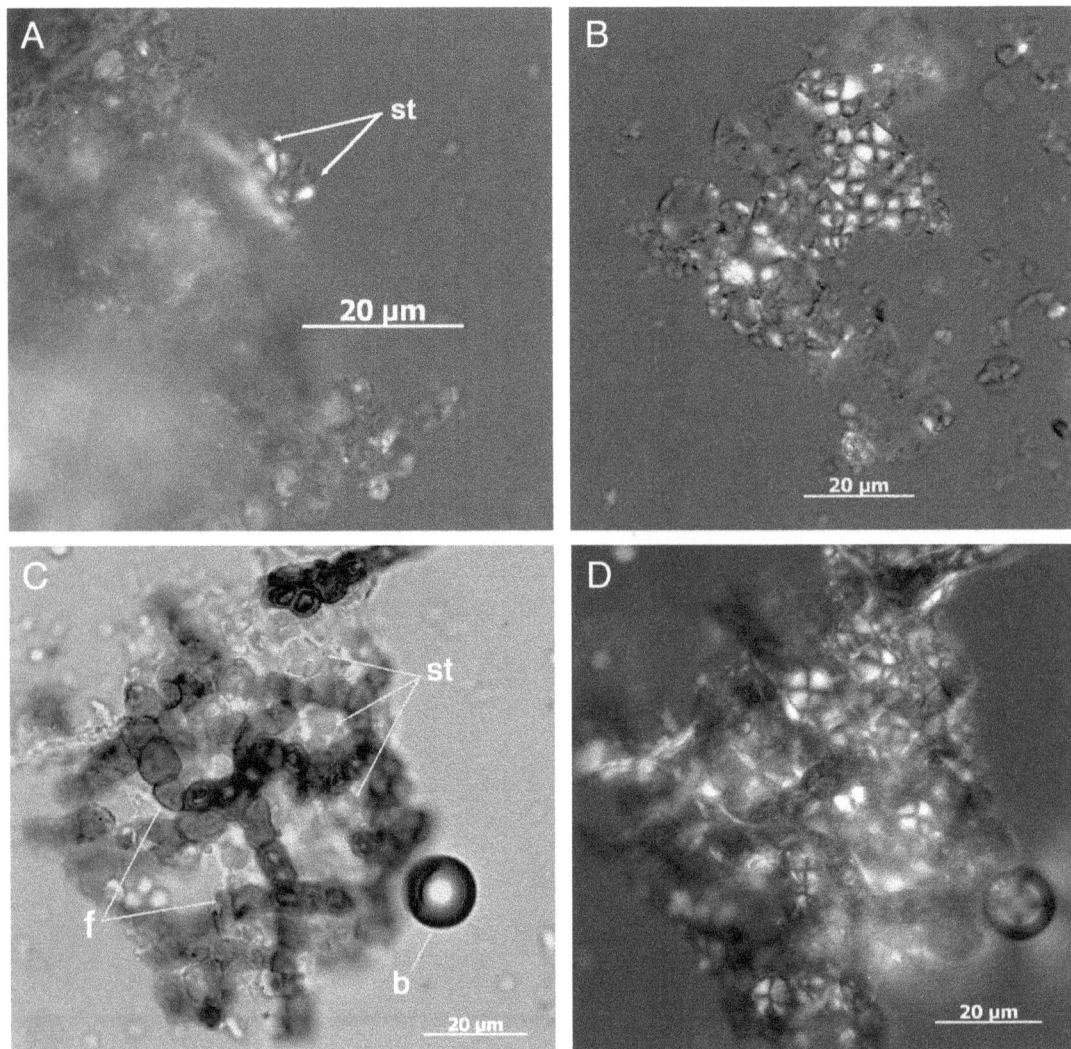

Figure 4. (A) Starch granules (st) from Tool #6 viewed in cross-polarised light. (B) Mass aggregation of granules from Tool #20. (C) Mass aggregation of starch granules (st) surrounded by fungal bodies (f). (D) Same view as C but in cross-polarised light.

experiment and that in fact there appears to be little difference in the nature of starch degradation when tools are buried or left on the surface over a period of 16 weeks.

Granule frequency

A total of 328 starch granules were recovered from the surface and subsurface tools in this experiment, which includes counts from reflected light observations of the tool edges and the single aqueous extraction from the tool surface. The aqueous extractions targeted those areas identified to have the most abundant starch residues on the tool. A total of 312 granules were recorded from the 16 week sample (Group 1), which included surface and subsurface tools, and a total of 16 granules were recorded in the 108 week tool sample (Group 2). This represents a large variation in starch counts between these two samples. No sample was initially taken at the time the tools were deposited, but it might be assumed that initial sampling would recover many hundreds if not thousands of granules. Only a single extraction was taken at one spot on the tool; multiple samples would be expected to increase counts in both experimental groups.

The overall trend is clear; a substantial removal of starch from all tools occurs relatively soon after deposition. A further sample of artefacts selectively analysed in a series of time intervals would confirm if this decay path is exponential as found with other studies of organic decay (e.g. Haslam 2004:1721). The recovery rates of starch from the 108 week sample are already representative of the types of counts likely to be encountered in many archaeological contexts (e.g. Barton 1998, 2005; Cosgrove *et al.* 2007; Pearsall *et al.* 2004; Perry 2004, Piperno *et al.*

2000; Piperno *et al*. 2004) with single to frequently less than 100 granules recovered per tool; though the method of extraction is an important factor here. Variation in the numbers of recovered granules from each tool in surface and subsurface contexts is very high, possibly reflecting small-scale processes affecting organic preservation at the scale of each tool. To generate enough data for reasonable statistical observations to be made a far larger experimental sample than this one would be necessary (see Haslam, this volume).

Physical properties of the starch granules

An important aspect of this study is that it provides an opportunity to view the physical condition of starch granules exposed to two different depositional contexts. Starch has been shown to remain in good physical condition for up to 100 years when kept as dried tuber or processed flour (Barton 2007). That study also showed that starch had retained the biochemical and physical properties of modern starch, though in some cases granules had been slightly altered internally, as indicated by poorer birefringence and a disjointed extinction cross (Barton 2007). The starch granules from this study show a range of physical alterations and in some cases provide some clues as to the mechanism(s) that may have aided their preservation.

Granules from all tools generally occur in good condition (Figure 3C) but some display a range of physical alterations including cracking (Figure 5A and B), complete breaks fragmenting the granule, and a degraded or fuzzy extinction cross (Figure 3D). As found in a previous study (Barton 2007) some starch granules from a single artefact may appear in very good condition while others from the same sample may be partially or badly degraded. I argued that this may have resulted from differential preservation of the starches within a single residue deposit on a tool surface. Some granules may lie in situations close to or near the surface of a residue and thus be available to attack by microorganisms whereas other granules may be sealed within the residue and shielded from biological decay.

Granules enclosed in sediment were common in this study. In some cases granules appeared, when slide mounted, to be partially surrounded by very small particulates (e.g. Figure 5, B-D). Sometimes the granule appeared fully enclosed, as shown in Figure 5, C-D. It is suggested that this sediment layer may be enough to prevent microbes attacking starch granules forming a protective barrier around the granule. This observation is what would be expected if Particulate Organic Matters (POMs) also play a role in organic preservation of residues at the micro scale (see Barton and Matthews 2006, Golchin *et al*. 1998, Waters and Oades 1991). This shield of micro-particulates is likely an important aspect of a general mechanism of long-term organic preservation of all residues, not just starch granules (see figures in Piperno and Holst (1998) for similar examples of starch granules surrounded by small particulate matter).

Given that organic breakdown and decay in soils is argued to be relatively rapid, occurring over timescales of days rather than months (see Haslam 2004), unless some other process can retard either microbial activity or chemical breakdown, the persistence of starch after two years is counter to many general expectations of organic decay.

Phytoliths were present on several tools. The dumbbell form was most common and is associated with grasses (Piperno 2006). Given the depositional context within a grassed paddock, this is highly unsurprising – but again indicates caution with any interpretation of adhering matter on a tool surface. The presence of organics alone (especially isolated occurrences) only indicates the presence of organic matter in the deposit, not the presence of a use-derived residue. For each tool a contextual case must be made as to why this adhering matter was or was not derived from its use.

CONCLUSIONS

Starch granules were documented on six of the seven artefacts used for this study; the eighth piece could not be found at the conclusion of the experiment. This experiment confirms that modern starch granules can survive up to 108 weeks in an open field site, subject to the attentions of

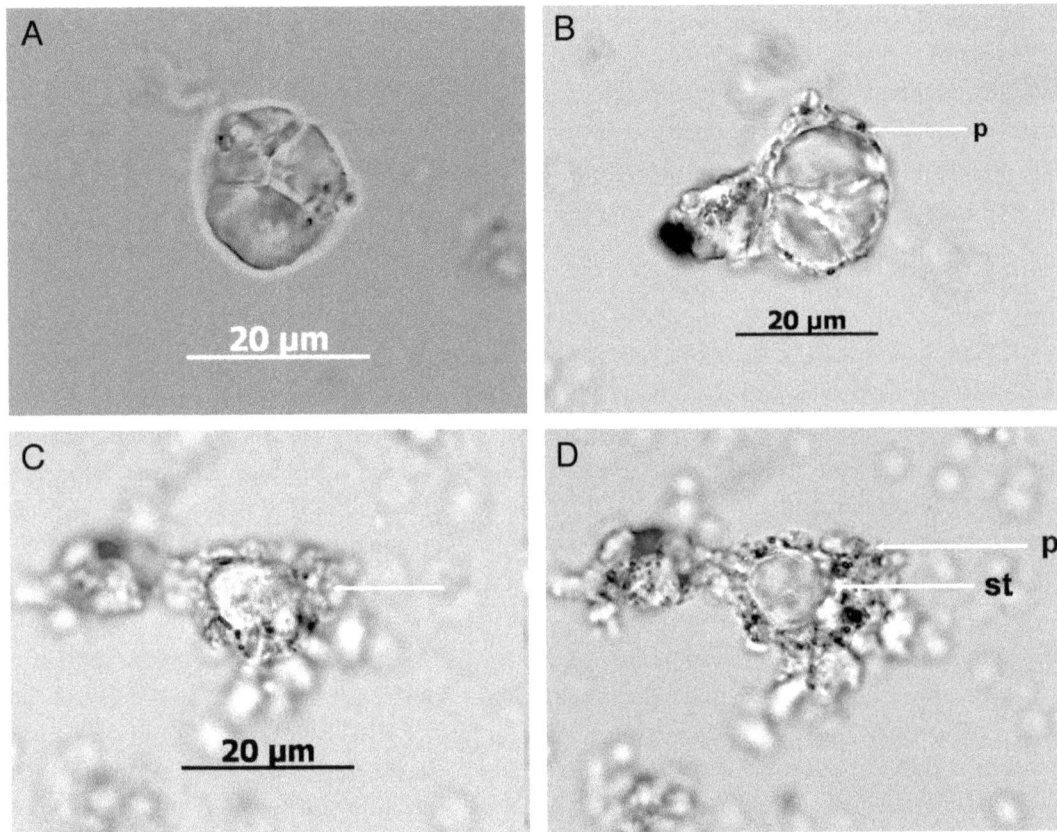

Figure 5. (A) Damaged starch granule from Tool #19 (4 months, surface sample). (B) Damaged starch granule from Tool #5 (4 months, subsurface sample), cracked and surrounded by particulate matter (p). (C) and (D) consecutive views of starch granule from Tool #20 (4 months, subsurface sample) completely encased in particulate matter (p). View in C focussed on top of encased granule (indicated by arrow) and view in D after focussing downwards into the object, revealing starch granule (st) within.

the meso-fauna (ants in particular) and micro-fauna (fungi were visible on nearly all recovered artefacts). Recovery rates from a single aqueous extract range from two granules to 151 granules. Preservation is variable for each artefact, suggesting that the nature of preservation may be peculiar to each piece, even within the same soil matrix.

A major finding was the discovery that tools left on the surface apparently had far better starch preservation than those buried in the A-horizon soils. This was unexpected (though see Haslam 2004 for a counter view) as I had anticipated that the act of burial would provide a more secure environment for long-term starch preservation (e.g Lu 2003). However, these tools were buried at shallow depth, between 10 to 20 cm, well within the upper organic-rich and well aerated soil layer where the activity of microorganisms would be high (Barton and Mathews 2006:79-83).

Further experiments with starch granules will help us understand the complex nature of organic preservation at the microscopic scale and are essential if we are to fully understand the mechanisms and conditions of organic residue preservation. The results of this and other experiments reveal the complexities of residue formation but should increase confidence in the utility of this class of archaeological information to make important contributions to current scholarship.

ACKNOWLEDGEMENTS

I would most like to thank Dr Laila Haglund for her enormous patience waiting for the results of this study to actually come to light in a publication and for allowing me to participate in her own experimental program. I would also like to thank Warkworth Mining Limited for permission to undertake the work on lands under their management. The Mine Management, and notably their environmental and geological staff, was very helpful with advice and in ensuring that the experimental location remained as undisturbed as possible. The project was initiated when the author was working in contract archaeology but all analyses were carried out while funded by the Wellcome Trust at the University of Leicester, UK.

REFERENCES

Barton, H. 2005. The case for rainforest foragers: the starch record at Niah Cave, Sarawak. *Asian Perspectives* 44:56-72.

Barton, H. 2007. Starch residues on museum artefacts. *Journal of Archaeological Science* 34: 1752-1762.

Barton, H. and P.J. Matthews 2006. Taphonomy. In R.Torrence and H. Barton (eds) *Ancient Starch Research*, pp.75-94. Walnut Creek: Left Coast Press Inc.

Barton, H., Torrence, R., and Fullagar, R. 1998. Clues to stone tool function re-examined: comparing starch grain frequencies on used and unused obsidian artefacts. *Journal of Archaeological Science* 25: 1231-1238.

Barton, H. and J.P. White 1993. Use of stone and shell artefacts at Balof 2, New Ireland, Papua New Guinea. *Asian Perspectives* 32: 169-181.

Brown, P. 1988. Residue analysis of stone artefacts from Yombon, West New Britain. BA Hons Thesis. Department of Anthropology, University of Sydney.

Bruier, F.L. 1976. New clues to stone tool function: plant and animal residues. *American Antiquity* 41: 478-484.

Cosgrove, R., J. Field and A. Ferrier 2007. Environmental history of the humid tropics region of north-east Australia: the archaeology of Australia's tropical rainforests. *Palaeogeography, Palaeoclimatology, Palaeoecology* 251: 150-173.

Eglinton, G and G.A. Logan,1991. Molecular preservation. *Philosophical Transactions of the Royal Society of London* B 333: 315-328.

Fullagar, R. 2006. Starch on artefacts. In R.Torrence and H. Barton (eds) *Ancient Starch Research*, pp.177-203. Walnut Creek: Left Coast Press Inc.

Golchin, A., A.J.Baldock and J.M. Oades 1998. A model linking organic matter decomposition, chemistry, and aggregate dynamics. In R.Lal, J.M. Kimble, R.F. Follett and B.A. Stewart (eds) *Soil Processes and the Carbon Cycle*, pp.245-266. New York: CRC Press.

Gurfinkel, D.M. and U.M. Franklin 1988. A study of feasibility of detecting blood residues on artefacts. *Journal of Archaeological Science* 15: 83-97.

Haglund, L. 2002. Archaeological Investigations within Warkworth Mining Lease: Aboriginal Sites along Sandy Hollow Creek, NSW. Vol.V: Taphonomic experiments and observations. Report to Warkworth Mining Limited, Haglund and Associates.

Hall, J., S. Higgins and R. Fullagar 1989. Plant residues on stone tools. In W.Beck, A.Clarke and L. Head, (eds) *Plants in Australian Archaeology*, pp.136-160. St Lucia: University of Queensland, Tempus 1.

Haslam, M. 2004. The decomposition of starch grains in soils: implications for archaeological residue analyses. *Journal of Archaeological Science* 31: 1715-1734.

Horrocks, M., G. Irwin, M. Jones and D. Sutton 2004. Starch grains and xylem cells of sweet potato (Ipomoea batatas) and bracken (Pteridium esculentum) in archaeological deposits from northern New Zealand. *Journal of Archaeological Science* 31: 251-258.

Jahren, A. H., N. Toth, K. Schick, J.D. Clark and R.G. Amundson 1997. Determining stone tool use: chemical and morhpological analyses of residues on exprimentally manufactured stone tools. *Journal of Archaeological Science* 24: 245-250.

Loy, T.H. 1987. Recent advances in blood residue analysis. In W.R. Ambrose and J.M.J. Mummery (eds) *Archaeometry: Further Australasian studies*, pp.57-65. Canberra: Australian National University.

Loy, T.H. 1990. Prehistoric organic residues: recent advances in identification, dating, and their antiquity. In E. Pernicka (ed.), *Archaeometry '90*, pp.645-656. Basel and Boston: Springer Verlag.

Loy, T.H. 2006. Optical properties of potential look-alikes. In R.Torrence and H. Barton (eds) *Ancient Starch Research*, pp.123-124. Walnut Creek: Left Coast Press Inc.

Loy, T.H., M. Spriggs and S. Wickler 1992. Direct evidence for human use of plants 28,000 years ago: starch residues on stone artefacts from the northern Solomon Islands. *Antiquity* 66: 898-912.

Lu, T. 2003. The survival of starch in a subtropical environment. In D.M. Hart and L.A. Wallis (eds), *Phytolith and Starch Research in the Australasian-Pacific-Asian Regions: The State of the Art*, pp.119-126. Terra Australis 19. Canberra: Pandanus Books.

Pearsall, D.M., K.Chandler-Ezell and J.A. Zeidler 2004. Maize in ancient Ecuador: results from residue analysis of stone tools from the Real Alto site. *Journal of Archaeologcial Science* 31: 423-442.

Perry, L. 2004. Starch analyses reveal the relationship between tool type and function: an example from the Orinoco valley of Venezuela. *Journal of Archaeological Science* 31: 1069-1081.

Piperno, D.R. and I. Holst 1998. The presence of starch granules on prehistoric tools from the humid Neotropics: indicators of early tuber use and agriculture in Panama. *Journal of Archaeological Science* 25: 765-776.

Piperno, D.R., A.J. Ranere, I. Holst and P. Hansell 2000. Starch granules reveal early root crop horticulture in the Panamanian tropical forest. *Nature* 407: 894-897.

Piperno, D.R. 2006. *Phytoliths. A Comprehensive Guide for Archaeologists and Paleoecologists*. Oxford: AltaMira Press.

Piperno, D.R., I. Weiss, I. Holst, and D. Nadel 2004. Processing wild cereal grasses in the Upper Palaeolithic revealed by starch grain analysis. *Nature* 430: 670-673.

Shafer, H.J. and R.G. Holloway 1979. Organic residue analysis in determining stone tool function. In B. Hayden, (ed.) *Lithic Use-wear Analysis*, pp.385-399. London: Academic Press.

Therin, M. 1998. The movement of starch grains in sediments. In Fullagar, R. (Ed.) *A Closer Look: Recent Australian Studies of Stone Tools*, pp.61-72. Sydney: Sydney University Archaeological Methods Series 6.

Ugent, D., S. Pozorski and T. Pozorski 1981. Prehistoric remains of the sweet potato from the Casma Valley of Peru. *Phytologia* 49: 401-415.

Ugent, D., S. Pozorski and T. Pozorski 1982. Archaeological potato tuber remains from the Casma Valley of Peru. *Economic Botany* 36: 182-192.

Waters, A.G. and J.M. Oades 1991. Organic matter in water-stable aggregates. In W.S. Wilson (ed.), *Advances in Soil Organic Matter Research: The Impact on Agriculture and the Environment*, pp.163-174. Cambridge: Royal Society of Chemistry.

Williamson, B. 2006. Investigation of potential contamination on stone tools. In R.Torrence and H. Barton (eds) *Ancient Starch Research*, pp.89-90. Walnut Creek: Left Coast Press Inc.

10

Toward using an oxidatively damaged plasmid as an intra- and inter-laboratory standard in ancient DNA studies

Loraine Watson[1†], Julie Connell[2†], Angus Harding[3] and Cynthia Whitchurch[4]

1. Molecular Aquatic Research Group
Department of Genetics, University of Stellenbosch
South Africa
Email: 15734331@sun.ac.za

2. Queensland Health Forensic and Scientific Services
Cooper Plains QLD 4108 Australia

3. Queensland Institute of Medical Research
Herston Rd, Herston QLD 4006 Australia

4. Department of Microbiology
Monash University
Melbourne VIC 3800 Australia

† Authors have contributed equally to the production of this manuscript

PROLOGUE

The following paper was originally presented by Dr Thomas H. Loy at the 6th International Conference on Ancient DNA and Associated Biomolecules held in Israel, July 2002. It is included here with editorial and formatting changes with the intention of demonstrating the passion and lateral thinking that underpinned Tom's approach to the field of Molecular Archaeology. The paper represents research from three honours projects conducted during the late 1990s and early 2000s. Building a modern model for ancient DNA that could be used during routine procedures was a concept that Tom had long held as an important step forward for the burgeoning discipline. With the equipment and technology that was available at the time, the Damaged Plasmid Model concept was completely viable and worthy of detailed validation. As with all historical accounts, an understanding of more recent developments in molecular techniques and equipment will highlight the need for considerable optimisation of the model before it can be used as an inter-laboratory standard for ancient DNA.

ABSTRACT

For some years laboratories working with ancient DNA have been optimising extraction methods using either undamaged modern DNA or authentic ancient DNA. This approach is unsatisfactory for a number of reasons, chief being the inherent variability from sample to sample. In addition, quantitative comparison of methods is generally impossible using typically small samples of ancient DNA, as well as being ethically questionable. We have now perfected a method whereby we can oxidatively damage the plasmid pUC19 using copper sulfate, ascorbic acid and hydrogen peroxide to create artificially damaged DNA that mimics the behaviour of ancient DNA. We have used this damaged plasmid to assay our extraction methods to quantitatively monitor the yield and degree of damage induced during the extraction, purification, and storage of DNA. We suggest that the damaged plasmid can be used as a monitoring standard within the one laboratory, as we have done, and more importantly to compare yields and efficiencies between different laboratories.

KEYWORDS
ancient DNA, plasmid, model, laboratory standard

INTRODUCTION
The analysis of degraded, forensic and ancient (archaeological, palaeontological) DNA (aDNA) using polymerase chain reaction (PCR) amplification has demonstrated significant applications to such diverse areas as origins of domestication, palaeoepidemiology, evolution and population studies, and forensic and archaeological identification of taxa of origin from tissue, bones and blood residues. Despite these achievements, however, considerable obstacles remain to be overcome. One such obstacle is the state of degradation of aDNA samples that has prohibited the amplification of DNA fragments larger than a few hundred base pairs - making the elucidation of genetic sequences a time-consuming, and at times, impossible assignment. Damage and typically small sample sizes significantly reduce amplification yield and create a situation where extremely low-levels of copurified inhibitors and contaminants have significant detrimental effects. Evidence suggests that the major culprit in DNA degradation, especially in aDNA, is oxidative damage. This type of damage causes weaknesses in the chemical bonds that hold the nucleotides together, leading to fragmentation of DNA upon the application of heat – the first (denaturing) step in any PCR amplification (Pääbo 1989; Rogan and Salvo 1990a, 1990b).

Here, we propose that the use of deliberately oxidatively damaged modern DNA that mimics the degradation seen in ancient samples provides one avenue for increasing the reliability of conclusions drawn from aDNA studies, by presenting a standard for assessing extraction, amplification and verification of aDNA protocols and results.

DNA DAMAGING AGENTS
To create a model system of artificially damaged DNA, an understanding of the degradative process is needed. Studying aDNA creates a problem because it is difficult to determine the amount or type of degradation exhibited by the DNA sample. A way around this issue can be through using analogies drawn from studies of modern DNA that has undergone chemical/physical alteration (i.e. damage; Rogan et al. 1990b). Since the issue of damaged DNA is of major concern to those studying the effects of carcinogens and mutagens, a considerable body of research exists on the types of chemical alterations to DNA and the agents that induce them.

OXIDATION
Damage caused through oxidation is the most common form recognised. A large proportion of oxidative damage in living systems occurs through the action of hydroxyl radicals ($\cdot$OH) and this also appears to be the major contributing factor in the oxidative damage of aDNA (Rogan et al. 1990b). $\cdot$OH attacks the integrity of the DNA molecule by adding to either the C5 or C6 of pyrimidines via their shared double-bond or the C4, C5 and C8 of purines (Dizdaroglu 1993). Radicals thus created are unstable and undergo further reactions in the presence of oxygen (O_2) to form other reactive radicals. For example thymine (T), once attacked by $\cdot$OH, goes through a series of reactions in the presence of O_2 to produce thymine peroxyl radicals. The presence of radicals has the net effect of causing strand breaks in the DNA molecule (Halliwell and Gutteridge

1989). The sugar backbones of DNA can also be attacked by •OH through removal of hydrogen (H) atoms from any of the five carbons (Dizdaroglu 1993).

HYDROLYSIS

DNA is also hydrolysed by water both contained within the DNA matrix and surrounding the molecule. Again, a large body of work exists on the effects of hydrolysis on DNA (Demple and Linn 1980; Levin and Demple 1990; Lindahl 1993; Lindahl and Andersson 1972; Lindahl and Nyberg 1972). However, it has also been shown that DNA can bind to other substrates, for example, the inorganic hydroxyapatite matrix of bone (Tuross 1994), thus reducing the effects of hydrolysis, extending the period of time in which DNA may still be recovered. It is generally believed that hydrolytic attack affects the base-sugar bonds of DNA causing the production of abasic sites (also known as apurinic/apyrimidinic (AP) sites) (Doetsch and Cunningham 1990; Lindahl 1993).

IONISING RADIATION AS AN AGENT OF •OH

A considerable number of papers discuss the effects of ionising radiation on DNA in living cells. A large proportion of the damage is thought to be caused by free radicals - particularly •OH (Dizdaroglu 1991; Steenken 1987). Because of this, the damage can be classified as oxidative (Halliwell and Aruoma 1991). Types of damage caused by radiation include alkali- and heat-labile sites (Lafleur *et al.* 1979), various base modifications, single-strand and double-strand breaks and AP sites (Breimer 1988). The alkali/heat-labile lesions are generally brought about by chemical reactions between •OH and the sugar moiety of the nucleotide (Lafleur *et al.* 1979). DNA containing such lesions, when subjected to alkaline and/or high heat environments (as in a PCR), dissociates at these positions. This phenomenon is particularly relevant and most pronounced in single stranded DNA (Lafleur *et al.* 1979). The most common forms of base modifications are guanine converted to 8'-hydroxy guanine and thymine converted to thymine glycol (Halliwell 1993; Lindahl 1993).

CO-FACTOR DEPENDENCE

Research has also been conducted into the effects of certain chemicals upon the structure of DNA. Generally, the chemical action occurs by way of an oxidation reaction (e.g. potassium bromate [$KbrO_3$]) (Ballmaier and Epe 1995). One substance mentioned throughout the literature is hydrogen peroxide (H_2O_2) - a significant source of •OH. However, research also shows that co-factors aid in the production of free radicals. Importantly, certain transition metal ions, particularly iron (Fe^{2+}) and copper (Cu^{2+}) (Halliwell *et al.* 1989), are necessary to catalyse these oxidative reactions (Halliwell 1993; Halliwell *et al.* 1989; Meneghini and Martins 1993). The presence of ascorbic acid in Cu^{2+}-catalysed oxidative reactions has also been shown to increase the extent of damage.

DAMAGE AFFECTS REPLICATION

Studies have been carried out both *in vivo* and *in vitro* to determine the effects of lesions on DNA replication. One important question centres upon whether the various types of damage cause miscoding or non-coding events during replication. That is, does damage cause the incorporation of the wrong nucleotide or does copying stop completely at such lesions? Evidence suggests, for example, that the presence of a thymine glycol causes the termination of copying *in vitro* and is lethal *in vivo*. On the other hand, the presence of 8'-hydroxy guanine residues is thought to cause miscoding but does not inhibit replication (Breimer 1988).

Under normal conditions (room temperature and pressure) damaged strands of DNA are held together because the complementary strand is often undamaged. Heat denaturation during the course of amplification causes nicked or AP portions to break into smaller single strand fragments. The result is that both template length and DNA polymerase activity are reduced. The presence of long DNA fragments extracted from ancient sources (e.g. Nielsen *et al.* 1994) cannot guarantee successful amplification of long products – a well-known phenomenon that reflects the chemical alteration of DNA, especially in regions not protected by histone or histone-like proteins within the chromatin structure (Wolffe 1995).

DEVELOPING A DAMAGED PLASMID MODEL

Over the past ten years we have been using a combination of modern and ancient sources of DNA to optimise DNA extraction, purification, amplification and sequencing reactions. The advantage of using modern DNA is that usually it can be accurately quantified, allowing losses during purification and the efficiency of PCR processes to be closely monitored. The use of modern DNA, however, even in very dilute concentrations (1 fg/µl) does not provide a real comparison for aDNA because the modern DNA is largely intact and undamaged.

On the other hand, using authentic aDNA sources for methods development and optimisation is problematic for several reasons. First, yield is often variable and difficult to quantify – even from the same sample. Second, because of the unique relationship between taphonomic processes and sample material (soft tissue, bone and blood residue on stone tools), no two samples can be said to have undergone the same history of degradation. A general trend of increasing DNA damage can be extrapolated from our experiences with extraction, purification and amplification with ancient samples. Blood residues generally provide longer sequences of intact DNA, followed by bone, while mummified tissue often contains the most nucleic acid modification (damage). Third, archaeological and forensic samples have intrinsic value that introduces a serious ethical concern regarding their use for purposes such as optimisation and quantification experiments. Finally, issues revolving around the detection and prevention of contamination events have cast doubts on the veracity of many aDNA analyses. Current practice dictates the repetition of key experiments at independent laboratories. While admirable in intention, the requirement necessitates the production of both identical laboratory conditions and utilisation of common methodologies, requirements that are often beyond the capacity of isolated research teams.

Developing and using a modern damaged DNA model that is representative of aDNA provides a multi-locational means for testing protocols and experimental hypotheses that could not, and indeed should not, be undertaken on ancient specimens. The current paper presents results on the production and use of a Damaged Plasmid Model (DPM). Our aim was to produce a reliable mimic for authentic degraded DNA samples that could be easily substituted into current laboratory procedures.

METHODS

The plasmid pUC19 was chosen as the model to better understand and characterise the damage to aDNA because its sequence is known, it can be produced in abundance at high purity, and amplification primers are easily designed.

DNA DAMAGE REACTION PROTOCOL

Linearised plasmid DNA is oxidatively damaged in 20 µl reactions. The •OH that damage the DNA are generated by a combination of H_2O_2, $CuSO_4$ and ascorbic acid. The reaction is stopped on ice and with the addition of EDTA (Aruoma *et al.* 1991; Prutz 1990). The extent of damage can be controlled by altering the concentration of H_2O_2 and the incubation time.

PURIFICATION PROCEDURES

1. Ethanol precipitation

A solution containing the DNA sample to be purified, 1/10 volume 3 M sodium acetate and 2 volumes 100% EtOH was incubated at -70°C for 30 min. The solutions were centrifuged at 15,000 X g in a cool room (4°C) for 30 min. The supernatant was removed and pellets were washed in 100 µl 70% EtOH. A further centrifugation was performed for 15 min. The second supernatant was removed, pellets were air-dried and resuspended in Milli-Q water to an equivalent volume as the original DNA sample.

2. Sephadex G50 columns

DNA in solutions was purified using Amersham G-50 Sephadex columns, in accordance with the manufacturer's directions.

POLYMERASE CHAIN REACTION

1. Enzymes and Primers

AmpliTaq DNA Polymerase, Stoffel Fragment (10 U/µl) was used for PCR amplification. Five plasmid primers were designed using both MacVector and Amplify 1.2 software. The sequences are:

pUC19A forward	5'-GGCCGAGCGCAGAAGTGGTC-3'
pUC19B reverse	5'-GCAATGGCAACAACGTTGCG-3'
pUC19C reverse	5'-ACAACATGGGGGATCATGTA-3'
pUC19D reverse	5'-CGCCGGGCAAGAGCAACTCG-3'
pUC19E forward	5'-GCAAGCAGCAGATTACGCGCAG-3'

Amplification of the following products is possible; (i) 823 bp with primers pUC E and pUC D; (ii) 420 bp with primers pUC A and pUC D; (iii) 217 bp with primers pUC A and pUC C; and (iv) 122 bp with primers pUC A and pUC B.

2. Amplification protocol

PCR was performed at 50 µl final volumes. To 10 µl volumes of DNA in solution (variable concentrations) 2.25 mM MgCl$_2$, 1x PCR buffer, 0.2 mM combined dNTPs, 30 pM primers (5' and 3'), 0.03 U/µl Taq DNA polymerase and milliQ H$_2$O were added. PCR amplification for plasmid DNA consisted of 36 cycles with an initial cycle of 95°C for 3 min, 58°C for 20 seconds and 72°C for 2 min. The remaining cycles consisted of 95°C for 15 seconds, 58°C for 10 seconds and 72°C for 2 min. A final extension period (72°C) was 4 min before holding at 4°C.

EXTRACTION PROCEDURES:

The extraction buffer III (EBIII) extraction method was developed by one of the authors and has been described previously (Matheson and Loy 2001). The sample is incubated at 57°C in a solution of EBIII (10 mM Tris-HCl, pH 8.0; 100 mM NaCl, 250 mM EDTA, pH 8.0; 1:20 10% SDS; 0.45% NP-40; 0.45% Tween-20; 0.45% Triton X-100; 100 mM DTT) and proteinase K (25 µg/ml) for 8 hours. The DNA is then purified using a SpinBind silica membrane column.

The EBIIIβ extraction method is based on the same buffer used in the above extraction. However, this buffer contains β-mercaptoethanol in addition to the other ingredients. The presence of this chemical facilitates the dissociation of DNA from proteins by breaking the di-sulfide bonds. The sample and buffer are incubated at 90°C for 45 minutes. The DNA is then purified using SpinBind silica membrane columns, as in the EBIII extraction method.

RESULTS AND DISCUSSION

The Damaged Plasmid Reaction

Using the techniques described above, the plasmid can be damaged to varying degrees. Published data on the extraction of degraded DNA consistently describe samples that contain a large proportion of low molecular weight fragments but also contain a small proportion of high molecular weight sequences (Cherry 1994; DeSalle and Grimaldi 1994; Handt *et al.* 1996; Pääbo 1989, 1990). The current Damaged Plasmid Model successfully mimics both the range and the relative proportions of DNA fragments found in ancient and forensic DNA samples (Figure 1). The fragment sizes present in the plasmid (Figure 1, lane 3) range from 2000 bp to less that 100 bp. Given that the total size of pUC19 is roughly 3000 bp, the damaged DNA size range is similar to the range of fragment sizes seen in other published aDNA extraction data (e.g. Nielsen *et al.* 1994).

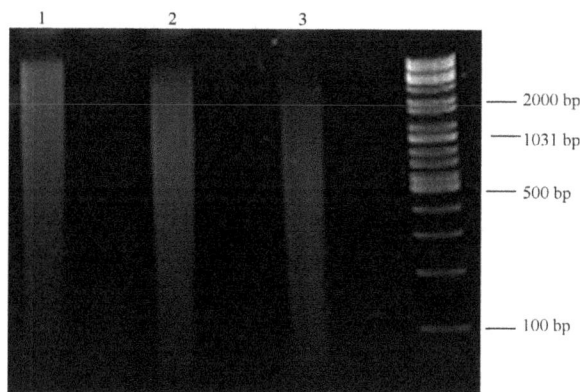

Figure 1. Effects of changing H_2O_2 concentration and/or time of incubation on the extent of damage to the plasmid. The increased extent of damage seen in lanes 2 and 3 when compared with that in lane 1 reflects the effects that different reaction conditions can have on the damage reaction. Polyacrylamide 4-20% gradient gel.

Plasmid digestion prior to the damage reaction was an important innovation that removed the possibility of accidental retention of supercoiled DNA. An accurate model required the retention of a long smear of DNA including fragments of ≥ 1000 bp. The presence of supercoiled DNA in the smear would have introduced an unidentifiable structural variable into the reaction, complicating the results and reducing the model's merit.

Increasing the concentration of H_2O_2 in the damage reaction or lengthening the period of incubation has the effect of altering the ratio of long versus short fragments in the resultant damaged DNA (Figure 1). Altering these two variables allows the extent of damage to be controlled. Decreasing the concentration of DNA has a similar effect (data not shown).

The effect of heat on damaged DNA

The first step of any PCR reaction is heating to denature the DNA. This step causes DNA with single strand breaks to fragment. Oxidatively damaged DNA (like that present in the DPM and in aDNA) contains many heat-labile sites that form single strand breaks upon heating. Therefore the first heating step of PCR causes greater fragmentation of the already damaged aDNA. Heating of the damaged plasmid results in a shift of the range of fragment sizes present (Figure 2). Many PCR reactions use an extended heating step at the beginning of the reaction. As can be seen in Figure 2, extended initial heating decreases the amount of longer DNA fragments. The damaged plasmid was also used to assess the effect of longer PCR cycles on the template DNA. The amount of high molecular weight fragments is significantly reduced following a longer PCR reaction.

The effect of concentration on amplification

Dilution series PCR analysis provides a striking illustration of the amplification differences between undamaged and damaged DNA. The results indicate that while low molecular weight fragments can be amplified from degraded DNA samples - even at very low concentrations (0.1 x 10^{-6} µg/µl), high molecular weight sequences are only amplified from undamaged DNA. The precise reason for this phenomenon is not known but is thought to be derived from the types of oxidative damage known to occur in aDNA. Single strand nicks in the degraded, double stranded

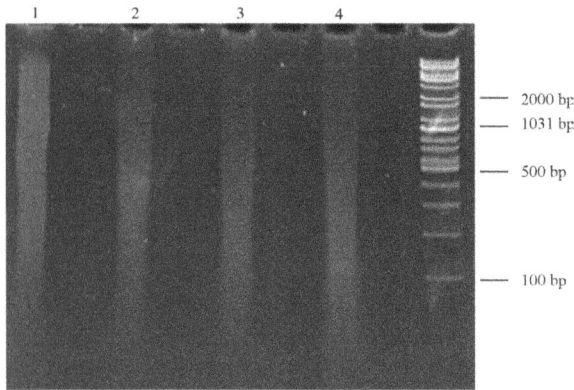

Figure 2. Effects of heating/PCR on damaged DNA. Heating the damage reaction to 95°C results in denaturation and strand breaks at heat-labile sites caused by oxidative damage. The unheated, damaged DNA in lane 1 has a relatively high molecular weight compared with the damaged DNA in lane 2 that has been heated to 95°C for 5 minutes or that in lanes 3 and 4 that have been subjected to the heating and cooling cycles of a short and long PCR reaction (respectively). Polyacrylamide 4-20% gradient gel.

DNA are 'invisible' after extraction since the short lengths of DNA between nicks are held together by their complementary base pairs. denaturing the DNA during PCR causes the strands to 'fall apart' at the nicked sites, leaving only short strands available for amplification (see Figure 2). Compounding the amplification problem is the ratio of long to short fragments of extracted aDNA. Since shorter pieces are in relative abundance, they generally will be preferentially amplified during any PCR, wihch dramatically reduces the likelihood of detecting longer DNA sequences. The compilation of amplification results for damaged and undamaged plasmid DNA at 0.02 pg/μl (Figure 3) can be compared directly with Pääbo's (1990) amplifications of mitochondrial DNA from 7,000 year old and modern human brains. The same pattern can be seen in his results as in both the Damaged Plasmid Model amplifications, where low molecular weight fragments can be retrieved from degraded DNA but not higher molecular weight sequences. The similarity between these results is striking and is further support for the Damaged Plasmid Model as a valid tool for understanding damaged DNA.

USING THE DAMAGED PLASMID MODEL TO ASSESS EXTRACTION TECHNIQUES

As an example of how the DPM can be used to assess the damage caused to DNA by routine laboratory procedures, a modification to a routine extraction method used in the University of Queensland Archaeological Sciences Laboratory for many years was assessed. Recently β-mercaptoethanol has been used as an additional ingredient in the EBIII (see methods) extraction buffer due to its ability to dissociate proteins and disulfide bonds. The modified method has produced successful results in extracting DNA from ancient samples (Loy, unpublished data).

To fully characterise the new buffer (EBIIIβ) and the effect it has on DNA, side-by-side extractions were carried out on both damaged and undamaged pUC19. Each extraction used 8 ng of substrate DNA. The extractions were assessed via PCR. While the original EBIII buffer allows amplification of up to 400 bp of damaged DNA following extraction, the EBIIIβ buffer and method appear to cause more damage to the DNA (Figure 4). In comparison, undamaged DNA subjected to either extraction buffer could be amplified. The results indicate that while the addition of β-mercaptoethanol produces a buffer that is effective for separating aDNA from its encompassing organic matrix, care must be exercised in order to minimise further damage to the already fragile DNA.

CONCLUSIONS

Using techniques designed to cause predictable oxidative damage to modern DNA, it has been possible to create a Damaged Plasmid Model that accurately and repeatably mimics extraction and amplification results commonly obtained from aDNA. Damage present in the DPM is consistent with damage caused by hydroxyl radical attack that produces single strand nicks and AP sites in DNA.

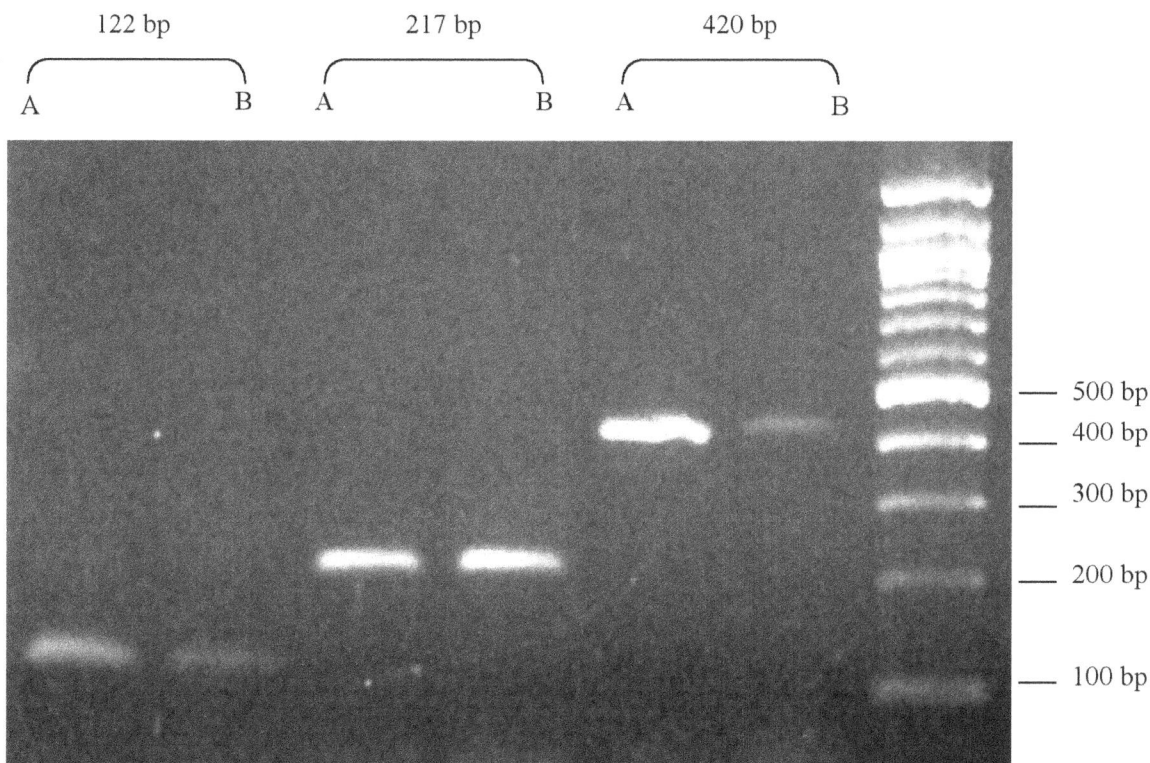

Figure 3. Amplification of the damage reaction. The damaged plasmid can display the same characteristics of aDNA when it comes to PCR. Like most aDNA samples (e.g. Pääbo 1990) the fragment of higher molecular weight cannot be amplified from the damaged plasmid. Here 20 fg of undamaged (A) or damaged (B) DNA was amplified using primers targeting fragments of increasing size.

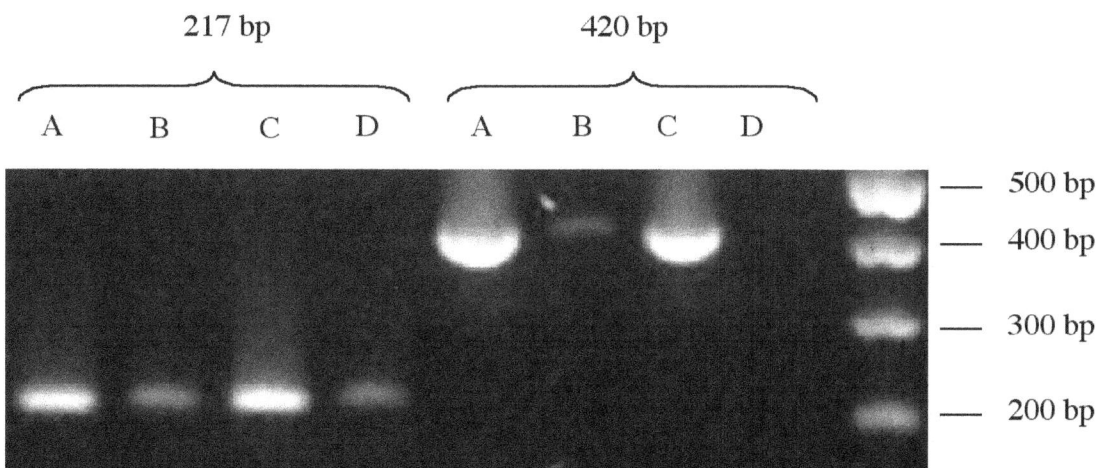

Figure 4. Using the DPM to Assess Extraction Techniques. 8 ng of either damaged or undamaged pUC19 was extracted and purified with two extraction methods (EBIII and EBIIIß) commonly used in the UQ Archaeological Sciences Laboratory. DNA fragments of sizes up to and including 420 bp could be amplified from undamaged extracted plasmid when either EBIII (A) or EBIIIß (C) was used as the extraction method. However, only the EBIII method allowed amplification of all fragment sizes when the damaged plasmid was used as substrate (B). When EBIIIß is used to extract damaged plasmid DNA only the 217 bp fragment could be amplified (D).

Distinct advantages are conveyed by using the DPM, including (i) the original size and sequence of the plasmid are known; (ii) the types of damage are controlled; (iii) any quantity of damaged plasmid DNA can be produced and used; and (iv) the plasmid DNA has no forensic or cultural significance and is thus available for fundamental experimentation, unlike archaeological or forensic material.

We propose that the Damaged Plasmid Model presents an important step forward in providing feasible tools for inter- and intra-laboratory assessments of current damaged DNA extraction and amplification protocols. We also suggest that with further development the model could potentially be used as a verification standard against which authentic degraded DNA results can be measured.

ACKNOWLEDGMENTS

The authors would like to acknowledge the late Dr Thomas H. Loy for his visionary influence in the conception and implementation of the projects that arose from his concept of a model for ancient DNA.

REFERENCES

Aruoma, O.I., Halliwell, B., Gajewski, E. and Dizdaroglu, M. 1991. Copper-ion-dependent damage to the bases in DNA in the presence of hydrogen peroxide. *Biochemistry Journal* 273(Pt 3):601-604.

Ballmaier, D. and Epe, B. 1995. Oxidative DNA damage induced by potassium bromate under cell-free conditions and in mammalian cells. *Carcinogenesis* 16(2):335-342.

Breimer, L.H. 1988. Ionizing radiation-induced mutagenesis. *British Journal of Cancer* 57:6-18.

Cherry, M.I. 1994. Ancient DNA and museums. *South African Journal of Science/S. Afr. Tydskr. Wet.* 90:8-9.

Demple, B. and Linn, S. 1980. DNA N-glycosylases and UV repair. Nature 287:203-208.

DeSalle, R. and Grimaldi, D. 1994. Very old DNA. *Current Opinion in Genetics & Development* 4(6):810-815.

Dizdaroglu, M. 1991. Chemical determination of free radical-induced damage to DNA. *Free Radicals in Biology and Medicine* 10(3-4):225-242.

Dizdaroglu, M. 1993. Chemistry of free radical damage to DNA and nucleoproteins. In B. Halliwell and O.I. Aruoma (eds) *DNA and Free Radicals*, pp.19-39. Chichester: Ellis Horwood.

Doetsch, P.W. and Cunningham, R.P. 1990. The enzymology of apurinic/apyrimidinic endonucleases. *Mutation Research* 236(2-3):173-201.

Halliwell, B. 1993. Oxidative DNA damage: meaning and measurement. In B. Halliwell and O.I. Aruoma (eds) *DNA and Free Radicals*, pp.67-79. Chichester: Ellis Horwood.

Halliwell, B. and Aruoma, O.I. 1991. DNA damage by oxygen-derived species. Its mechanism and measurement in mammalian systems. *FEBS Letters* 281(1-2):9-19.

Halliwell, B. and Gutteridge, J.M.C. 1989. *Free Radicals in Biology and Medicine*. Oxford: Oxford University Press.

Handt, O., Krings, M., Ward, R.H. and Pääbo, S. 1996. The retrieval of ancient human DNA sequences. *American Journal of Human Genetics* 59(2):368-376.

Harding, A. 1997. The development of a modern plasmid DNA damage model which mimics the damaged state of ancient DNA. Unpublished ms on file. Brisbane: Department of Microbiology, The University of Queensland

Lafleur, M.V., Woldhuis, J. and Loman, H. 1979. Alkali-labile sites and post-irradiation effects in gamma-irradiated biologically active double-stranded DNA in aqueous solution. *International Journal of Radiation Biology and Related Studies in Physics, Chemistry and Medicine*. 36(3):241-247.

Levin, J.D. and Demple, B. 1990. Analysis of class II (hydrolytic) and class I (beta-lyase) apurinic/ apyrimidinic endonucleases with a synthetic DNA substrate. *Nucleic Acids Research* 18(17):5069-5075.

Lindahl, T. 1993. Instability and decay of the primary structure of DNA. *Nature* 362:709-715.

Lindahl, T. and Andersson, A. 1972. Rate of chain breakage at apurinic sites in double-stranded deoxyribonucleic acid. *Biochemistry* 11(19):3618-3623.

Lindahl, T. and Nyberg, B. 1972. Rate of depurination of native deoxyribonucleic acid. *Biochemistry* 11(19):3610-3618.

Matheson, C. and Loy, T.H. 2001. Genetic sex identification of 9,400 year old human skull samples from Çayönü Tepesi, Turkey. *Journal of Archaeological Science* 28:569-575.

Meneghini, R. and Martins, E.L. 1993. Hydrogen peroxide and DNA damage. In B. Halliwell and O.I. Aruoma (eds) *DNA and Free Radicals*, pp.83-93. Chichester: Ellis Horwood.

Nielsen, H., Engberg, J. and Thuesen, I. 1994. DNA from arctic human burials. in B. Herrmann and S. Hummel (eds) *Ancient DNA*, pp. 122-140. New York: Springer-Verlag.

Pääbo, S. 1989. Ancient DNA: extraction, characterization, molecular cloning, and enzymatic amplification. *Proceedings of the. National Academy of Sciences* 86:1939-1943.

Pääbo, S. 1990. Amplifying ancient DNA. In M.A.Innis, D.H.Gelfand, J.J. Sninsky and T.J.White (eds) *PCR Protocols: A Guide to Methods and Applications*, pp. 159-166. San Diego: Academic Press.

Prutz, W.A. 1990. The interaction between hydrogen peroxide and the DNA-Cu(I) complex: effects of pH and buffers. *Z Naturforsch* [C] 45(11-12):1197-1206.

Rogan, P.K. and Salvo, J.J. 1990a. Molecular genetics of pre-Columbian South American Mummies. *Molecular Evolution* 122:223-234.

Rogan, P.K. and Salvo, J.J. 1990b. Study of nucleic acids isolated from ancient remains. *Yearbook of Physical Anthropology* 33:195-214.

Steenken, S. 1987. Addition-elimination paths in electron-transfer reactions between radicals and molecules. *The Royal Society of Chemistry Journal. Faraday Trans.* 1(83):113-124.

Tuross, N. 1994. The biochemistry of ancient DNA in bone. *Experientia* 50:530-535.

Wolffe, A. 1995. *Chromatin Structure and Function*. San Diego: Academic Press.

11

Method validation in forensics and the archaeological sciences

Vojtech Hlinka, Iman Muharam and Vanessa K. Ientile
DNA Analysis, Forensic and Scientific Services, Queensland Health
39 Kessels Rd
Coopers Plains
QLD 4108 Australia
Email: vojtech_hlinka@health.qld.gov.au

ABSTRACT

With the development and application of scientific methods in both archaeology and forensics, it is important to test or validate these methods to ensure they can withstand rigorous scrutiny from scientific and other communities and perform as expected. Approaches to method validation within archaeological and forensic science are discussed in this paper, with a focus on outlining current forensic validation procedures. An example of how a validation process has been applied in assessing technology for quantifying human DNA from forensic samples is provided. We propose that guidelines for method validation in archaeological science could be based on those currently utilised in forensics.

KEYWORDS

validation, scientific methods, archaeology, forensics, PCR, DNA

INTRODUCTION

An increasing number of archaeological and forensic laboratories are becoming service providers for public and private institutions. With this role comes the responsibility to provide accurate scientific data using an accepted methodology that has been scrutinised and reviewed effectively through a rigorous validation procedure. This procedure is of particular importance if results are to be compared with those obtained elsewhere, for example from another laboratory, or if the results are to be included in an official report such as a contracted report from an archaeological laboratory or a statement for a court of law as is often the case in forensics. Validation ensures that results have been obtained using a proven method that is suitable for answering a specific scientific question.

WHAT IS VALIDATION?

Validation is a process by which a method is primarily assessed for:

* its adequacy to suit its intended purpose (NATA 2004);
* its reliability (SWGDAM 2004);
* whether it has suitable operational conditions for obtaining results (SWGDAM 2004); and
* its limitations (SWGDAM 2004).

Validation is useful for achieving a number of desired outcomes. It minimises re-invention of methods in different laboratories. Methods that have been validated are more readily accepted, more easily standardised, and can be compared internationally between different laboratories. Validation also helps to identify potential limitations specific to a method or laboratory.

Many laboratories follow validation processes in order to meet specific quality standards (either informal or set at the national or international level). A validated scientific protocol will have an associated report detailing the validation process for the specific protocol that can be reviewed by external quality assessors. Because of their role in the criminal justice system, and specifically in the identification of people through analysis of DNA material, forensic laboratories in Australia are required to meet guidelines (based on the international ISO 17025 standard) set by the National Association of Testing Authorities (NATA). They must meet all quality standards that are listed in the guidelines. Non-compliance with the NATA guidelines may result in loss of accreditation and also risks loss of credibility of the laboratory with peers. Although forensic science is related to archaeological science in certain aspects such as the identification of human remains based on analysed DNA, not all archaeological laboratories operate under a set of national or international guidelines. Such guidelines may be beneficial to some archaeological science disciplines, through demonstration of standards in a universal and clearly understood manner.

Validation should be distinguished from other method-assessment processes such as verification or evaluation. Verification is the process by which collaborating lines of evidence are collected in order to determine if a method is working as expected within a specific laboratory's own conditions (operators, equipment, environment) (adapted from Hedges 2003:667). During verification, results from a few samples are compared with results obtained from other evidence. In the forensic field, this evidence is usually validation data, typically in the form of publications or reports that detail the performance characteristics of the standard method. The outcomes of the verification process are closely linked to the quality and reliability of the validation process. However, validation is a more intensive and rigorous process than verification.

Evaluation is a process by which the suitability of a method is assessed. Typically, results from a number of methods are compared to determine which is more suitable for an intended task. However, the detailed criteria encountered in a validation are not a prerequisite in this process and are only recommended for application to the method selected at the end of an evaluation.

The focus of this article is the use of the validation process as distinct from the verification and evaluation processes with which archaeological scientists may be more familiar. Validation ensures a rigorous and high standard of maintenance during any form of scientific analysis by guaranteeing that samples will be used economically and maintaining sample integrity and continuity. Integrity maintenance has two aspects: the first is avoiding contamination or deleterious handling or processing that could damage, make void or devalue evidence; the second is the avoidance of unnecessary destruction or reduction of the sample size. Both archaeological and casework forensic samples are finite and it may be necessary to use samples for other lines of evidence or for reworking, such as independent evaluation of the evidence.

For casework samples in the field of forensic science, scientific evidence needs to be admissible in court. The expert witness or reporting officer may be questioned about any obtained results and therefore must be able to convince the court of the applicability of the chosen scientific method and significance of the results in context. It is not unusual for an expert witness to be challenged on specific scientific protocols that were used by the testing laboratory, in order to scrutinise the chosen protocols and unveil any breakdowns in quality. An example of a challenge to an established forensic protocol occurred recently over the applicability of low copy number (LCN) analysis methods for the Omagh bombing case in the United Kingdom (The Crown Prosecution Service Press Release 14 January 2008). The presiding judge expressed concerns because, although the testing laboratory had internally validated and published scientific papers on the technique, there was an alleged lack of external validation by the wider scientific community. As a result, the use of LCN analysis was suspended in the UK for a short period and a full review of forensic cases involving LCN technology was ordered.

Forensic anthropological investigation methods have also been scrutinised by courts in the past (Ubelaker 2008). For example, in the trial of The Prosecutor of the Tribunal against Radislav Krstic (Case No. IT-98-33), the validity of methods used to estimate the age at death of human remains from the Balkans was questioned because the methods were based on American standards. Because these standards were developed on the basis of a reference population dataset not necessarily representative of the Balkans skeletal population, it was argued that the methods had not been validated appropriately for the case in question (Ubelaker 2008).

FORENSIC AND ARCHAEOLOGICAL SCIENCE IN AUSTRALIA

In Australian forensic science, NATA set validation and verification guidelines and specify the minimum acceptable standards for a laboratory. DNA Analysis, Forensic and Scientific Services in Queensland Health is accredited under NATA based on ISO/IEC 17025 standards, while Quality and Management is certified ISO9001. NATA (2004) recommended components of a chemical validation include determining selectivity, linearity, sensitivity, accuracy [trueness and precision (repeatability and reproducibility)], limit of detection, limit of quantitation, range, ruggedness, and measurement uncertainty (see Table 1). Method validation is required for laboratory accreditation by ISO/IEC 17025.

Table 1. NATA (2004) recommended components of a chemical validation

Component	Description
Linearity	Linearity or the linear response range is the range of values between which the method produces a linear calibration line. Linearity can only be applied to methods where a linear relationship can be established. For non-linear systems, relationships can also be established in equation form and expressed as a range (see range). Linearity is determined by graphing results from the replicate analysis of a reference material of known composition to produce a linear equation $y=mx+c$. In this equation m is the slope and c the y-intercept, x is the value on the x-axis or the independent variable, and y is the value obtained from the instrument.
Sensitivity	Sensitivity is a measure of the response of measurements to only slight changes in the sample being tested. For linear systems, sensitivity is the slope (m) of the linearity graph ($y=mx+c$). A mean slope with a high positive (e.g. >1) or low negative linear value (e.g. <-1) indicates that on average the method is highly sensitive.
Accuracy-Trueness	Accuracy-trueness is a measure of how near a result is to the accepted or 'true' value for the method being assessed. Accuracy-trueness is measured by determining the bias or systemic error. Trueness is determined by replicate analysis of a reference material of known composition. It is calculated as a fraction or percentage of the measured result to the accepted or 'true' value and can be assessed with a t-test.
Accuracy-Precision-Repeatability	Precision is the closeness of agreement between independent replicate test results. Repeatability is a measure of the maximum acceptable difference between two test results obtained at the same time by the same analyst under identical conditions on the same material. Normally, repeatability is calculated at the 95% confidence level (and two correctly obtained results will not differ from one another by more than the repeatability value in more than 1 in 20 cases).
Accuracy-Precision-Reproducibility	Reproducibility is a measure of the maximum acceptable difference between two test results obtained on the same material by different analysts at different times. This value of reproducibility is the one generally used to estimate the limits of uncertainty of a result. Inter-laboratory reproducibility is most conveniently determined in collaborative trials.
Selectivity	Selectivity is a measure of accuracy in the presence of interference. It is tested by comparing results for samples containing impurities with results for samples without impurities. Single point tests are acceptable but various points with varying amounts of inhibitor will add more data about the interference of different amounts of a substance.

Limit of detection	The lower limit of detection is the lowest value for a sample that can be reliably distinguished from zero, but not necessarily quantified, by the test method. It is the lowest value that is greater than the uncertainty associated with it and therefore is an indication of the value at which detection becomes problematic.
Limit of quantitation	The limit of reporting or quantitation is the lowest value that can be determined with acceptable repeatability and accuracy by the test method. Depending on the level of certainty required (e.g. whether or not the analysis is for legal purposes) this is usually taken as three times or ten times (for greater certainty) the limit of detection.
Range	The working range is defined as the minimum to maximum acceptable working values. The minimum acceptable working value beyond any reasonable doubt is the limit of quantitation. The maximum acceptable working value is the maximum value that can be determined with acceptable repeatability and accuracy by the test method.
Ruggedness	Ruggedness is a measure of how robust a method is, and therefore an assessment of how changes in the normal course of performing a method can affect the results. Ruggedness is tested by measuring the effects of one variable at a time or a combination of variables identified as the most likely to affect results. Testing is performed with controls.
Measurement uncertainty	Measurement uncertainty is a measure of the range of values that can reasonably be attributed to the specific quantity being measured. It takes into account all the recognised effects on a result such as overall long-term precision, bias and calibration effects and uncertainties, and other effects. Measurement uncertainty is usually expressed as a standard deviation or confidence interval.

In addition to the basic NATA validation guidelines for accreditation in forensics, DNA Analysis, Forensic and Scientific Services in Queensland occasionally utilises Scientific Working Group on DNA Analysis Methods (SWGDAM) guidelines (SWGDAM 2004). Internal Standard Operating Procedures (SOPs) within Queensland Health ensure that validation guidelines are followed to acceptable standards. The SOPs are used and maintained internally through an intranet Quality Information System.

An example of an internal validation undertaken at DNA Analysis, Forensic and Scientific Services, Queensland Health was that of the Quantifiler Human DNA Quantitation Kit (Applied Biosystems 2003). Run on the ABI Prism 7000 SDS instrument from Applied Biosystems, it forms a real-time quantitative PCR system with in-built inhibition detection. The method determines the human DNA concentration of an extracted DNA sample and the approximate level of PCR inhibitors present. Thus, an estimate for the approximate sample volume for the amplification of DNA via PCR can be calculated from the resulting concentration. The same reaction also determines whether a sample needs further purification prior to amplification . Although the method was validated in the United States to SWGDAM guidelines (see Applied Biosystems 2003), an in-house validation at Queensland Health with external collaboration was necessary to comply with NATA guidelines.

We assessed the accuracy of the Quantifiler system in DNA Analysis, Forensic and Scientific Services as a component of the validation process. In assessing the long-term accuracy of the system in-house it was found that specific lots of the DNA standard on average displayed a two-fold bias in concentration values. Originally the average bias was corrected for and eventually, the Quantifiler standards were replaced with other standards certified at known concentrations by another manufacturer. This change resulted in a lower and more acceptable bias. Controls for the acceptance and rejection of Quantifiler runs as well as for monitoring changes in them were introduced. The lesson learnt from this process was that although a method may pass an original validation, monitoring systems in the form of suitable quality controls must be in place to demonstrate that a method continues to work acceptably over time and that conditions encountered during the original validation have not changed.

Forensic scientists occasionally misinterpret validation guidelines, leading to some incorrect views about how validations should be performed (Butler 2006). One common misinterpretation of the SWGDAM guidelines for conducting an internal validation is the belief that it is necessary to have 50 samples per experiment or 50 samples of the same type per experiment (Butler 2006:4). The 50 samples recommended in the guidelines actually refer to the minimum total sample size for an entire internal validation (Butler 2006:4). In summary, the sample size has to be large enough to produce statistically valid results with methods such as the Student's t-test for assessing accuracy (Butler 2006). Butler (2006) considers between five and ten samples to be sufficient for a validation experiment in forensics.

A potential difference in method validation between forensic and archaeological sciences is that in forensic science NATA guidelines state that a standard method needs to be validated by collaborative studies before its use can be verified in a laboratory. Such a method is more likely to be deemed valid in a court of law. In archaeological science, either validation or verification processes may be applied.

Archaeological method verification is used to provide evidence of whether a method is performing satisfactorily within a laboratory when compared with its published performance characteristics or a similar standard method. It is generally considered suitable when its performance exceeds or equals that of a published method. A method is therefore deemed valid if it produces the desired results and has been checked or verified internally against external published standards.

Radiocarbon dating is one example where researchers have paid significant attention to ensuring the method produces valid results in terms of accuracy. It is also an example of a scientific method that was applied to archaeological samples before a validation of the accuracy of the method was obtained. Libby first described the radiocarbon method for dating organic material in 1949 (Libby 1955). Because questions arose over the accuracy of the method (Libby 1963), Libby then tested the method against well-dated materials to validate its accuracy. Assumptions in the original method such as the constant concentration of radiocarbon in the atmosphere were subsequently shown to be incorrect, with readings from dendrochronologically-dated bristlecone pines being younger before about 1200 BC (Fagan 1991:123). It took several decades of cooperation between different laboratories to produce validated methods for obtaining reproducible calibrated radiocarbon dates (Fagan 1991:123-124). Verification of the radiocarbon method in different laboratories identified localised problems with radiocarbon dates such as the marine, estuarine and freshwater reservoir effects (e.g. Fischer and Heinemeier 2003; Ulm 2006). Over a few decades of validation and verification processes, it was demonstrated that the radiocarbon method was suitable for its intended purpose when the appropriate calibrations can be performed.

For DNA projects in the Archaeological Sciences Laboratories (ASL) at the University of Queensland in the years between 1995 to 2005, the validity of a method was commonly determined through trials on modern reference samples (e.g. control samples) and comparison of results with those from a published or accepted method. The method had to be tested on the same samples or at least samples of the same type. If successful, the test was then applied to mock archaeological samples or limited archaeological samples (e.g. unprovenanced samples). If this was successful, then the final stage involved application of the method with an aim of answering a question of archaeological significance. Successful results from subsequent independent use of the methods by researchers and students, usually in-house, meant the methods went through an archaeological 'verification' process.

An example of a method that went through this verification process was a silica and chaotropic-based DNA extraction method. This method was developed by Boom *et al.* (1990) and tested in other laboratories (Höss and Pääbo 1993). In the ASL, it was subsequently tested and optimised by several researchers on modern samples before being applied to archaeological samples (e.g. Hlinka 2003; Matheson 2001). A major component of testing in the ASL was comparison of results using this particular method with results obtained from other methods running comparable sample controls. Slightly modified versions of the method continue to be used by other laboratories and have been verified as suitable for archaeological samples such as bones and teeth (Rohland and Hofreiter 2007a, 2007b). However, a validation of the method performed to set criteria such as those outlined in Table 1, combined with a comparison of different methods validated by the same criteria, would be a more efficient system. In this way some of the comparisons made using finite archaeological samples as well as repetition of work in method testing by different researchers could have been minimised.

As a further control, Dr Tom Loy pioneered a Laboratory Information Management System called MARS (Multirelational Archaeological Research Sciences database) at the University of Queensland. MARS contained a list of standard reference methods used routinely by DNA researchers in the ASL and tracked day-to-day laboratory work. Methods deemed suitable for use through the verification process, such as the silica and chaotropic-based DNA extraction method, were listed in the database and considered validated. Methods were further verified with continued successful use by other researchers and students, both internally and externally.

In Australian archaeology, current guidelines for laboratories not bound by biohazard, quarantine, internal or other quality control regulations still include moral and ethical standards and codes such as the Code of Ethics of the Australian Archaeological Association (AAA). The AAA Code of Ethics states that methods in archaeological science used to investigate archaeological sites and materials should be utilised in ways that "conserve the archaeological and cultural heritage values of the sites and materials" (AAA 2008). The Code also advises that "… members will advocate the conservation, curation and preservation of archaeological sites, assemblages, collections and archival records" (AAA 2008). By extension, the AAA Code of Ethics therefore encourages archaeologists to utilise methods that have minimal impact on the integrity of archaeological evidence and methods where continuance of archaeological evidence can be maintained. Similarly, while research guidelines such as the Joint NHMRC/AVCC Statement and Guidelines on Research Practice (NHMRC and AVCC 1997) and the Australian Code for the Responsible Conduct of Research (NHMRC 2006) do not contain any specific validation guidelines, they do discuss peer review mechanisms that are directly applicable to determining if a method is valid.

RECOMMENDATIONS

The degree to which a method should be validated in archaeology or forensics is a balance of costs, risks and technical possibilities and depends on the status of the method and the needs and requirements relating to its application (Butler 2006). Often a method will appear complex because it can comprise several sub-methods, and it may be necessary then to validate the individual sub-methods as well as the whole system. A new in-house method will require rigorous validation whereas minor modifications to a validated in-house method may only require a few checks before implementation. The basic rule is that every method evaluated as suitable for use first requires validation, then verification.

We propose that guidelines for method validation in some fields within archaeological science could be based on those currently utilised in forensic science (e.g. NATA 2004; SWGDAM 2004), although these are intended primarily for chemical methods. Current archaeological practice for evaluating scientific methods for their suitability could be enhanced by the use of standardised criteria or parameters that can be compared intra- and inter-laboratory. Where applicable, the use of the NATA criteria would ensure a minimum standard and provide scope for more rigorous comparisons and evaluations. In addition, the publication of more rigorous methods testing in archaeological science would help in assessing the suitability of specific methods.

The forensic validation criteria presented here are skewed towards quantifiable analytical methods such as DNA research. However, archaeological excavation or processing methods could also be validated with these criteria by analysing quantifiable characteristics such as the number and types of artefacts recovered from mock and archaeological sites. The criteria would need to be adapted in cases such as the validation of reconstruction methods (e.g. pottery, bone and site reconstructions).

In addition to the validation criteria, we also recommend the creation of a database of standardised methods that have been validated externally and that are recommended for in-house validation and verification to avoid 're-inventing the wheel'.

CONCLUSION

Validation is a necessary process in both forensics and archaeological science. Adoption of the forensic validation model presented here could provide a framework for application in many areas of archaeological science. The validation process ensures that methods are standardised and comparable and hence more readily acceptable internationally.

ACKNOWLEDGEMENTS

We would like to dedicate this article to the memory of Dr. Tom Loy. We thank the members of the Archaeological Sciences Laboratories, the School of Social Science and the Institute for Molecular Bioscience at the University of Queensland, and the members of Forensic and Scientific Services, Queensland Health.

REFERENCES

AAA (Australian Archaeological Association) 2008. *www.australianarchaeologicalassociation.com.au/ethics.htm* Accessed 29th May, 2008.

Applied Biosystems 2003. Quantifiler™ Kits Quantifiler™ Human DNA Quantification Kit and Quantifiler™ Y Human Male DNA Quantification Kit User's Manual, pp.175.

Boom, R., Sol, C.J., Salimans, M.M., Jansen, C.L., Wertheim-van Dillen, P.M., and J. van der Noordaa 1990. Rapid and simple method for purification of nucleic acids. *Journal of Clinical Microbiology* 28:495-503.

Butler, J. 2006. Debunking some urban legends surrounding validation within the forensic DNA community. *Profiles in DNA* September 2006:3-6.

Fagan, B.M. 1991. *In The Beginning: An introduction to archaeology.* 7th edition. New York: Harper Collins Publishers.

Fischer, A. and Heinemeier, J. 2003. Freshwater reservoir effect in 14C dates of food residue on pottery. *Radiocarbon* 45(3):449-466.

Hedges, R. 2003. Puzzling out the past. *Nature* 422:667.

Hlinka, V. 2003. *Genetic Speciation Of Archaeological Fish Bones.* Unpublished PhD thesis. School of Social Science, University of Queensland.

Höss, M. and Pääbo 1989. DNA extraction from Pleistocene bones by a silica-based purification method. *Nucleic Acids Research* 21:3913-3914.

Libby, W.F. 1955. *Radiocarbon Dating.* Chicago: University of Chicago Press.

Libby, W.F. 1963. The accuracy of radiocarbon dates. *Antiquity* 37:213-219.

Matheson, C. 2001. *Genetic Analysis Of Human Population Groups And Subgroups From Samples Of Degraded DNA.* Unpublished PhD thesis. Department of Biochemistry, University of Queensland.

NATA (National Association of Testing Authorities) 2004. Guidelines for the Validation and Verification of Chemical Test Methods. *Technical Note 17*.

NHMRC and AVCC 1997. Joint NHMRC/AVCC Statement and Guidelines on Research Practice. *www.nhmrc.gov.au/funding/policy/researchprac.htm* Accessed 31st January, 2007.

NHMRC 2006. Australian Code for the Responsible Conduct of Research *www.nhmrc.gov.au/funding/policy/code.htm*. Accessed 31st January, 2007.

Rohland, N. and Hofreiter, M. 2007a. Ancient DNA extraction from bones and teeth. *Nature Protocols* 2(7): 1756-1762.

Rohland, N. and Hofreiter, M. 2007b. Comparison and optimization of ancient DNA extraction. *BioTechniques* 42(3):343-352.

SWGDAM 2004. Scientific Working Group on DNA Analysis Methods (SWGDAM) 2004 Revised Validation Guidelines *Forensic Science Communications* 6(3):1-6.

The Crown Prosecution Service (UK) 2008. Press Release 14 January 2008: "Review of the use of Low Copy Number DNA analysis in current cases". *www.cps.gov.uk/news/pressreleases/101_08.html*. Accessed 10 July 2008.

Ubelaker, D.H. 2008. Issues in the global applications of methodology in forensic anthropology. *Journal of Forensic Sciences* 53(3): 606-607.

Ulm, S. 2006. Marine and estuarine reservoir effects in central Queensland: determination of ΔR values. In *Coastal Themes: An Archaeology of the Southern Curtis Coast, Queensland*. Terra Australis 24: 47-64.

12

Mesolithic stone tool function and site types in Northern Bohemia, Czech Republic

Bruce L. Hardy[1] and Jiří A. Svoboda[2]

1. Department of Anthropology
Olof Palme House
Kenyon College
Gambier, OH 43022 USA
Email: hardyb@kenyon.edu

2. Institute of Archaeology
Academy of Sciences of the Czech Republic
Brno, Czech Republic

ABSTRACT

The sandstone plateaus cut by canyons with rockshelters in Luxembourg, Germany and the northern Czech Republic remain relatively unexplored archaeologically. In northern Bohemia, Czech Republic, these were mainly settled during the Mesolithic. Recent surveys have revealed a network of Mesolitihic sites representing short and long-term occupations in canyon rockshelters. One long-term (Pod zubem) and one short-term site (Pod křídlem) were selected for functional analysis of stone tools, including microscopic use-wear and residues analyses, in order to investigate the uses of different site types in terms of subsistence, economic, and seasonal activities. Results indicate that stone tools were used on a variety of materials at the long-term occupation site of Pod zubem. At Pod křídlem, where the record is more sparse, the analyses suggest that plant processing was the primary activity. Faunal, macrobotanical, and lithic functional anlayses suggests specialised use of tools and sites in the Mesolithic of Northern Bohemia.

KEYWORDS

Mesolithic, residue analysis, use-wear analysis, plant processing, hafting, northern Bohemia, Czech Republic

INTRODUCTION

The Mesolithic in the Czech Republic
One of the typical but still little explored landscape features in western Central Europe are the restricted areas of sandstone plateaus cut by canyons with rockshelters that are scattered from Luxembourg over western and central Germany to the northern part of the Czech Republic. Archaeological research within the sandstone rockshelters shows that the surrounding lands, being unattractive for agriculture, were mainly settled during the Mesolithic period. In this manner, these sites contribute substantially to the knowledge of a relatively little known period of central European prehistory (cf. Street *et al.* 2001).

In the Czech Republic (Northern Bohemia) this type of research is of a relatively recent date. Systematic survey and excavation projects, held between 1997-2005, and partly supported by National Geographic Society grant 98/6330 'Last Hunter-gatherers of Northern Bohemia', revealed a network of key Mesolithic rockshelter sites across the landscape (Figure 1). Surveying the whole area followed several strategies. At the first stage, creating a geographic network of sites was the main aim, so that representative rockshelters in the various microregions, valleys and canyons were selected for exploration. At the second stage, we concentrated on the most promising areas where the individual sites created interrelated systems with internal hierarchy in size and function. One of them is the microregion on the Robecsky Brook, presented in this paper, another one is the Kamenice River canyon further north, explored later (Svoboda *et al.* 2007). During this project, the selected sites were approached in a complex manner, from viewpoints of geology and sedimentology, paleobotany, archaeozoology, archaeology, and anthropology (Svoboda 2003). At certain sites, special attention was paid to use-wear and residue analyses as well.

A closer look at the map that resulted from this survey and a comparative analysis of the individual sites reveal a certain hierarchy in the size and location of the rockshelters, complexity of archaeological features such as hearths and pits, and the quantity and variability of artefacts. In

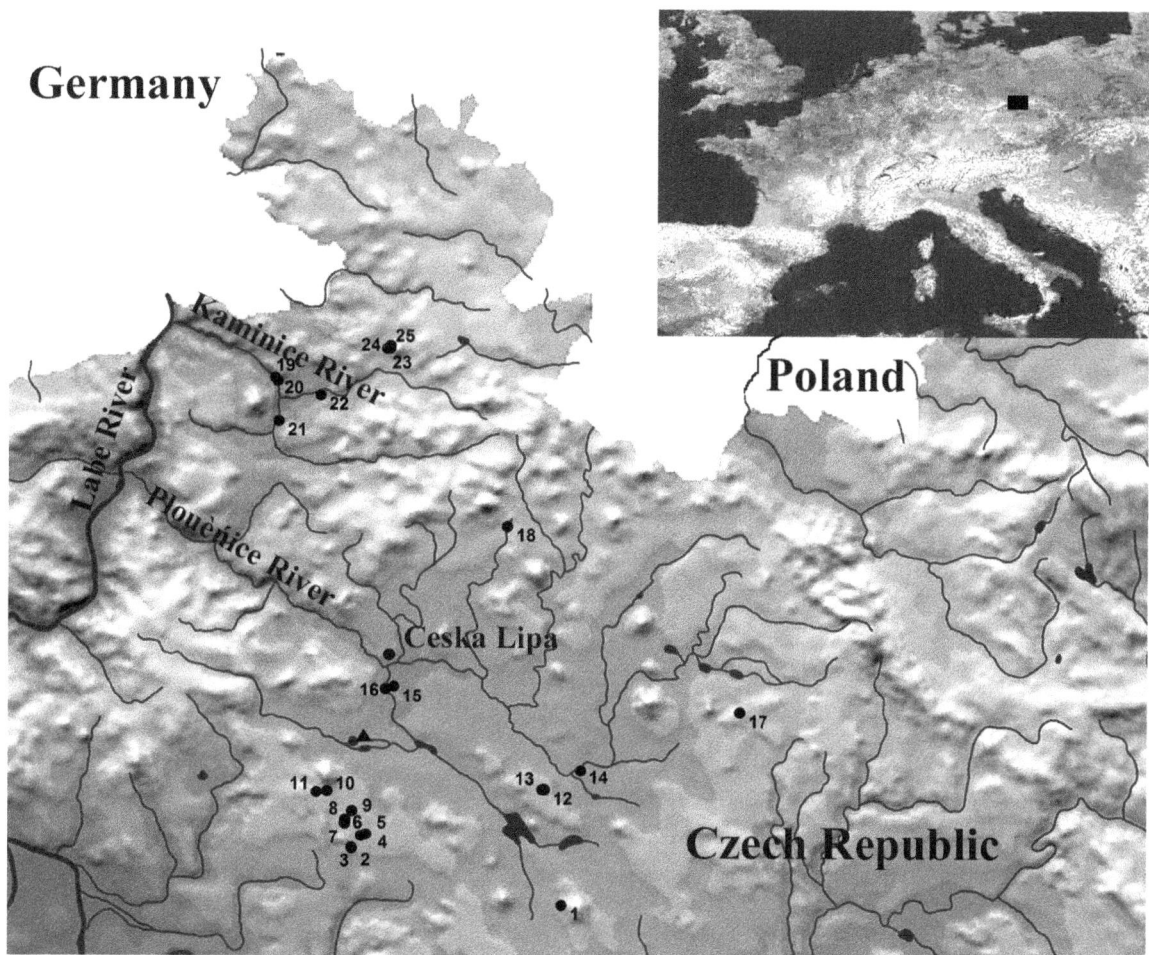

Figure 1. Map of the North Bohemian rockshelter sites with Mesolithic occupations (points) and location of the North Bohemian region in Europe. The wide area corresponds to present-day national border against Germany and Poland. The main crossing river (from south to north) is Labe (Elbe); the plateau lies in elevations between 250-300 m a.s.l., with single elevations reaching to 500-750 m a.s.l., while highest elevations (over 1000 m a.s.l.) occur only in westernmost part of the area. - 1: Bezděz, 2: Vysoká Lešnice, 3: Nízká Lešnice, 4: Strážník, 5: Stará skála, 6: Máselník, 7: Černá Louže, 8: Pod Černou Louží, 9: Šídelník, 10: Heřmánky, 11: Hvězda, 12: Uhelná rokle, 13: U obory, 14: Donbas, 15: Pod zubem, 16: Pod křídlem, 17: Černá Novina, 18: Údolí Samoty, 19: Dolský Mlýn, 20: Okrouhlík and surrounding sites, 21: Arba, 22: Sojčí rockshelter, 23: Jezevčí rockshelter, 24: Nosatý kámen, 25: Švédův rockshelter. Triangle: open air sites of the Stvolínky – Holany area.

addition, potentials for preservation of organic materials, which naturally add more data on past activities performed at a site, vary from one site to another.

From this viewpoint, the most complex site clusters excavated were in the deep canyons of larger brooks, such as the Kamenice river canyon in the very north (sites of Okrouhlík and Dolský Mlýn) and the Robečský Brook canyon ('Peklo' or 'Hellegrund' valley) in the center of the explored area (Svoboda et al. 1999). Because the Robečský brook canyon area offers the best conditions for organic preservation within the whole area, we have chosen the sites of Pod zubem ('under the tooth') and Pod křídlem ('under the wing') for the analysis presented in this paper (Figure 1).

The Pod zubem and Pod křídlem sites

The Pod zubem site, cadastre (county) of Česká Lípa, is located under a large, isolated rockhelter in a shallow side valley adjacent to the Robečský brook canyon, next to an active water spring, and 260 m a.s.l. This large but shallow rockshelter was explored in 1997 by three trenches adjacent to its northern wall, all of them 2 m wide and 2-3 m long. Trench A recorded a sequence of cultural layers not deeper than 1 m, whereas in trenches B and C we explored a large, oval-shaped depression reaching a total depth of 1.7 m (Figure 2).

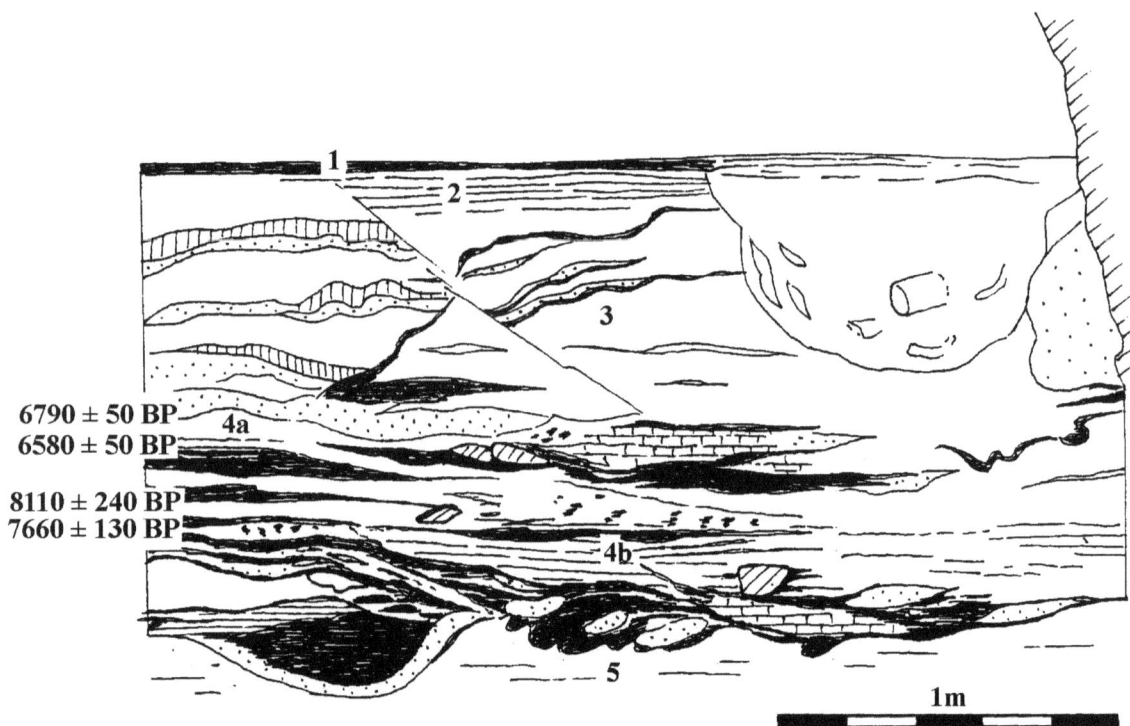

Figure 2. The Pod zubem rockshelter, section in trench C. 1: grey, dusty layer; 2: brown-greyish, stripped layers of sand (both layers subrecent); 3: yellowish-grey, coarse-grained sand (Bronze Age, Neolithic); 4a-c: brownish-grey, fine-grained sand, with hearths and other anthropogenic features (Mesolithic); 5: yellow, coarse-grained sand, with silty bands (archaeologically sterile). Position of the ¹⁴C samples from Layers 4a and 4b is indicated.

The upper three layers provided subrecent and Bronze Age occupations, especially Stroked pottery materials. Mesolithic Layer 4 was complex, with sandy-to-clayish layers interstratified by lenses of charcoal, burnt sand, and calcareous material. Thus, it was subdivided into stratigraphic units 4a-c, all providing Mesolithic lithic industries, with microlith types diagnostic for middle and late Mesolithic periods, bone tools (a chisel-shaped artifact and awls), associated faunal remains, and one human tooth. Charcoal for the ¹⁴C dates (Table 1) was sampled from hearths in Layers 4a and 4b and yielded dates ranging from 7461 to 9025 cal B.P.

Table 1. ¹⁴C dates for the Pod zubem and Pod křídlem rockshelters (all from charcoal). Calibration after the CALIB.REV.4.3. programme (Stuiver and Reimer 1993)

Rockshelter	Context	Depth (cm)	Layer	Sample #	Date (BP)	Interval (2 sigma)	Date (cal. BP)
Pod zubem	hearth	75	4a	GrN 23332	6790 ± 70	7510-7785	7656
Pod zubem	charcoal deposit	80	4a	GrN 23333	6580 ± 50	7421-7570	7461
Pod zubem	charcoal deposit	115-120	4b	GrN 23335	7660 ± 130	8182-8748	8412
Pod zubem	charcoal deposit	115	4b	GrN 23334	8110 ± 240	8408-9545	9025
Pod křídlem	charcoal deposit	50-70	4	GrN 23331	8160 ± 80	8815-9401	9124

Whereas the upper Mesolithic Layer 4a extended through all three trenches, and a remarkable circular hearth, plastered with sandstone blocks, was recovered in trench A, the middle and lower Mesolithic Layers (4b,c) were limited to filling of the depression in trenches B and C (Figure 3). In terms of paleobotany, this site yielded charcoal of pine, oak, and hazel, including hazelnut shells (Opravil 2003). The rich malacofauna (invertebrates with shells) corresponds to predominantly woodland communities, although a single find of Unionidae (freshwater mussels) may have been a contribution to the Mesolithic diet (Ložek 2003). The vertebrate fauna represents a wider variety of environments such as open habitats, woodland, and water environment of the nearby Robečský brook valley. Part of the faunal assemblage (Table 2) evidently results from hunting (hare, deer and red deer, elk, birds), and the relative abundance of certain species seems to reflect specialised hunting for fur animals (marten, squirrel, beaver, wild cat, fox; Horáček 2003).

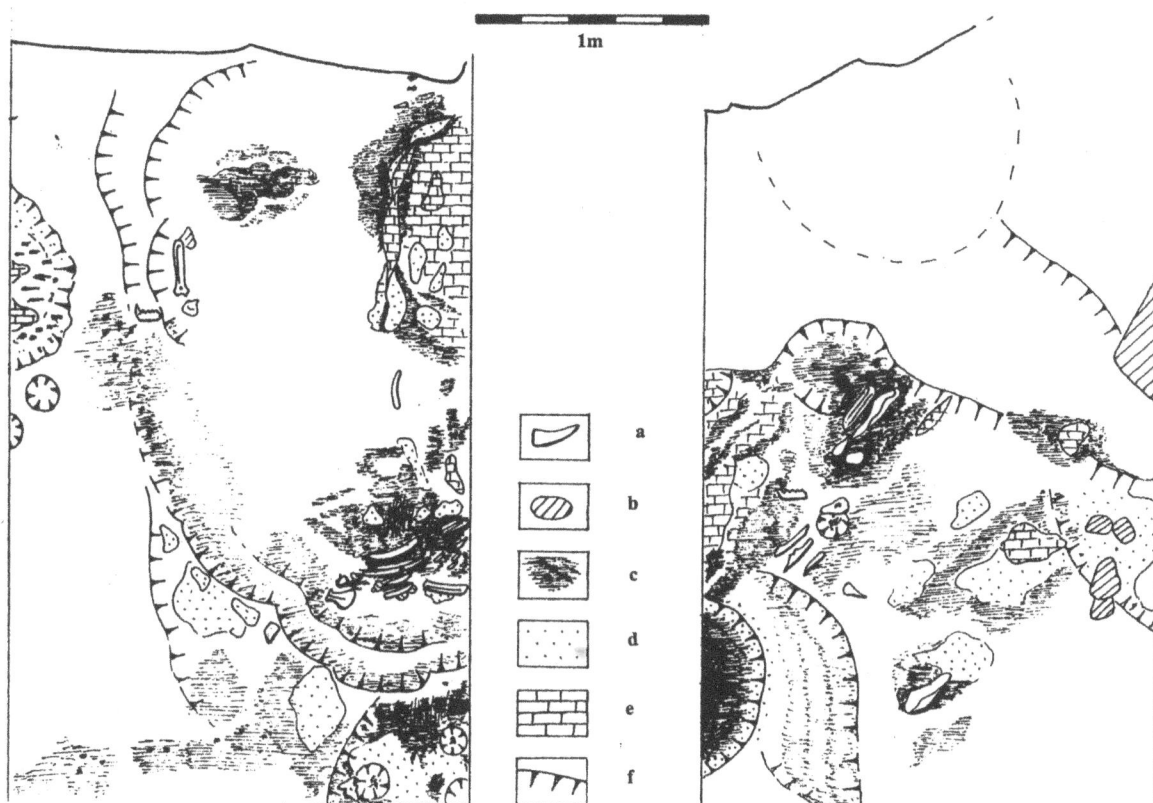

Figure 3. The Pod zubem rockshelter, 1997 excavation, lowermost layer. a) white triangle – bone; b) diagonally hatched – sandstone blocks; c) charcoal; d) dotted - red-burnt sand; e) 'brick-like' hatched – calcareous deposits with organic material; f) limits of excavation.

Table 2. Fauna potentially hunted by Mesolithic humans at the Pod zubem rockshelter (MNI). Data by I. Horáček (2003)

Fauna	Pod zubem, Layer 4
Eurasian red squirrel (Sciurus vulgaris)	3
European beaver (Castor fiber)	3
Brown hare (Lepus europaeus)	12
European roe deer (Capreolus capreolus)	2
Red deer (cf. Cervus elaphus)	2
Eurasian elk (Alces alces)	1
Wild boar (Sus scrofa)	3
Auroch/Bison (Bos/Bison)	1
Pine marten (Martes martes)	10
Wild cat (Felis sylvestris)	4
Red fox (cf. Vulpes vulpe)	5
Wolf (Canis lupus)	2
Perching birds (Aves, Passeriformes)	3
Non-perching birds (Aves, non Passeriformes)	3

The Pod křídlem site, cadastre (county) of Kvítkov, is located at the foot of the canyon about 3 m above the Robečský brook water level and 247 m a.s.l. A trench of 2 x 3 m was excavated to explore the sediments. In the uppermost part of the deposits, we revealed a sequence of several 20th century hearths, underlain by hearths with subrecent and medieval pottery. The underlying Mesolithic Layer 4 included charcoal layers with burnt stones from slightly redeposited hearths, the largest of which provided a ^{14}C date of 9124 cal B.P. (Table 1). The lithic industry comprised only 54 artefacts, one of which was a microlithic triangle diagnostic of early or middle Mesolithic, consistent with the ^{14}C date. This layer also provided charcoal of pine and fragments of hazelnut shells. The base of this sequence was formed by sterile sandy deposits (Figure 4).

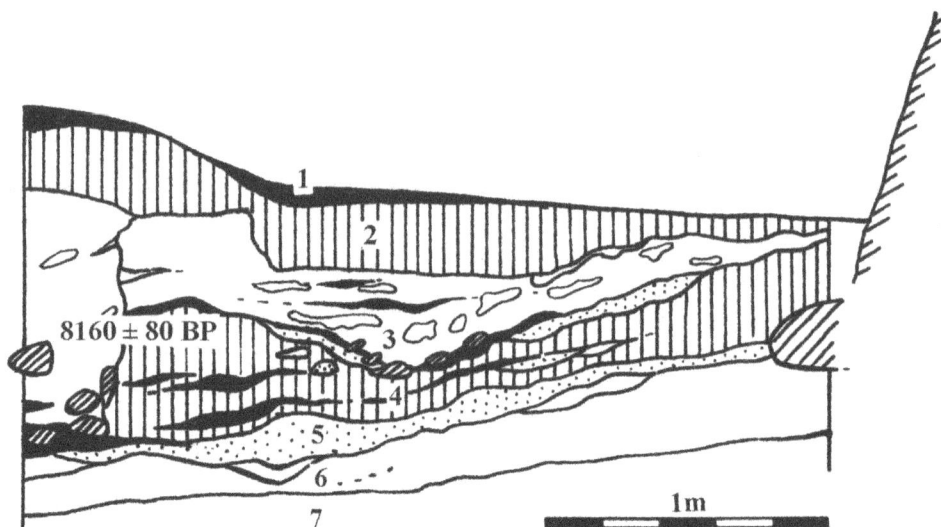

Figure 4. The Pod křídlem rockshelter section showing location of the ^{14}C dated charcoal sample. 1: recent layer; 2: brownish, humus sand (subrecent); 3: light grey, clayish layer with spots of white sand (subrecent); 4: brown-reddish to brown-grayish sand (Mesolithic); 5: rusty sand horizon; 6: yellow-brownish sand; 7: basal, coarse-grained sand (Layers 5-7 archaeologically sterile). Position of the ^{14}C sample from Layer 4 is indicated.

In summary, the two sites excavated in the Robečský brook canyon are each of a different type. Pod zubem represents a large and complex settlement with elaborate hearths, pits, abundant artefacts of stone and bone, faunal remains, and a multi-layered stratigraphy documenting a sequence of long-term occupations. In contrast, Pod křídlem is a small satellite site with simple hearths accompanied by a only a few artefacts, which (based on typology and ^{14}C date) may correspond to short episodes within the more complex settlement sequence documented at Pod zubem. Judging from the location just above the brook, Pod křídlem would be an ideal fishing post, even if no fish bones were recovered (in Northern Bohemia, fish bones are recorded only from the rockshelter of Dolský Mlýn on the Kamenice river).

These two sites were selected for detailed functional analyses of stone tools (both residue and use-wear analyses) because they represent cases of a large and long-term site and a small, rather episodic site that were partly contemporary. The excavations contribute to our understanding of hunting and other subsistence strategies, including aspects of seasonality. The faunal composition at Pod zubem shows a dominance of forest species, especially animals hunted for furs, whereas the morphology of bone industry suggests working with hides. The finds of hazelnuts complete the evidence on subsistence, but may also indicate a season of occupation during late summer and/or early fall. This observation would fit with the evidence of fur working, possibly in expectation of the approaching winter. With this in mind, we undertook functional analyses of a sample of stone tools from the upper and lower Mesolithic Layers (4a,b) of Pod zubem and a comparative sample from Mesoltihic Layer 4 of Pod křídlem, where almost half of the small assemblage was examined, in order to gain a more complete picture of economic and subsistence activities.

STONE TOOL FUNCTION IN THE MESOLITHIC: PREVIOUS RESEARCH

Just as excavations in the Czech Republic have been more focused on the Upper Paleolithic than the Mesoltihic, so functional studies of stone tools in Europe have also tended to neglect the Mesolithic. The small number of studies that have been undertaken have primarily used the high-power method of use-wear analysis with magnifications up to 500x (Keeley 1980) although low-power magnification (50-200x) has also been used (e.g. K. Hardy 2004). The high-power method primarily relies on the recognition of polishes for identification of use-material, with supporting evidence provided by striations and edge damage. Microliths (including crescents, triangles, lunates, and backed bladelets) commonly show evidence of hafting and are generally assumed to have been part of composite tools with an emphasis on use as armatures or projectile points (Crombé et al. 2001; Dumont 1988). Other tool types, including blades, scrapers, denticulates, etc., are used for a variety of purposes such as wood, plant, hide, meat or antler/bone processing (e.g. Cahen et al. 1979; Dumont 1988; Finlayson 1990; Finlayson and Mithen 1997; K. Hardy 2004; Thorsberg 1983; Uwe-Heussner 1989).

The current study was described preliminarily by Hardy (1999) and has been updated subsequently with further microscopic analysis (including scanning electron microscopy, see below) and more specific identification of residues. Another type of analysis showing mechanical blow impacts on tips of microlithic projectiles is currently being undertaken by Yaroshevich (pers. comm. 2006) on materials from two Mesolithic sites (Okrouhlík and Dolský Mlýn) within the same region of northern Bohemia.

METHODS

Microscopic residue analysis attempts to elucidate stone tool function. In this case, function is a broad term encompassing a combination of how a tool was used (e.g. to cut or to scrape, also referred to as use-action) and the material on which a tool was used (e.g. hide, wood, etc.; also referred to as use-material). When it is possible to ascertain both use-action and use-material, it is possible to reconstruct tool function (e.g. whittling wood, scraping hide, etc.). Use-actions are inferred from the distribution of wear patterns (polishes, striations, edge rounding) and residues on tool's surface. Residues that show distinct patterning or co-occurrence with use-wear are referred to as use-related. Residues with no clear patterning (even distribution over the entire surface) and no associated wear patterns are not considered to be use-related.

A total of 70 artefacts (Pod zubem, n=46; Pod křídlem, n=24) were analysed microscopically for the presence of use-wear traces and use-related residues (Hardy 1999). Minimally handled, unwashed artefacts greater than 0.5 mm in size were placed in new resealable plastic bags as they were removed from the excavation in order to reduce the potential loss of residues through washing (Hardy and Garufi 1998; Hardy et al. 2001). The opportunistic sample included all artefacts removed from in situ excavations at either site during the two weeks of analysis. The artefacts were examined using incident light microscopy at magnifications ranging from 50-500x using an Olympus BH microscope. Observed residues and use-wear were photographed and their locations recorded on line drawings of each artifact. Sediment samples taken from all levels at each site were also examined for the presence of residues.

Residues were identified based on comparison with published and comparative experimental material and included hair, feathers, plant and woody remains, resins, and starch grains (Anderson-Gerfaud 1990; Beyries 1988; Brom 1986; Brunner and Coman 1974; Catling and Grayson 1982; Fullagar 1991; Hardy 1994; Hardy and Garufi 1998; Hather 1993; Hoadley 1990; Kardulias and Yerkes 1996; Lombard 2004; Pearsall 2000; Teerink 1991; Williamson 1996). Residues were interpreted as use-related based on their patterning on the artifact surface, concordance with use-wear patterns, and their absence from sediment samples. Residues found on both artefacts and in the surrounding sediments were not classed as use-related (e.g. small rootlets). The typological categories represented in the sample included blades, bladelets, flakes, unifacial points and a crescent. Blades and bladelets are here defined as flakes twice as long as they are wide (Whittaker 1994) with blades having a breadth >0.7 cm. For the purposes of this analysis, the two categories are grouped together since they represent different sizes of the same artifact shape.

A small number of identifiable residues were removed from artifact surfaces for further analysis with scanning electron microscopy (SEM). These residues were removed by applying double-sided adhesive tape to the tool surface, then peeling the tape away and sticking the other side to an SEM stub. The stubs were sputter-coated with gold and examined with a JEOL 840a SEM at magnifications up to 2000x.

Use-wear observations included the assignation of polishes to two broad categories indicating the relative hardness of the use-material, hard/high silica or soft (B. Hardy 2004; Hardy *et al.* 2001). The hard/high silica category does not distinguish between hard materials such as bone, antler and wood from high silica materials such as grasses since high silica materials can produce polishes similar in nature to those produced by hard materials (Fullagar 1991). The direction and orientation of striations, edge damage, and edge rounding were also recorded.

RESULTS

The combined sample from both sites consisted of 40 blades and bladelets, 28 flakes, one crescent and one unifacial point, for a total of 70 artefacts. Of these, 45 (64.3%) had some form of functional evidence (Table 3). Specific materials identified include plant, wood (particularly conifers/gymnosperms), feathers, hair, resin, and starch grains.

Pod Zubem

A total of 46 stone artefacts were analysed from the long-term occupation at Pod zubem including 30 blades/bladelets, 15 flakes, and a single unifacial point (Table 3). The residues observed include plant fragments, wood fragments, starch grains, feathers, hair and resin (mastic). In addition, use-wear polishes and edge damage contribute to our understanding of tool function and hafting.

Blades and Bladelets

Twenty-three of 30 blades and bladelets (76.7%) preserved some type of functional evidence. Residues included plant and wood tissue, and single instances of hair and feathers, suggesting use on a wide range of materials. The most common pattern of residue/use-wear includes combinations of plant/wood residues with hard/high silica polishes. In several cases, it was possible to identify the wood residue as gymnosperm (conifers) in origin due to the presence of bordered pits (Figure 5-6) (Hoadley 1990). In one case, tracheids are characterised by paired bordered pits indicating either larch (*Larix* sp.) or spruce (*Picea* sp.) as the source wood (Figure 5) (Hoadley 1990). Differentiation between larch and spruce is not possible based on the preserved anatomy. Furthermore, nearly half (14/30, 46.7%) of the blades/bladelets also show evidence for hafting. The distribution of functional evidence often suggests that a blade or bladelet was hafted

Table 3: Summary of functional evidence by tool type*

Pod zubem/Pod křídlem combined					
Evidence	Blades and Bladelets	Flakes	Crescent	Unifacial Pt.	Total
Plant	9/40	3/28	0/1	0/1	12/70
Wood	3/40	1/28	0/1	0/1	4/70
Starch grains	2/40	1/28	0/1	0/1	3/70
Feathers	1/40	1/28	0/1	0/1	2/70
Hair	1/40	1/28	0/1	0/1	2/70
Soft	4/40	1/28	0/1	0/1	5/70
Hard/high silica	12/40	4/28	0/1	0/1	16/70
Hafted	14/40	3/28	0/1	1/1	18/70
Striae only	2/40	1/28	0/1	0/1	3/70
None	9/40	16/28	0/1	0/1	27/70

Pod zubem				
Evidence	Blades and Bladelets	Flakes	Unifacial Pt.	Total
Plant	2/30	1/15	0/1	3/46
Wood	3/30	1/15	0/1	4/46
Starch grains	0/30	1/15	0/1	1/46
Feathers	1/30	1/15	0/1	2/46
Hair	1/30	1/15	0/1	2/46
Soft	4/30	1/15	0/1	5/46
Hard/high silica	9/30	2/15	0/1	11/46
Hafted	14/30	2/15	1/1	17/46
Striae only	2/30	1/15	0/1	3/46
None	7/30	7/15	0/1	14/46

Pod křídlem				
Evidence	Blades and Bladelets	Flakes	Crescent	Total
Plant	7/10	2/13	0/1	9/24
Wood	0/10	0/13	0/1	0/24
Starch grains	2/10	0/13	0/1	2/24
Feathers	0/10	0/13	0/1	0/24
Hair	0/10	0/13	0/1	0/24
Soft	0/10	0/13	0/1	0/24
Hard/high silica	3/10	2/13	0/1	5/24
Hafted	0/10	1/13	0/1	1/24
Striae only	0/10	0/13	0/1	0/24
None	2/10	9/13	0/1	11/24

*Note: categories of evidence are not mutually exclusive. Totals in columns may exceed the number in the sample.

longitudinally. One half of the tool shows hafting evidence while the other half shows evidence for use. Hafting evidence includes striations confined to one area of an artifact as well as additive resin residue which may have served as a mastic (Figure 5).

Flakes

Eight of the 15 flakes (53%) analysed showed evidence of use. Flakes appear to have been used on a wide range of substrates with some flakes exhibiting use on multiple materials. PZ 824 (Figure 6) was used for cutting both gymnosperm wood and avian tissue. Feathers are potentially identifiable to Order level based on the patterning of nodes and internodes as well as the shape and arrangement of prongs at internodes (Brom 1986). The patterning on the feathers in Figure 6C-D is consistent with the Order Procellariiformes, which includes fulmars, shearwaters, and petrels. Given the fragmentary nature of the feather sample and the lack of certain diagnostic elements, however, this identification should be considered tentative, particularly since Procellariiformes are generally highly pelagic and Pod zubem is far from the ocean. Figure 7 shows a flake that was hafted using resin as a mastic. The worked edge shows numerous fragments of plant tissue (not specifically identifiable).

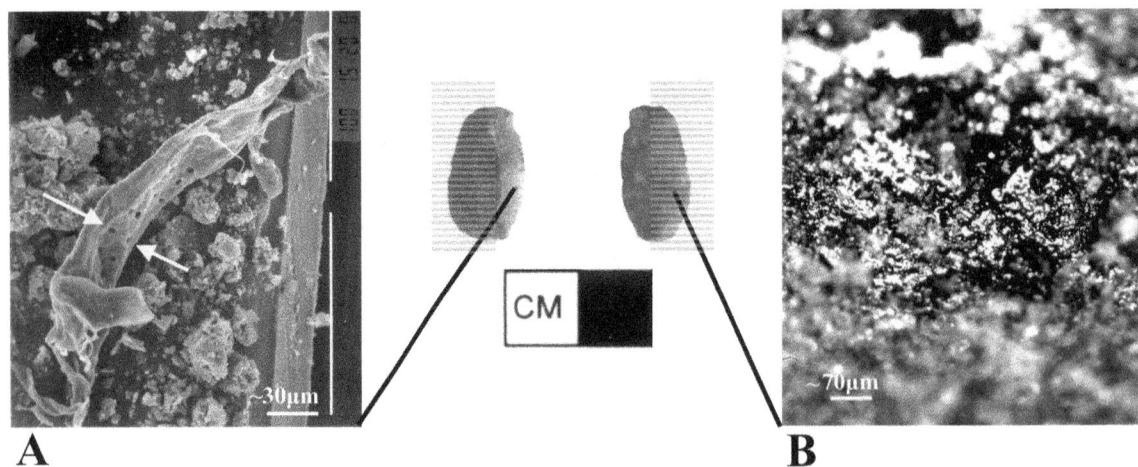

Figure 5. PZ 503: Bladelet with evidence of hafting and woodworking, A) SEM photo of softwood tracheid with paired bordered pits indicated by arows, B) resin. Shading indicates area under the haft.

Unifacial Point

The one unifacial point in the sample shows evidence for hafting in the form of striations on the proximal portion of the tool caused by movement of the artifact in the haft. No evidence was found on the distal end to indicate the material on which the tool was used.

Pod křídlem

Twenty-four artefacts were examined from the short-term occupation site of Pod křídlem, 12 (50%) of which showed functional traces (Table 3). Eight of ten blades and bladelets had signs of use, all with evidence of plant or hard/high silica material. The remaining two showed no evidence of use. Of the 13 flakes examined, only four (30.7%) showed signs of use, all on plant or hard/high silica material. One microlithic crescent examined preserved no evidence of use.

Figure 8 shows a bladelet that has been used for processing a starchy plant material. The plant tissue on the bladelet's margin consists of storied parenchyma (energy storage tissue) with numerous starch grains within it, identifiable by a characteristic extinction cross visible under cross-polarised light (Haslam 2004). This type of tissue is typically found in storage organs such as roots and tubers (Hather 1993). Nuts are also high in starch and contain storage (parenchyma) tissue (Decke 1982). Hazelnut remains were found at both Pod zubem and Pod křídlem. If this tissue does represent hazelnut, then the bladelet may have been used to peel the outer skin. However, once the hard outer shell has been removed, the skin on hazelnuts is soft, thin and edible, making this type of processing unnecessary. A more likely scenario would involve using a stone tool to pry open the hard outer shell. Paz (2001) documented this process on *Beilschimiedia* nuts from archaeological deposits in Sulawesi. In the process of prying open the shell of the nut with a stone tool, the underlying flesh of the nut was cut. This process would potentially produce a residue similar to that in Figure 7. Unfortunately, no further diagnostic anatomy was found to distinguish between nuts and roots or tubers.

DISCUSSION

Pod zubem and Pod křídlem were chosen for analysis because they represent two distinct types of occupation (long-term and short-term) and are in close proximity to one another. Thus, they have the potential to provide information about differences in activities at different site types, at least in terms of stone tool function. Based on the current analysis, the two sites show marked differences in terms of tool use.

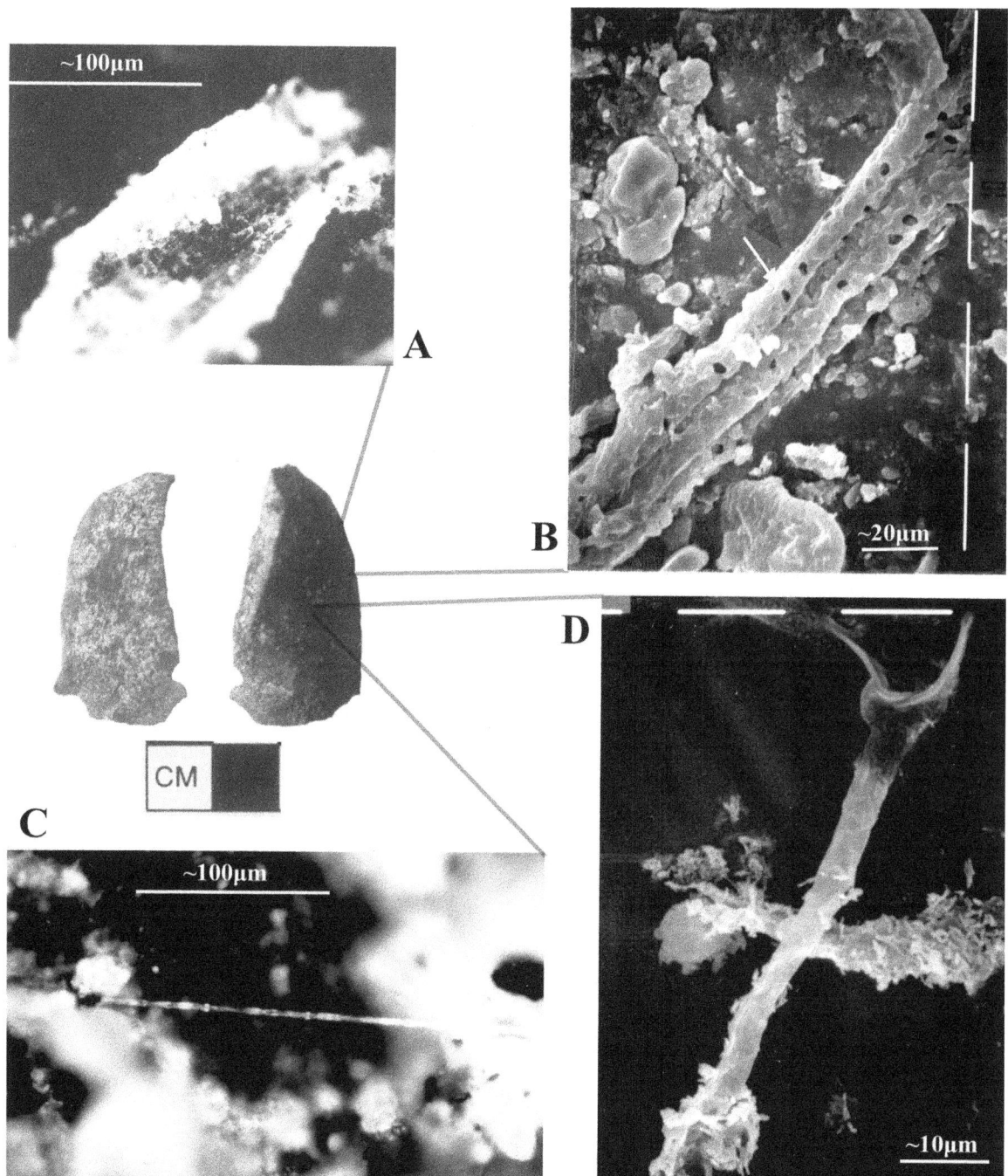

Figure 6. PZ 824: Flake used for woodworking and cutting avian tissue, A) wood cells in oblique section, B) SEM photo of softwood tracheids with bordered pits indicated by arrow, C) feather barbule, D) SEM photo of feather barbule with two projecting prongs.

On the basis of the environmental and faunal analyses, Pod zubem, the larger settlement, was within a predominantly woodland environment with pine, oak, and hazel, not far from a larger brook. The Mesolithic inhabitants concentrated on collecting hazelnuts and hunting hares, deer, elks and birds, with special emphasis on fur animals such as squirrel, marten, wild cat, beaver, and fox. Hazelnuts are available seasonally in late summer and early fall. The evidence for skin-working and the presence of fur-bearing animals suggests a late fall or winter occupation. In contrast, the Pod křídlem site represents short-term and probably task-specific stays deeper inside the valley, directly above the brook.

Parallel to the longer occupation with numerous hearths, faunal remains, and macrobotanical remains, stone tool use at Pod zubem is more varied than that at Pod křídlem. Stone tools were used on a wide range of materials including plant, wood, mammals, and birds. In some cases, a single tool was used on many different materials. Hafting is common at Pod zubem and was often accomplished by the use of a resin as a mastic. While it has not been possible to specifically

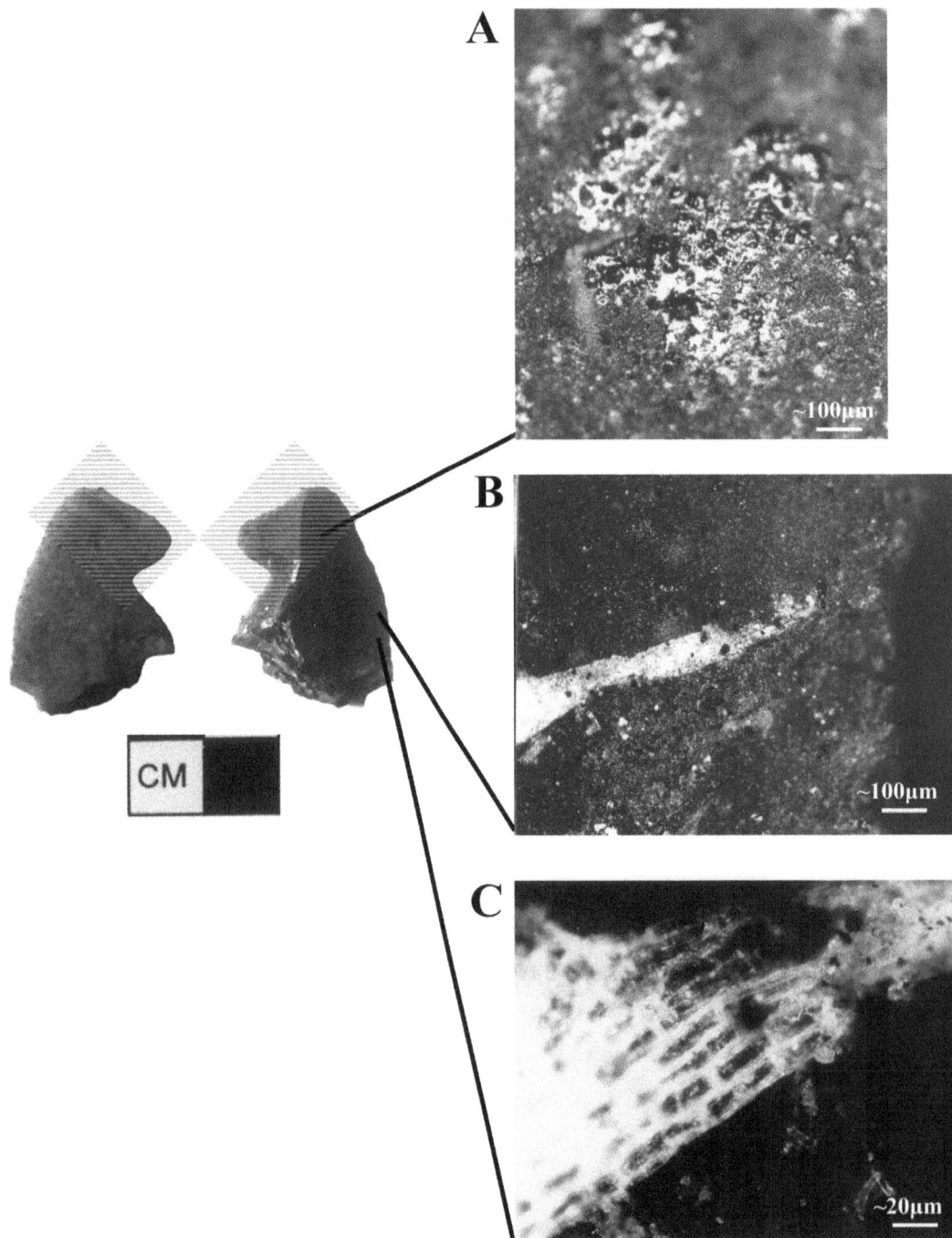

Figure 7. PZ 475: Hafted flake used for plant processing, A) resin, B and C) plant tissue on edge. Shading indicates area under the haft.

identify the source of the resin, the evidence for processing of gymnosperms suggests various conifers (juniper, pine, spruce, larch, etc.) as possiblities. These findings are consistent with archaeological data that suggest that the site represents a longer term, repeated occupation.

By contrast, the more ephemeral site of Pod křídlem demonstrates a more specialised use of tools. All artefacts with signs of use were utilised on plants or hard/high silica material. Furthermore, there is little evidence of hafting. These data fit a pattern of hand-held tool use focused on plant material. The starchy storage tissue found on PK 251 can be interpreted as evidence of food processing, either of nuts or roots and tubers. If this is indeed nut processing, this may suggest a late summer/early fall occupation similar to Pod zubem. If instead this tissue derives from roots or tubers, it does not provide insight into seasonality.

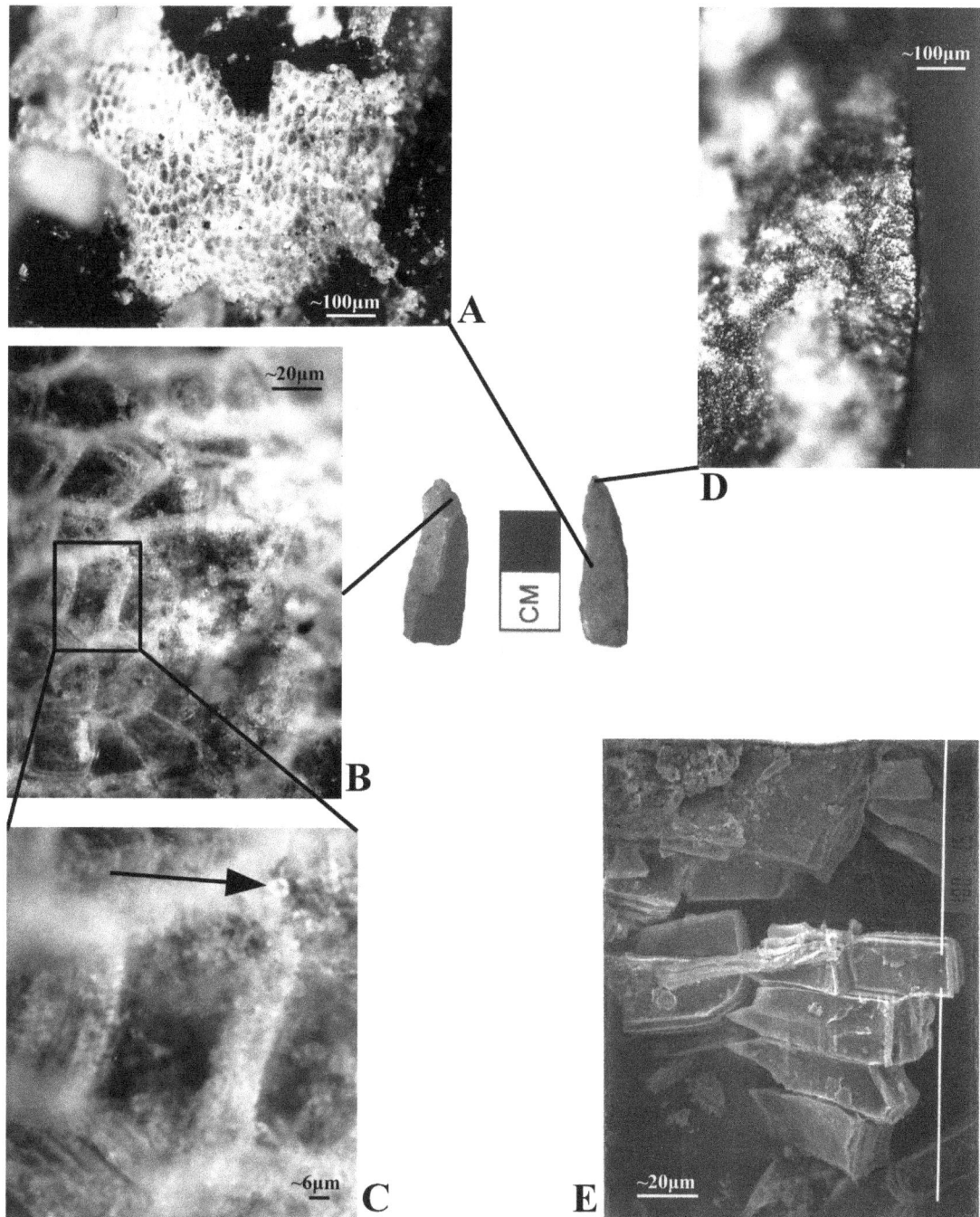

Figure 8. PK 251: Bladelet used for processing starchy plant material. A) storied parenchyma cells, B and C) parenchyma cells with starch grains, arrow indicates characteristic extinction cross, D) hard/high silica polish, E) SEM photo of storied parenchyma.

Another possible explanation for the differences between these two sites is differential residue preservation. Often, it appears that animal residues are underrepresented, at least in comparison with the archaeozoological assemblages (Hardy *et al.* 2001; Hardy, 2004). Microscopic residue analysis can, however, detect animal residues when they are present (e.g. Hardy *et al.* 2001; Lombard 2004; Loy and Dixon 1998; Williamson 1996). Furthermore, when preservation conditions are favourable, animal residues may be found in large amounts on numerous artefacts as has been demonstrated on Aurignacian artefacts from southwestern Germany (Hardy *et al.* in press). In this case, the sites are very similar in geological makeup (sediment, depositional environment) and differential preservation between the two is unlikely. Therefore, the lack of animal residues at Pod křídlem is not likely due to taphonomic bias.

Plants processed at Pod zubem and Pod křídlem include wood (specifically softwood/ gymnosperm) and possibly roots, tubers or nuts. Evidence for root or tuber exploitation has been documented at numerous Mesolithic sites across Europe. Root and tuber remains include arrowhead, knotweed (Calowanie, Poland: Kubiak-Martens 1996), cattail, wild beet, bulrush, wood fern (NP3 and S51, Northern Netherlands: Perry 1999), wild garlic, pignut (Halsskov, Denmark: Kubiak-Martens 2002), and non-specific vegetative storage tissue (Roc del Migdia, Catolonia: Holden *et al.* 1995). Hazelnuts are quite common at many Mesolithic sites (see Regnell *et al.* 1995; Zvelebil 1994; among others), although it is difficult to estimate exactly how important they were in the Mesolithic diet (Mithen *et al.* 2001).

Wood processing definitely occurred at Pod zubem, particularly of gymnosperms and in one case larch or spruce. However, what the wood was being used for is unclear. There is evidence that conforms to hafting traces (striations, bright spots, resin) seen on experimental and archaeological material (e.g. Crombé *et al.* 2001; Rots 2005) and hafting requires the preparation of a handle. Thus, the woodworking could relate to the preparation of a haft. Other uses are certainly possible but lack any supporting evidence on which to form a hypothesis.

CONCLUSIONS

Clarke (1976) anticipated the importance of plant use in the Mesolithic many years ago and subsequent research has confirmed his hypothesis (e.g. Mason 2000; Mithen *et al.* 2001; Zvelebil 1994). While macrobotanical remains were found at the long-term occupation site of Pod zubem, the application of microscopic residue analysis has yielded additional evidence for plant processing. At the short-term occupation of Pod křídlem, few macrobotanical remains were recovered; thus, residue analysis has demonstrated an activity that was otherwise hardly visible archaeologically. The evidence for processing of wood at Pod zubem fits well with the faunal assemblage which is dominated by forest species. The macrobotanical identification of hazelnuts along with the microscopic identification of roots, tubers, or nuts adds to the growing body of literature demonstrating the importance of plant foods in the Mesolithic.

The evidence presented here also points to the exploitation of mammals and birds at Pod zubem, especially the smaller fur animals (cf. Charles 1997). Evidence for frequent hafting at the site may relate to the construction of composite points or barbs similar to those recreated experimentally (Crombé *et al.* 2001), whereas the impact blows identified by Yaroshevich (pers. comm. 2006) on some of the geometric projectiles complete the picture of hunting by projectiles.

In addition to providing information about subsistence and economic activities, we can see a clear distinction in site function. Stone tools were used on a wide variety of materials at Pod zubem, reflecting the longer-term occupation and the richer archaeological assemblage. The archaeological record at Pod křídlem is sparser, but can be supplemented by the functional analyses that point to plant processing as an important activity. The combination of faunal, macrobotanical, and stone tool functional anlayses suggests specialised use of tools and sites in the Mesolithic of Northern Bohemia.

ACKNOWLEDGEMENTS

We would like to thank Kenyon College for supporting this research, the Institute of Archaeology, Academy of Sciences of the Czech Republic, Brno, Czech Republic for allowing access to the materials, and the staff of the SEM Laboratory at Miami University, Ohio for assistance with scanning electron microscopy. The research program 'Last Hunter-Gatherers in Northern Bohemia' was funded by National Geographic Grant 98/6330.

REFERENCES

Anderson-Gerfaud, P. 1990. Aspects of behaviour in the Middle Paleolithic: Functional analysis of stone tools from southwest France. In P. Mellars (ed.) *The Emergence of Modern Humans*, pp. 389-418. Ithaca, New York: Cornell University Press.

Beyries, S. 1988. *Industries Lithiques: Traçeologie et Technologie*. London: British Archaeological Report, International Series 411 (1 and 2).

Brom, T. 1986. Microscopic identification of feathers and feather fragments of Palearctic birds. *Bijdragen tot de Dierkunde* 56:181-204.

Brunner, H. and B. Coman 1974. *The Identification of Mammalian Hair*. Melbourne: Inkata Press.

Cahen, D., L. Keeley and F.L. Van Noten 1979. Stone Tools, toolkits, and human behavior in prehistory. *Current Anthropology* 20: 661-683.

Catling, D. and J. Grayson 1982. *Identification of Vegetable Fibres*. New York: Chapman and Hall.

Charles, R. 1997. The exploitation of carnivores and other fur-bearing mammals during the north-western european Late Upper Paleolithic and Mesolithic. *Oxford Journal of Archaeology* 16: 253-277.

Clarke, D. 1976. Mesolithic Europe: the economic basis. In G. Sieveking, I.H. Longworth and K. Wilson (eds) *Problems in Economic and Social Archaeology*, pp. 449-481. London: Duckworth.

Crombé, P., Y. Perdaen, J. Sergant and J.-P. Caspar 2001. Wear analysis on early Mesolithic microliths from the Verrebroek site, East Flanders, Belgium. *Journal of Field Archaeology* 28: 253-269.

Decke, U. 1982. Mikroskopische Untersuchung an geschälten und zerkleinertenölsamen und Nußkernen. *European Food Research and Technology* 174: 187-194.

Dumont, J. 1988. *A Microwear Analysis of Selected artifact Types from the Mesolithic Sites of Star Carr and Mount Sandel*. BAR British Series 187. Oxford: B.A.R.

Finlayson, B. 1990. The Function of Microliths, Evidence from Smittons and Starr SW, Scotland. *Mesolithic Miscellany* 11: 2-6

Finlayson, B. and S. Mithen 1997. The microwear and morphology of microliths from Gleann Mor, in H. Knecht (ed.) *Projectile Technology*, pp. 107-129. New York: Plenum Press.

Fullagar, R. 1991. The role of silica in polish formation. *Journal of Archaeological Science* 18: 1-24.

Hardy, B.L. 1994. Investigations of stone tools function through use-wear, residue and DNA analyses at the Middle Paleolithic site of La Quina, France. Unpublished Ph.D. thesis. Bloomington: Indiana University.

Hardy, B.L. 1999. Preliminary results of residue and use-wear analyses of stone tools from two Mesolithic sites, northern Bohemia, Czech Republic. *Archaeologicke Rozhledy* 51: 274-279.

Hardy, B.L. 2004. Neanderthal behaviour and stone tool function at the Middle Palaeolithic site of La Quina, France. *Antiquity* 78:547-565.

Hardy, B.L. and G.T. Garufi 1998. Identification of woodworking on stone tools through residue and use-wear analyses: experimental results. *Journal of Archaeological Science* 25: 77-84.

Hardy, B.L., M. Kay, A.E. Marks, and K. Monigal 2001. Stone tool function at the Paleolithic sites of Starosele and Buran Kaya III, Crimea: behavioral implications. *Proceedings of the National Academy of Sciences, USA* 98: 10972-10977.

Hardy, B.L., M. Bolus and N. J. Conard. In press. Hammer or crescent wrench? Stone tool form and function in the Aurignacian of southwest Germany. *Journal of Human Evolution*.

Hardy, K. 2004. Microwear analysis of a sample of flaked stone tools. In C. Wickham-Jones and K. Hardy (eds) *Camas Daraich: A Mesolithic Site at the Point of Sleat, Skye*. Scottish Archaeological Internet Report 12 [URL http://www.sair.org.uk/sair12].

Haslam, M. 2004. The decomposition of starch grains in soils: implications for archaeological residue analyses. *Journal of Archaeological Science* 31: 1715-1734.

Hather, J. 1993. *An Archaeobotanical Guide to Root and Tuber Identification, Volume I, Europe and South West Asia*. Oxford: Oxbow Books.

Hoadley, R.B. 1990. *Identifying Wood: Accurate Results with Simple Tools*. Newtown, CT: Taunton Press.

Holden, T., J. Hather, and J.P.N. Watson 1995. Mesolithic plant exploitation at the Roc del Migdia, Catalonia. *Journal of Archaeological Science* 22: 769-778.

Horáček, I. 2003. Obratlovčí fauna z pískovcových převisů severních Čech. In J. Svoboda (ed.) *Mezolit Severních Čech*, pp. 48-57. The Dolní Věstonice Studies 9. Brno: Institute of Archaeology.

Kardulias, N. P. and R. W. Yerkes 1996. Microwear and metric analysis of threshing sledge flints from Greece and Cyprus. *Journal of Archaeological Science* 23:657-666.

Keeley, L. 1980. *Experimental Determination of StoneTool Uses*. Chicago: University of Chicago.

Kubiak-Martens, L. 1996. Evidence for possible use of plant foods in Paleolithic and Mesolithic diet from the site of Calowanie in the central part of the Polish plain. *Vegetation History and Archaeobotany* 5:33-38.

Kubiak-Martens, L. 2002. New evidence for the use of root foods in pre-agrarian subsistence recovered from the late Mesolithic Site at Halsskov, Denmark. *Vegetation History and Archaeobotany*11:23-31.

Lombard, M. 2004. Distribution patterns of organic residues on Middle Stone Age points from Sibudu Cave, Kwazulu-Natal, South Africa. *South African Archaeological Bulletin* 59:37-44.

Ložek, V. 2003. Fosilní měkkýši ve výplních pískovcových převisů a jejich význam pro poznání pravěkého prostředí. In J. Svoboda (ed.) *Mezolit Severních Čech*, pp.43-47. The Dolní Věstonice Studies 9. Brno: Institute of Archaeology.

Loy, T.H. and E.J. Dixon 1998. Blood residues on fluted points from Eastern Beringia. *American Antiquity* 63: 21-46.

Mason, S.L.R. 2000. Fire and Mesolithic subsistence – managing oaks for acorns in northwest Europe? *Palaeogeaography, Palaeoclimatology,Palaeoecology* 164: 139-150

Mithen, S., N. Finlay, W. Carruthers, S. Carter and P. Ashmore 2001. Plant use in the Mesolithic: Evidence from Staosnaig, Isle of Colonsay, Scotland. *Journal of Archaeological Science* 28: 223-234.

Opravil, E. 2003. Rostlinné makrozbytky. In Jiří Svoboda (ed.), *Mezolit Severních Čech*. The Dolní Věstonice Studies 9. Brno: Institute of Archaeology, 38-42.

Paz, V. 2001. Cut not smashed: A new type of evidence for nut exploitation from Sulawesi. *Antiquity* 75:497-498.

Pearsall, D. 2000. *Paleoethnobotany: A Handbook of Procedures*. 2nd Edition. New York: Academic Press.

Perry, D. 1999. Vegetative tissues from Mesolithic sites in the northern Netherlands. *Current Anthropology* 40:231-237.

Regnell, M., M-J. Gaillard, T.S. Bartholin and P. Karsten 1995. Reconstruction of environment and history of plant use during the late Mesolithic (Ertebólle culture) at the inland settlement of Bökeberg III, southern Sweden. *Vegetation History and Archaeobotany* 4: 67-91.

Rots, V. 2005. Wear traces and the interpretation of stone tools. *Journal of Field Archaeology* 30: 61-73.

Street, M., M. Baales, E. Cziesla, S. Hartz, M. Heinen, O. Jöris, I. Koch, C. Pasda, T. Terberger, and J. Vollbrecht 2001. Final Paleolithic and Mesolithic research in reunified Germany. *Journal of World Prehistory* 15: 365-453.

Svoboda, Jiří A. (ed.) 2003. *Mesolithic of Northern Bohemia. The Dolní Věstonice Studies 9*. Brno: Institute of Archaeology.

Svoboda, J., V. Cílek, L. Jarošová, and V. Peša 1999. Mezolit z perspektivy regionu. Výzkumy v ústí Pekla. *Archeologické rozhledy* 51: 243-264.

Svoboda, J., M. Hajnalová, I. Horáček, M. Novák, A. Přichystal, A. Šajnerová, and A. Yarosevich 2007. Mesolithic settlement and activities in rock shelters of the Kaminice River Canyon, Czech Republic. *Eurasian Prehistory* 5: 95–127.

Stuiver, M. and P.J. Reimer 1993. Extended ^{14}C data base and revised CALIB 3.0 ^{14}C age calibration program. *Radiocarbon* 35: 215-230.

Teerink, B.J. 1991. *Hair of West European Mammals: Atlas and Identification Key*. Cambridge: Cambridge University Press.

Thorsberg, K. 1983. Bruksskadeanalys av valda artefakter fran tva tidigmesolitiska boplatser vid Hornborgasjon (Use-wear Analysis of Selected Artefacts from Two Early Mesolithic Sites at Hornborgasjon). *Tor* 20:11-44.

Uwe-Heussner, K.1989. Gebrauchsspurenuntersuchungen an Flintgeraten aus den mesolithischen Grabern von Schopsdorf, Kr. Hoyerswerda. *Veroffentlichungen des Museums fur Ur- und Fruhgeschichte* 23: 55-57.

Whittaker, J. 1994. *Flintknapping: Making and Understanding Stone Tools*. Austin: University of Texas.

Williamson, B.S. 1996. Preliminary stone tool residue analysis from Rose Cottage Cave. *Southern African Field Archaeology* 5:36-44.

Zvelebil, M. 1994. Plant use in the Mesolithic and its role in the transition to farming. *Proceeding of the Prehistoric Society* 60: 35-74.

13

Chloroplast DNA from 16th century waterlogged oak in a marine environment: initial steps in sourcing the Mary Rose timbers

Alanna K. Speirs[1], Glenn McConnachie[2] and Andrew J. Lowe[3,4]

1. School of Social Science
The University of Queensland
Brisbane QLD 4072 Australia
Email: alanna.speirs@gmail.com

2. Conservation Manager, Mary Rose Trust
College Road, HM Naval Base
Portsmouth PO1 3LX United Kingdom

3. The School of Integrative Biology
The University of Queensland
Brisbane QLD 4072 Australia

4. School of Earth and Environmental Science
University of Adelaide
Adelaide SA 5000 Australia

ABSTRACT

This paper reports initial results of a palaeogenetic analysis of timbers from the hull of the English Tudor flagship *Mary Rose*. The study is the first step in assessing the feasibility of extracting and amplifying chloroplast DNA (cpDNA) from these timbers, which were preserved in a marine environment for more than four centuries. The ultimate goal of this research is to determine the provenance of oak (*Quercus* spp.) used in the ship's manufacture, following previous work demonstrating that the chloroplast genome of modern European oak populations exhibits a strong phylogeographic structure. Experimental trials revealed that extraction methods developed for modern oak wood were inadequate owing to the presence of polymerase chain reaction (PCR) inhibitors in the *Mary Rose* timbers. A series of treatments were tested to develop a new extraction protocol, resulting in cpDNA recovery from one archaeological sample. These results represent the first successful extraction and amplification of cpDNA from waterlogged archaeological oak wood from a marine environment.

KEYWORDS

oak wood, DNA, palaeogenetics, PCR, chloroplast, *Mary Rose*

INTRODUCTION

This paper reports initial results of a palaeogenetic analysis of timbers from the hull of the English Tudor flagship *Mary Rose*. The study is the first step towards an ultimate goal of determining the geographic provenance of oak (*Quercus* spp.) used in the ship's manufacture. Stage I of the research program was designed to test the viability of extracting and amplifying DNA from the waterlogged *Mary Rose* timbers, which were submerged in a marine environment for more than four centuries.

PALAEOGENETICS AND OAK

Palaeogenetics, the study of ancient DNA (aDNA), offers direct genetic evidence of extinct and extant species from preserved samples (Capelli *et al.* 2003; Gugerli *et al.* 2005; Hofreiter *et al.* 2001; Pääbo *et al.* 2004). After the death of an organism, hydrolytic and oxidative processes cause fragmentation and modification of DNA (Lindahl 1993; Pääbo *et al.* 2004). These processes reduce the size of DNA fragments that may be successfully amplified using the polymerase chain reaction (PCR), resulting in average aDNA fragment lengths of 50-500bp (Hofreiter *et al.* 2001; Hoss *et al.* 1996; Pääbo 1989; Pääbo *et al.* 2004). However, using PCR primers designed to target small fragments, aDNA up to tens of thousands of years old has been successfully amplified from the tissues of animals and plants (Pääbo *et al.* 2004).

Wood is one of the most common archaeological plant remains (Gugerli *et al.* 2005), and the application of palaeogenetics to wooden archaeological artefacts may reveal a wealth of information about the genetic structure of past natural populations. Such information has potential applications in studies of wood traceability, historical shipbuilding, resource management and trade. Other potential applications include modern forest management, species conservation and monitoring climate change and evolution of species (Capelli *et al.* 2003; Deguilloux *et al.* 2002; Dumolin-Lapegue *et al.* 1999; Gugerli *et al.* 2005; Tani *et al.* 2003).

European oaks of the genus *Quercus* provide a suitable case study for genetic determination of provenance as oak cpDNA remains unaltered through many generations (Ferris *et al.* 1993). Present oak populations are likely therefore to reflect variation that has been established for hundreds of years (Hewitt 1999; Petit *et al.* 1993). Analysis of cpDNA from fresh bud and leaf samples from 2613 *Quercus* populations Europe-wide identified polymorphisms that distinguish 32 distinct genotypes, referred to as haplotypes (Dumolin-Lapegue *et al.* 1997; Ferris *et al.* 1993, 1995; Petit *et al.* 1993; Petit *et al.* 2002). Maps constructed from these data demonstrate patchy haplotype distributions that tend to run along a longitudinal gradient and are differentiated by latitude (Hewitt 1999; Petit *et al.* 2002). Chloroplast DNA haplotype maps corroborate patterns of oak post-glacial colonisation deduced from palynological studies that identify Iberia, Italy and the Balkans as regions of refugia during the last glacial period (~115ka - 15kyr BP). Therefore, oak cpDNA variation exhibits a strong phylogeographic structure that reflects patterns of oak post-glacial migration northwards during the Holocene (10kyr BP – present) (Brewer *et al.* 2002; Huntley and Birks 1983; Petit *et al.* 2002).

Although all early oak haplotype work used fresh leaf and bud tissues as a DNA source, recent studies have demonstrated that is possible to extract and amplify chloroplast DNA from oak wood. Between 2002 and 2004, Deguilloux and colleagues developed molecular tools for provenancing modern European oak wood using DNA extracted from sawdust. Deguilloux *et al.* (2003) optimised oak wood PCR amplification techniques by designing a set of nine primer pairs that target short cpDNA sequences (50-200 bp). Collectively, these targets distinguish nine common European cpDNA haplotypes, including those found in Britain. The authors successfully genotyped nine of 22 oak samples and compared these haplotypes with cpDNA maps to differentiate wood of western and eastern European origin. Using this method, Deguilloux *et al.* (2004) determined that haplotypes confirm the region of origin of wood used to manufacture oak barrels in the French cooperage industry.

Molecular investigations have identified that cpDNA in oak wood is rapidly degraded within the first few years after felling (Deguilloux *et al.* 2002; Dumolin-Lapegue *et al.* 1999). Deguilloux *et al.* (2002) characterised cpDNA degradation in oak logs as seen in amplifiable fragment lengths, and found that approximately 11 years after felling only segments less than 250 bp could be amplified from the sapwood of logs. Despite these results, Dumolin-Lapegue *et al.* (1999) report amplification of 400 bp of cpDNA from 600 year old oak wood from a dam, while Tani *et al.* (2003) amplified up to 600 bp of nuclear DNA from 3600 year old *Cryptomeria japonica* wood that had also been buried in a dam. A review of previous wood DNA studies (see Table 1) indicates that DNA is increasingly degraded with age, resulting in smaller amplifiable fragments. It appears though that the level of DNA degradation becomes relatively stable after about two years after felling, with amplified fragments of 50 to 500 bp in length. Our research aimed to build on the methods of these previous studies to extract and amplify cpDNA from waterlogged archaeological oak material from the *Mary Rose*, as part of a program to determine the geographic source of the timbers used in the ship's construction.

Table 1. Summary of previous oak cpDNA studies, indicating plant material, age and maximum fragment size amplified

Tissue	Age (years)	Size (bp)	Reference
Leaf/Bud	Fresh	1688	Petit *et al.* 2002
Wood	1	1483	Deguilloux *et al.* 2002
Wood	2	566	Deguilloux *et al.* 2002
Wood	3	175	Deguilloux *et al.* 2002
Wood	11	187	Deguilloux *et al.* 2002
Wood	300	350	Dumolin-Lapegue *et al.* 1999
Wood	3600	500	Tani *et al.* 2003

Figure 1. The *Mary Rose* as depicted on the Anthony Roll. This is the only known contemporaneous image of the ship. (Image courtesy of The Mary Rose Trust).

THE *MARY ROSE*

The *Mary Rose* (Figure 1) was built in Portsmouth, England, between 1509 and 1511 and served as the flagship of King Henry VIII's fleet until 1545, when she sank in the Solent Channel during a battle with the French (Bridge and Dobbs 1996; Marsden 2003). Lying on its starboard side, the hull filled with sediments that preserved this portion in an anoxic environment, while the exposed port side of the ship decayed (Mouzouras *et al.* 1986; Rule 1983). Constructed almost entirely of oak (*Quercus* spp.), the remains of the ship were raised from the seabed in 1982 (Rule 1983; The Mary Rose Trust 2002) and are currently being conserved with a warm polyethylene glycol (PEG) spray solution (Figure 2). This treatment will prevent shrinkage when the timbers are dried out, starting in 2011 (The Mary Rose Trust 2002).

As the conservation program to stabilise the timbers nears completion, a number of important questions remain unanswered, including confirmation of the assumed source of the timber used in her construction and later refit. Contemporaneous documents relating specifically to timber supply for the *Mary Rose* have not been found. Clues to the source of the wood are restricted to documents recording that timber for contemporaneous ships was purchased local to Portsmouth (within 50km). The ship also had at least one major refit in 1536, probably in the River Medway at Chatham (Figure 3). Royal accounts indicate that timber used for ship rebuilding in the River Medway was purchased from southeast England (Bridge and Dobbs 1996).

Cottrell *et al.* (2002) and Lowe *et al.* (2004, 2006) have constructed a detailed oak cpDNA haplotype distribution map of Britain (Figure 4) that may be used to match haplotypes from oak timbers of the *Mary Rose* with their presumed region of origin. These authors found that the dominant haplotypes of Britain are those known as 10, 11 and 12. According to this map, oak populations around the Portsmouth region are dominated by haplotype 10, whereas the majority of southeast populations, near Chatham, are haplotype 12. Thus, timbers from the original

Figure 2. Salvaged starboard side of the *Mary Rose* on display in a dry dock in Portsmouth. The ship's remains were excavated in 1982 and are currently being conserved with a warm spray solution of the water-soluble wax polyethylene glycol (PEG). (Reproduced with permission from The Mary Rose Trust).

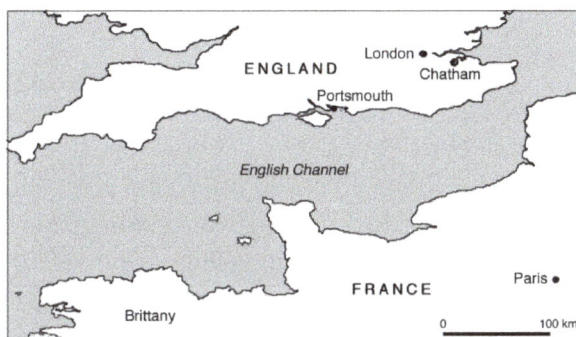

Figure 3. England and northern France identifying Portsmouth and Chatham. The *Mary Rose* was built in Portsmouth (1509-1511) and had at least one major refit (1536), probably in the River Medway near Chatham. The ship sank about 1 km from Portsmouth in the Solent. (Adapted from Bridge and Dobbs 1996).

construction of the *Mary Rose* would most likely be haplotype 10, whereas refit timbers would be haplotype 12. While positive correlation of the genetic signatures of the *Mary Rose* timbers with those of regional oak populations would not prove their source, failure to correlate would definitively exclude non-matching oak populations. As Stage I in a long-term project to genetically map the provenance of the *Mary Rose* timbers, existing cpDNA extraction protocols were assessed for their applicability to the specific archaeological circumstances surrounding the ship's taphonomic history, and new procedures were developed when necessary.

MATERIALS AND METHODS

DNA from the *Mary Rose* oak wood was expected to be extremely low in quantity and quality, therefore in this study production of target PCR product was considered primary proof of successful DNA extraction, while the production of primer dimer confirmed appropriate PCR conditions in the absence of DNA. Successfully amplifying cpDNA required a series of experimental modifications to established protocols. These steps are outlined in the methods and results sections below in the sequence in which

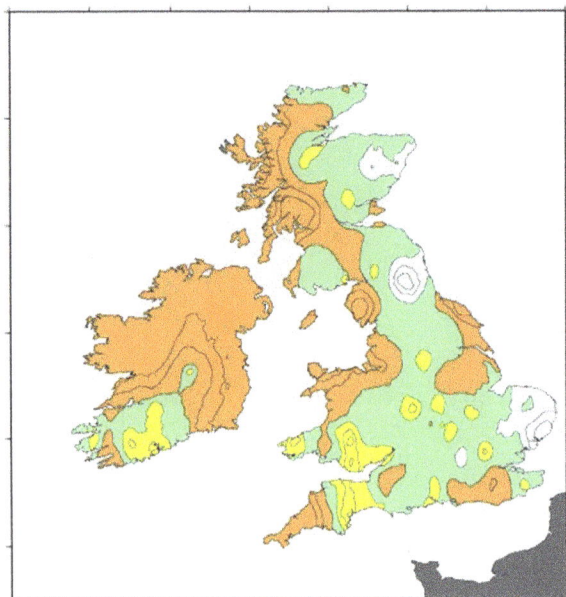

Figure 4. Distribution of three most common oak cpDNA haplotypes across the British Isles. Key: haplotype 10 = yellow, 11 = white and 12 = orange. A kriging average of cpDNA haplotype frequency is presented, where green are regions of no overall dominance and the outer circles of the three representative haplotype colours indicate where that type is found within neighbouring populations at a frequency greater that 60%; additional concentric lines within these areas represent haplotype dominances of 80 and 100%. Portsmouth and Chatham areas are dominated by haplotypes 10 and 12 respectively. (Modified from Lowe et al. 2006).

they were developed and are reviewed in the discussion section. In summary, the methods of Deguilloux *et al.* (2003) were replicated in an initial feasibility study using a modern oak sample; application of these methods to *Mary Rose* samples identified the presence of PCR inhibitors in these DNA extracts; ten *Mary Rose* samples were used in a series of experiments to develop a new protocol to remove inhibitors during the extraction process; and finally five resulting *Mary Rose* DNA extracts were used to optimise PCR conditions for amplifying cpDNA from this source.

SAMPLES

This study included samples of modern and archaeological oak wood. A modern wood sample was collected from Edinburgh, Scotland, in June 2005. The Mary Rose Trust provided ten *Mary Rose* samples from ten individual timbers of the salvaged hull, as indicated in Figure 5. Each timber sample is a core, approximately 10 cm long, 0.5 cm in diameter and weighing between 600-900 g. The labels in Figure 5 are used to refer to the samples throughout this study.

We originally used 180-grit silicon carbide wet/dry sandpaper to convert the samples to 20 mg or less of dry homogenous powder suitable for cell lysis. However, this method resulted in a loss of approximately two thirds of the core weight to the sandpaper grit matrix. An alternative method of producing sawdust using a mini hacksaw substantially reduced sample waste. Sawdust was transferred to a 1.5ml sealed plastic tube, weighed and stored at -20°C.

CONTAMINATION CONTROL

In aDNA research it is essential to avoid contamination of reagents with modern DNA (Cooper and Poinar 2000; Pääbo *et al.* 2004). Precautions applied in this study included performing all ancient and modern experiments in separate laboratories at the University of Queensland (UQ) and treating all equipment and workspaces with a bleach solution. All mobile equipment was irradiated with ultraviolet (UV) light to mutate contaminant DNA and we monitored reagent and laboratory contaminants by including negative controls (extractions and PCR amplifications performed without adding DNA). *Mary Rose* sample sawdust was prepared in the Archaeological Sciences Laboratory (ASL) UV room after irradiating the room for at least 10 minutes. Aliquots of reagents used in aDNA PCR amplifications were prepared in the ASL airflow displacement bubble. All modern oak work was performed in the UQ Life Sciences Laboratory, located in a different building to the ASL.

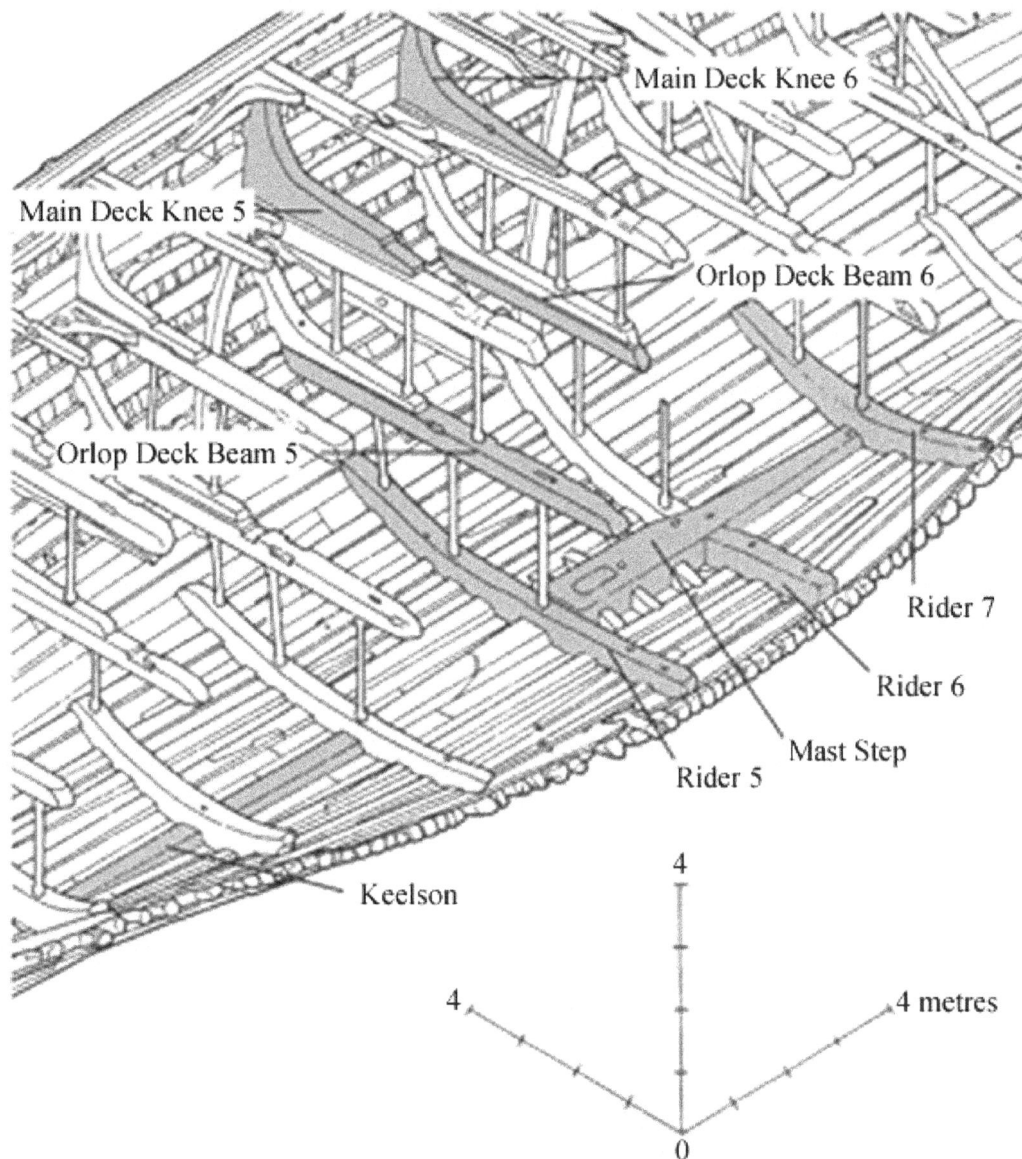

Figure 5. An isometric diagram of the *Mary Rose* hull showing the location of timbers (shaded) sampled for this study. The key lists the sampled timbers and their correlating reference label used in this study. As each timber has been sampled twice, individual cores from each timber are designated A and B. (Reproduced with permission from The Mary Rose Trust).

PREVIOUSLY PUBLISHED METHODS

DNA extractions were performed using the DNeasy Plant Minikit (Qiagen) as per Deguilloux *et al.* (2003). To increase the weight of the starting sample to a comparable 80-100 mg, we combined five separate 20 mg extractions through a single spin column.

DNA extracts were used in PCR amplifications according to the protocol of Demesure *et al.* (1995), which was also used for wood DNA by Deguilloux *et al.* (2003). PCR was conducted using five primer pairs designed by Deguilloux *et al.* (2003): tf42, µdt1, dt13, dt74 and dt74b. These primer pairs were chosen because they target cpDNA sequences less than 100 bp that contain informative polymorphisms and collectively differentiate amongst the four common British *Quercus* haplotypes 7, 10, 11 and 12 (after Cottrell *et al.* 2002). PCR products were separated by 3% agarose gel electrophoresis and PCR product was labelled with early termination (ET) dye for sequencing using the MegaBACE 1000 (Amersham Biosciences).

DETECTION OF PCR INHIBITORS IN *MARY ROSE* DNA EXTRACTS

Mary Rose sawdust samples extracted using the above methods failed to produce product or primer dimer following amplification. To verify that amplification failure was due to the presence of PCR inhibitors rather than the PCR conditions, we added modern DNA to the reactions (Bickley and Hopkins 1999). According to this method, modern DNA is added to the positive control and the *Mary Rose* extract. Successful amplification of the modern product in the positive control but not the *Mary Rose* extract indicates PCR inhibition.

Initially we used 2 µl pUC19, an *Escherichia coli* plasmid (Yanisch-Perron *et al.* 1985), to detect PCR inhibition in *Mary Rose* extracts. These PCR amplifications were performed using primers Af and Cr to amplify a 217 bp fragment (Connell 2002). We used modern oak wood DNA instead of the plasmid during subsequent PCR optimisation experiments,.

Extracts were assessed for nucleic acid quantity and purity with the NanoDrop ND-1000 spectrophotometer. Nucleic acids were measured at an absorbency of 260 nm and the quantity of DNA in each sample was calculated in ng/µl. Protein and phenols were measured at absorbencies of 280 and 230 nm. To assess the purity of DNA, the NanoDrop software calculates the ratio of absorbance at 260 and 280 nm (260/280) and at 260 and 230 nm (260/230), where pure DNA has a 260/280 ratio of ~1.8 and a 260/230 ratio within the range of 1.8-2.2. The information obtained was used to characterise potential PCR inhibitors in the *Mary Rose* extracts.

MODIFICATION OF PREVIOUSLY PUBLISHED METHODS

Two modifications of the lysis stage of the extraction method were trialled on samples HT4 and MD1. These changes were based on the protocol of Guy *et al.* (2003): (1) adding Proteinase K (Boehringer Mannheim) to the lysis stage; and (2) adding Proteinase K, Chelex 100 (BioRad) and polyvinylpyrrolidone (PVP) 360 (Sigma). Proteinase K digests proteins, PVP360 sequesters phenolics and Chelex 100 is an ion chelating resin (Guy *et al.* 2003).

NEW *MARY ROSE* EXTRACTION PROTOCOL

DNA was extracted from *Mary Rose* samples using a new protocol developed from the methods of Guy *et al.* (2003) and adapted to the DNeasy Plant Minikit protocol. 100 mg of dry sawdust from a single *Mary Rose* timber sample was divided into five extraction tubes of 20 mg each. To each of the five individual extractions, 10 µl of 20 mg/ml Proteinase K was added to the lysis stage incubation, which was lowered to 55°C and extended to 1 hr. Cellular debris was pelleted by centrifugation for 5 min at 14,000 rpm. The supernatant liquid was transferred to a sterile tube and incubated for 30 min at 55°C with 0.1 g of Chelex 100 and 125 µl of an 8% PVP 360 solution. The PVP 360 solution was made in lysis buffer AP1 (DNeasy Plant Minikit) and incubated for one hour at 55°C to dissolve.

Following the detergent precipitation stage, the sample was centrifuged for 10 min at 14,000 rpm and the supernatant liquid applied to the shredder column. After buffer AP3/E (DNeasy Plant Minikit) was added to each individual extract, the solution from all five tubes for a single timber sample was applied to a single spin column. The DNA was eluted in 100 µl buffer AE (DNeasy Plant Minikit) and stored overnight at -20°C before being used as template in PCR amplifications. This new method was performed on five *Mary Rose* samples (HT1, HT2, HT3, OD1 and OD2). PCR was performed using the *Mary Rose* extracts HT3 and OD2 in template volumes that ranged from 0.5 to 5 µl to assess the effect on amplification success.

The efficiency of the new *Mary Rose* DNA extraction protocol was tested using known quantities of modern oak wood DNA. A modern oak extract with a DNA concentration of 9.26 ng/µl (Table 2) was used to make a dilution series of 100, 50, 20, 10, 5, 2 and 1%. One hundred microlitres of each dilution, equating to 926, 463, 185.2, 92.6, 46.3, 18.52 and 9.26 ng of DNA respectively, was added to extractions instead of sawdust and tested with PCR.

MODIFIED *MARY ROSE* PCR PROTOCOL

The presence of primer dimer in reactions without DNA indicates appropriate PCR conditions. As the PCR protocol of Demesure *et al.* (1995) was unreliable in producing primer dimer, we adopted the ASL PCR protocol. This protocol includes excess primer to promote primer dimer formation in reactions without DNA, AmpliTaq DNA Polymerase Stoffel Fragment to reduce the production of false positive results (Loy 1997) and 5% dimethyl sulfoxide (DMSO), which has been shown to reduce PCR inhibition caused by some acidic plant substances (Bickley and Hopkins 1999). For a single 50 µl reaction with 2 µl DNA template, the ASL protocol includes 29 µl Milli-Q water, 8 µl $MgCl_2$, 5 µl 10x Stock Buffer, 2.5 µl 100% DMSO, 1 µl each 10 mM 5' primer, 3' primer and dNTP's and 0.5 µl AmpliTaq DNA Polymerase Stoffel Fragment.

Following the successful reduction of PCR inhibition from five *Mary Rose* DNA extracts, the ASL PCR protocol was optimised. Extracts were centrifuged for 15 min at 14,000 rpm and PCR template was extracted from the top aqueous layer to avoid residual Chelex 100, which inhibits polymerase. 10 µl Q-solution (Qiagen) was added, with a subsequent reduction in Milli-Q water. The 50-cycle PCR denaturation temperature was raised to 94°C and the elongation time was reduced to 45 sec to increase the efficiency of these stages. Also, the annealing temperature of primer pairs dt74 and dt13 was increased to 50°C and primer pairs tf42 and µdt1 to 55°C.

RESULTS

Modern wood
Using the methods of Deguilloux *et al.* (2003), all five primer pairs tested with modern oak wood DNA successfully produced a single band of expected fragment size.

Trial of previously published protocols on *Mary Rose* samples
The major cell wall components of wood tissue, such as carbohydrates (cellulose) and phenols (lignin) (Finney and Jones 1993; Levy 1977) are PCR inhibiting substances (Bickley and Hopkins 1999; Gugerli *et al.* 2005). Previous research has established that degradative processes have increased the iron content of the *Mary Rose* hull timbers, which also inhibits PCR (Finney and Jones 1993; Squirrell and Clarke 1987). Initial extraction of *Mary Rose* samples caused a strong yellow or brown discolouration of the DNeasy spin column membrane and elution. PCR amplification of these DNA extracts resulted in no detectable product or primer dimer. When tested with pUC19 DNA, the plasmid target fragment was amplified in the positive DNA control and primer dimer was amplified in the negative DNA control. In contrast, no product or primer dimer was produced in the presence of 2 µl *Mary Rose* extract, indicating the presence of PCR inhibitors.

Table 2 shows the results of NanoDrop assays for oak DNA extracted from modern wood and from *Mary Rose* samples. Sample HT4 Proteinase K had the largest amount of DNA at 129.39 ng/µl. Half of the samples had less than 45 ng/µl and three of those had less than 10 ng/µl. This latter category includes a modern DNA sample, which had 9.26 ng/µl. Most of the *Mary Rose* DNA samples had 260/280 and 260/230 absorbance ratios substantially below 1.8-2.2, indicating the presence of contaminants that absorb at 280 and 230 such as protein or phenol. The modern DNA extract had the highest 260/280 ratio of 1.42. A sample of HT5 Proteinase K had a 260/230 ratio of 1.76, however this sample also had an extremely low 260/280 ratio of <0.95, suggesting that contaminants were present in this sample and that the DNA purity is still very poor.

Modification of previously published methods
Both modifications based upon the method of Guy *et al.* (2003) were successful in removing the discolouration of the spin column membrane and subsequent DNA elution. However, only the second treatment that included Proteinase K, Chelex 100 and PVP360 successfully removed

Table 2. NanoDrop results for modern and archaeological oak wood DNA samples. Alternative extraction treatments are listed with Mary Rose sample labels. Nucleic acids are measured at an absorbance of 260 nm and the quantity of DNA is measured in ng/µl. The 260/280 and 260/230 ratio values indicate the purity of DNA in each sample, where pure DNA has values of ~1.8 and 1.8-2.2 respectively

Sample	DNA (ng/µl)	A260	260/280	260/230
Modern	9.26	0.18	1.42	0.46
HT2A	11.78	0.24	0.88	0.39
HT4A	104.14	2.08	0.86	1.5
HT5A	75.24	1.5	0.78	1.22
MD1A	57.73	1.15	0.75	1.65
HT5A Proteinase K	129.39	2.788	0.82	1.76
MD1A Proteinase K	32.63	0.65	0.94	0.89
MD1A Chelex 100/PVP360	2.6	0.05	0.88	-0.78
HT4A Chelex 100/PVP360	4.9	0.1	1.29	-3.32

inhibition from sample MD1A, as demonstrated by the presence of primer dimer after amplification. Despite this removal, no target DNA was amplified from this sample.

*New **Mary Ros**e extraction protocol*

Five *Mary Rose* samples (HT1, HT2, HT3, OD1 and OD2) were subjected to the new *Mary Rose* extraction protocol. Production of primer dimer in PCR amplifications of all five extracts indicated that the inhibiting substances had been removed. This was confirmed by the successful amplification of modern oak cpDNA 'spiked' into 5 µl of the *Mary Rose* extracts MD1 and HT4.

The efficiency of this new method was tested with a dilution series of modern oak wood DNA. Figure 6 illustrates that PCR product and primer dimer is visible in all test lanes in intensities inversely proportional to the quantity of DNA added. Primer dimer is also visible in the negative DNA extract and PCR controls. These results demonstrate that as little as 9.26 ng of DNA was extracted successfully using the new extraction protocol, amplified with primer pair dt74 and visualised on a 3% agarose electrophoresis gel.

Finally, volumes of *Mary Rose* DNA ranging between 0.5 and 5 µl were used in PCR amplifications with two sets of primers to assess whether template volume affected amplification success. As illustrated in Figure 7, primer dimer was produced in all test reactions, while PCR product dt74b was amplified from 0.5 and 1 µl of OD2 template. Subsequent PCR amplifications using 0.5 and 1 µl *Mary Rose* OD2 DNA template with each primer pair successfully amplified fragment dt74b again, and also fragment tf42.

DISCUSSION

Previously published methods

Application of the methods of Deguilloux *et al.* (2003) to *Mary Rose* samples resulted in a discoloured spin column membrane indicating the co-elution of PCR inhibitors. This conclusion was supported by the failure to produce product or primer dimer. To distinguish between inhibition and reaction failure, the ASL PCR protocol was adopted to encourage primer dimer production in successful PCR amplifications that do not contain DNA. The presence of PCR inhibitors in *Mary Rose* extracts was confirmed by the modern DNA inhibition test using pUC19 (Bickley and Hopkins 1999; Connell 2002; Yanisch-Perron *et al.* 1985). According to these results, the methods of Deguilloux *et al.* (2003) were not adequate for extraction and amplification of DNA

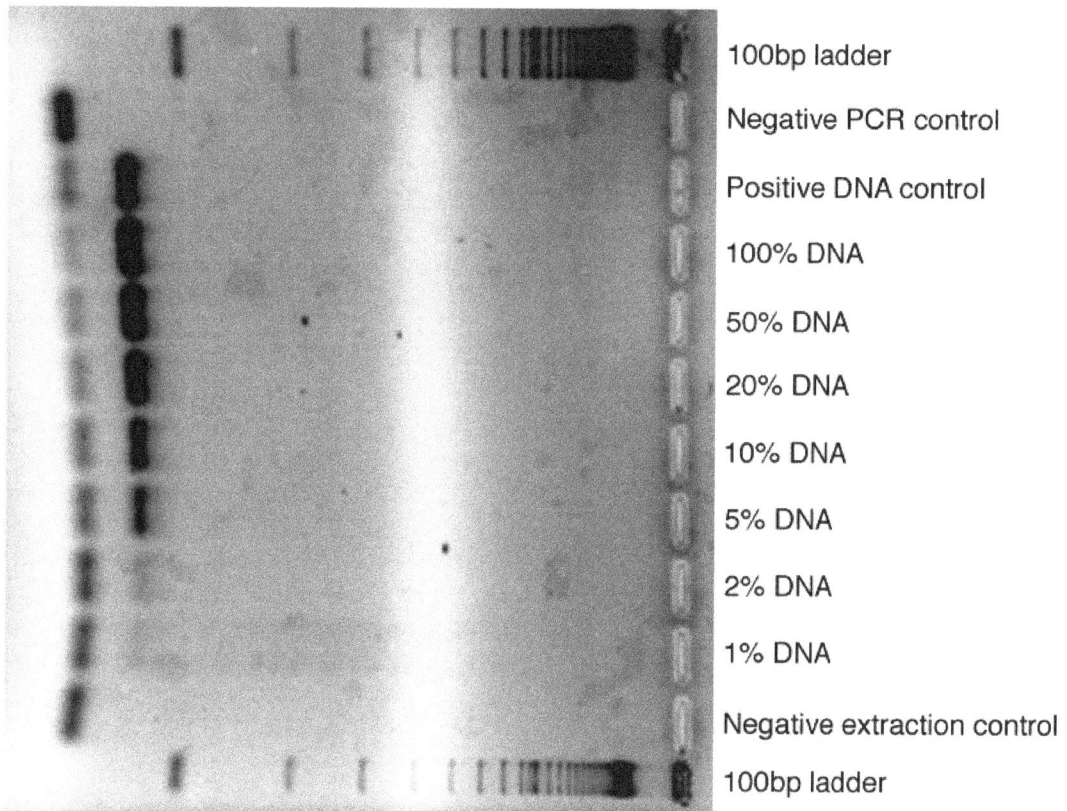

Figure 6. Inverse transillumination image of PCR products from the quantified test of the *Mary Rose* **DNA extraction protocol. This amplification was conducted with DNA extracted from a dilution series of modern oak DNA. Primer dimer is observed in all lanes. PCR product of the expected size can be seen in all test reactions, where it occurs in an inversely proportionate intensity to the primer dimer band, depending on the percentage of DNA added.**

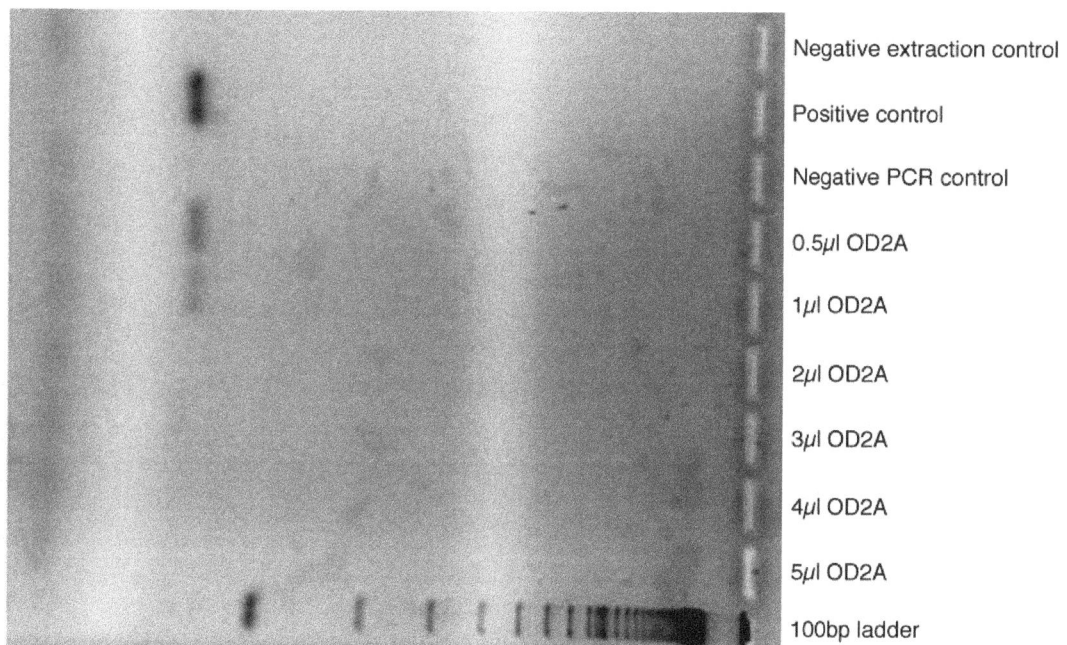

Figure 7. *Mary Rose* **DNA. Inverse transillumination image of PCR products from amplifications using** *Mary Rose* **template volumes ranging between 0.5 to 5µl. Primer dimer can be seen in all reactions except the positive DNA control. PCR products can be seen in the positive DNA control and in amplifications with 0.5 and 1 µl of OD2A template.**

Archaeological science under a microscope: studies in residue and ancient DNA analysis in honour of Thomas H. Loy ANCIENT CHLOROPLAST DNA FROM 16TH CENTURY WATERLOGGED OAK IN

A MARINE ENVIRONMENT: INITIAL STEPS IN SOURCING THE MARY ROSE TIMBERS

from the *Mary Rose* wood. Therefore, to progress the research it was necessary to characterise and determine how to remove the PCR inhibitors.

To characterise inhibitors we used the NanoDrop ND-1000 to determine the quantity and purity of DNA present in the *Mary Rose* extracts. Results from this analysis confirmed the expected low quantity and quality of DNA and indicated a high concentration of substances that absorb at wavelengths of 230 and 280 nm, most probably proteins and phenols.

Failure to amplify DNA due to inhibition does not necessarily mean that there is no DNA present. One approach to removing PCR inhibition is to dilute the extract prior to PCR. This method reduces the amount of inhibitors to a concentration tolerable to PCR, with a concurrent reduction in DNA template quantity (Bickley and Hopkins 1999). Although this method was successfully used by Tani *et al.* (2003) and Deguilloux *et al.* (pers. comm) when using aDNA samples it is imperative to maximise the concentration of PCR template in every reaction. Therefore, rather than reducing the amount of aDNA extract added to PCR, protocol modifications were designed to reduce PCR inhibitors such as iron and phenols during DNA extraction with the DNeasy Plant Minikit protocol.

Modification of previously published methods

Adding Proteinase K to the lysis stage to digest the proteins that appeared to be co-extracted with the *Mary Rose* DNA successfully removed the discolouration of the spin column membrane and elution, but did not remove PCR inhibition. These results suggested that the removal of extra protein made an improvement in the quality of the extraction, however proteins were not the sole cause of inhibition. Proteinase K treatment, followed by incubation with Chelex 100 and PVP360 was trialled and successfully produced primer dimer in the *Mary Rose* MD1 DNA reaction.

Chelex 100 was added as a multivalent ion chelating resin that targets metals such as iron and PVP360 was added to remove phenolics from the lysis solution by forming PVP/phenol complexes. These results indicate that the PCR inhibiting substances were a combination of protein, iron and phenolics. Furthermore, the enhanced amplification efficiency with these extracts after centrifugation suggested that the PCR may have been affected by residual Chelex 100. Successful amplification of modern oak DNA used to 'spike' 5 µl of these aDNA extracts demonstrated that PCR inhibiting substances had been removed.

New *Mary Rose* extraction protocol

The new protocol involves lysis stage treatments, as well as an increased starting sample size and optimised PCR conditions. A quantified test of the new protocol demonstrated that this method was efficient in extracting and amplifying as little as 9.26 ng of modern oak DNA from extracts.

This new protocol was tested on five *Mary Rose* samples (HT1, HT2, HT3, OD1 and OD2) and primer dimer was successfully amplified from 5 µl of each extract. Two of these extracts (HT3 and OD2) were then used in a series of amplifications with template volumes ranging from 0.5 to 5 µl. The target dt74b product was amplified from the smaller volumes (0.5 and 1 µl) of extract OD2 (Figure 7). This is the first extraction and amplification of cpDNA from waterlogged archaeological oak wood from a marine environment, and represents a significant advance towards the potential geographic provenancing of the *Mary Rose* timbers.

These results demonstrated that primer dimer and modern DNA products could be produced in the presence of up to 5 µl of *Mary Rose* extracts using this method. However, inhibiting substances were still present in the *Mary Rose* extracts, preventing amplification of product from more than 1 µl of these extracts. These findings suggest that some residual inhibitors may be closely associated with the DNA itself, however the concentration is tolerable to PCR at these lower volumes. Subsequent amplifications with 0.5 and 1 µl of three *Mary Rose* samples (HT3, OD1 and OD2) replicated the amplification of fragment dt74b from the OD2 extract, as well as fragment tf42. With the success of Stage I the next stages of this project can now proceed with sequencing of the amplified cpDNA fragments and comparison with geographic haplotype data.

*Contamination or **Mary Rose** DNA?*

Several arguments support the conclusion that the PCR products obtained were the target oak cpDNA fragments and not contaminating DNA from another source. First, throughout this study, PCR products of the expected size were consistently amplified from oak DNA from multiple extractions and using different reagents, suggesting that these fragments were in fact the target oak cpDNA products. This finding is supported by the use of primer pairs designed to be specific to oak cpDNA and used successfully in previous studies. It is highly unlikely that each extraction or PCR tube was affected by a contaminant that consistently amplified fragments of the expected size.

Second, *Mary Rose* sample OD2A PCR amplifications may have been contaminated with unknown DNA or modern oak wood DNA that were transferred into the ASL. In the first amplification of *Mary Rose* OD2 DNA (Figure 7), the 0.5 μl template was the first to be loaded into the PCR tube and the 1 μl template reaction was the last extract loaded. Thus, if the products were amplified from aerosol contaminants entering individual tubes, it is unlikely that it would demonstrate such a regular dispersal. It seems more likely that aerosol contamination would affect reactions in a mosaic pattern, as well as affecting at least one negative DNA control. Furthermore, in all amplifications using *Mary Rose* DNA extracts, the products only occur in the OD2A reactions.

It is more likely, given the pattern of occurrence, that if successful reactions were caused by contaminating DNA then it would have entered the OD2 extract itself. However, DNA was only amplified from *Mary Rose* extract OD2 when added at amounts of 0.5 and 1 μl, indicating that residual inhibitors were affecting the PCR amplification at larger template volumes. Alternatively, modern oak DNA added to reactions with 5 μl of *Mary Rose* extract still amplified fragments of the expected size. This supports the assumption that the PCR inhibitors may be linked with the *Mary Rose* DNA itself, and so modern contaminant DNA would have amplified in all the *Mary Rose* template volumes (up to 5 μl). These results support the conclusion that the products amplified from the OD2 extract are in fact fragments of *Mary Rose* cpDNA.

CONCLUSION

The aims of this study were to evaluate current methods of cpDNA analysis for their suitability to the specific archaeological case of the waterlogged *Mary Rose* timbers, and to develop procedures that would enable successful recovery of cpDNA from these samples. Chloroplast DNA was successfully extracted and amplified from modern and *Mary Rose* oak, demonstrating the viability of the new *Mary Rose* extraction protocol. These results are an encouraging start to the project aimed at determining the provenance of the flagship's timbers. The next stages will focus on sequencing the cpDNA fragments to verify the identity of the DNA. The extraction, amplification and sequencing results will then be replicated in an independent laboratory to support the authenticity of these fragments. Finally, the haplotype of each sample will be determined from these sequences for provenancing studies. The successful provenancing of these timbers would contribute significantly to our historical understanding of the archaeological remains of the *Mary Rose*.

On a broader scale, the successful extraction and amplification of DNA from archaeological oak wood offers the prospect of recovering unique and direct evidence of the genetic structure of *Quercus* spp. populations in the past. The *Mary Rose* DNA extraction protocol presents a technique for reducing PCR inhibitors from archaeological oak samples without dilution of the extract, thus maximising the concentration of DNA template available for analysis. Inhibition removal methods trialled in the development of this protocol have increased our understanding of different methods of reducing PCR inhibition in DNA extracts from waterlogged archaeological oak wood. This protocol and the lessons learnt in the current study broaden the range of tools and documented experience available to ancient DNA researchers, and expand the range of archaeological artefacts that can be examined for palaeogenetic information.

ACKNOWLEDGEMENTS

For permissions and assistance the authors would like to thank the Mary Rose Trust, the University of Queensland (UQ) Life Sciences Laboratory, the UQ Archaeological Sciences Laboratory and Dr Tom Loy.

REFERENCES

Bickley, J. and D. Hopkins 1999. Inhibitors and enhancers of PCR. In G.C. Saunders and H.C. Parkes (eds) *Analytical Molecular Biology Quality and Validation, Supplement 190*, pp. 81-102. Redwood Books Ltd.

Brewer, S., R. Cheddadi, J.L. de Beaulieu and M. Reille 2002. The spread of deciduous *Quercus* throughout Europe since the last glacial period. *Forest Ecology And Management* 156(1-3):27-48.

Bridge, M.C. and C. Dobbs 1996. Tree-ring studies on the Tudor warship Mary Rose. In Eds: J.S. Dean, D.M. Meko and T.W. Swetnam (eds) *Tree Rings, Environment and Humanity*, pp. 491-496. Tucson: Radiocarbon.

Capelli, C., F. Tschentscher and V.L. Pascali 2003. "Ancient" protocols for the crime scene? Similarities and differences between forensic genetics and ancient DNA analysis. *Forensic Science International* 131(1):59-64.

Connell, J. 2002. Towards the Repair of Ancient DNA and Using the Damaged Plasmid Model as an Intra- and Inter-Laboratory Control. Unpublished ms on file. Brisbane: Department of Biochemistry, University of Queensland.

Cooper, A. and H. Poinar 2000. Ancient DNA: do it right or not at all. *Science* 289:1139.

Cottrell, J.E., R.C. Munro, H.E. Tabbener, A.C.M. Gillies, G.I. Forrest, J.D. Deans and A.J. Lowe 2002. Distribution of chloroplast DNA variation in British oaks (*Quercus robur* and *Q-petraea*): the influence of postglacial colonisation and human management. *Forest Ecology And Management* 156(1-3):181-195.

Deguilloux, M.F., M.H. Pemonge, L. Bertel, A. Kremer and R.J. Petit 2003. Checking the geographical origin of oak wood: molecular and statistical tools. *Molecular Ecology* 12(6):1629-1636.

Deguilloux, M.F., M.H. Pemonge and R.J. Petit 2002. Novel perspectives in wood certification and forensics: dry wood as a source of DNA. *Proceedings Of The Royal Society Of London Series B* 269(1495):1039-1046.

Deguilloux, M.F., M.H. Pemonge and R.J. Petit 2004. DNA-based control of oak wood geographic origin in the context of the cooperage industry. *Annals Of Forest Science* 61(1):97-104.

Demesure, B., N. Sodzi and R.J. Petit 1995. A set of universal primers for amplification of polymorphic noncoding regions of mitochondrial and chloroplast DNA in plants. *Molecular Ecology* 4(1):129-131.

Dumolin-Lapegue, S., B. Demesure, S. Fineschi, V. LeCorre and R.J. Petit 1997. Phylogeographic structure of white oaks throughout the European continent. *Genetics* 146(4):1475-1487.

Dumolin-Lapegue, S., M.H. Pemonge, L. Gielly, P. Taberlet and R.J. Petit 1999. Amplification of oak DNA from ancient and modern wood. *Molecular Ecology* 8(12):2137-2140.

Ferris, C., R.P. Oliver, A.J. Davy and G.M. Hewitt 1993. Native oak chloroplasts reveal an ancient divide across Europe. *Molecular Ecology* 2(6):337-344.

Ferris, C., R.P. Oliver, A.J. Davy and G.M. Hewitt 1995. Using chloroplast DNA to trace postglacial migration routes of oaks into Britain. *Molecular Ecology* 4(6):731-738.

Finney, R.W. and A.M. Jones 1993. Direct analysis of wood preservatives in ancient oak from the *Mary Rose* by laser microprobe mass spectrometry. *Studies in Conservation* 38:36-44.

Gugerli, F., L. Parducci and R.J. Petit 2005. Ancient plant DNA: review and prospects. *New Phytologist* 166(2):409-418.

Guy, R.A., P. Payment, U.J. Krull and P.A. Horgen 2003. Real-time PCR for quantification of *Giardia* and *Cryptosporidium* in environmental water samples and sewage. *Applied And Environmental Microbiology* 69(9):5178-5185.

Hewitt, G.M. 1999. Post-glacial re-colonization of European biota. *Biological Journal Of The Linnean Society* 68(1-2):87-112.

Hofreiter, M., D. Serre, H.N. Poiner, M. Kuch and S. Pääbo 2001. Ancient DNA. *Nature Reviews Genetics* 2(5):353-359.

Hoss, M., P. Jaruga, T.H. Zastawny, M. Dizdaroglu and S. Pääbo 1996. DNA damage and DNA sequence retrieval from ancient tissues. *Nucleic Acids Research* 24(7):1304-1307.

Huntley, B. and H.J.B. Birks 1983. *An Atlas of Past and Present Pollen Maps of Europe, 0-13,000 Years Ago*. Cambridge: Cambridge University Press.

Levy, J.F. 1977. Degradation of wood. In S. McGrail (ed.) *Sources and Techniques in Boat Archaeology*, pp. 15-22. BAR Supplementary Series 29. Oxford:British Archaeological Reports.

Lindahl, T. 1993. Instability and decay of the primary structure of DNA. *Nature* 362(6422):709-715.

Lowe, A., R. Munro, S. Samuel and J. Cottrell 2004. The utility and limitations of chloroplast DNA analysis for identifying native British oak stands and for guiding replanting strategy. *Forestry* 77(4):335-347.

Lowe, A.J., C. Unsworth, S. Gerber, S. Davies, R.C. Munro, C. Kelleher, A. King, S. Brewer, A. White and J. Cottrell 2006. The route, speed and mode of oak postglacial colonisation across the British Isles: integrating molecular ecology, palaeoecology and modelling approaches. *Botanical Journal of Scotland* 57:59-82.

Loy, T. 1997. Ultrapure water, is it pure enough? *Ancient Biomolecules* 1:155-159.

Marsden, P. 2003. *Sealed by Time: The loss and recovery of the* Mary Rose. Trowbridge: Cromwell Press.

Mouzouras, R., E.B.G. Jones, R. Venkatasamy and S.T. Moss 1986. Decay of wood by micro-organisms in marine environments. *Record of the 1986 Annual Convention of the British Wood Preserving Association*: 27-45.

Pääbo, S. 1989. Ancient Dna - Extraction, Characterization, Molecular-Cloning, And Enzymatic Amplification. *Proceedings Of The National Academy Of Sciences* 86(6):1939-1943.

Pääbo, S., H. Poinar, D. Serre, V. Jaenicke-Despres, J. Hebler, N. Rohland, M. Kuch, J. Krause, L. Vigilant and M. Hofreiter 2004. Genetic analyses from ancient DNA. *Annual Review Of Genetics* 38: 645-679.

Petit, R.J., S. Brewer, S. Bordacs, K. Burg, R. Cheddadi, E. Coart, J. Cottrell, U.M. Csaikl, B. van Dam, J.D. Deans, S. Espinel, S. Fineschi, R, Finkeldey, I. Glaz, P.G. Goicoechea, J.S. Jensen, A.O. Konig, A.J. Lowe, S.F. Madsen, G. Matyas, R.C. Munro, F. Popescu, D. Slade, H. Tabbener, S.G.M. de Vries, B. Ziegenhagen, J.L. de Beaulieu and A. Kremer 2002. Identification of refugia and post-glacial colonisation routes of European white oaks based on chloroplast DNA and fossil pollen evidence. *Forest Ecology And Management* 156(1-3):49-74.

Petit, R.J., U.M. Csaikl, S. Bordacs, K. Burg, E. Coart, J. Cottrell, B. van Dam, J.F. Deans, S. Dumolin-Lapegue, S. Fineschi, R. Finkeldey, A. Gillies, I. Glaz, P.G. Goicoechea, J.S. Jensen, A.O. Konig, A.J. Lowe, S.F. Madsen, G. Matyas, R.C. Munro, M. Olalde, M.H. Pemonge, F. Popescu, D. Slade, H. Tabbener, D. Taurchini, S.G.M.de Vries, B. Ziegenhagen and A. Kremer 2002. Chloroplast DNA variation in European white oaks - phylogeography and patterns of diversity based on data from over 2600 populations. *Forest Ecology And Management* 156(1-3):5-26.

Petit, R.J., A. Kremer and D.B. Wagner 1993. Geographic structure of chloroplast DNA polymorphisms in European oaks. *Theoretical And Applied Genetics* 87(1-2):122-128.

Rule, M. 1983. *The* Mary Rose*: The Excavation and Raising of Henry VIII's Flagship*. London: Conway Maritime Press.

Squirrell, J.P. and R.W. Clarke 1987. An investigation into the condition and conservation of the hull of the Mary Rose. Part I: assessment of the hull timbers. *Studies in Conservation* 32(4):153-162.

Tani, N., Y. Tsumura and H. Sato 2003. Nuclear gene sequences and DNA variation of *Cryptomeria japonica* samples from the postglacial period. *Molecular Ecology* 12(4):859-868.

The Mary Rose Trust 2002. *The* Mary Rose *- Museum and ship hall*. Portsmouth: The Mary Rose Trust.

Yanisch-Perron, C., J. Vieira and J. Messing 1985. Improved m13 phage cloning vectors and host strains - nucleotide-sequences of the m13mp18 and pUC19 vectors. *Gene* 33(1):103-119.

14

Drawing first blood from Maya ceramics at Copán, Honduras

Carney D. Matheson[1], Jay Hall[2] and René Viel[3]

1. Anthropology Department
Lakehead University
Thunder Bay ON P7B 5E1 Canada
Email: cmatheso@lakeheadu.ca

2. School of Social Science
The University of Queensland
Brisbane, QLD 4072 Australia

3. Copán Formative Project
Copán Ruinas, Honduras CA

ABSTRACT

Residue analysis yielded trace blood residues on inside surfaces of four vessels of the Ventaron ceramic type from the Maya site of Copán, Honduras. Microscopy, biochemical detection of the heme subunit of hemoglobin/myoglobin and an immunological test all confirm the presence of blood. This identification supports the hypothesis that the Ventaron vessel type was used for blood ritual purposes during the Classic period.

KEYWORDS

blood residue, ceramics, Maya, Copán

INTRODUCTION

Molecular archaeology is a developing research field that analyses archaeological remains at the microscopic level for diverse ends relating to past human behavior. One of its main concerns is the identification of organic residues on stone artefacts in order to assess more accurately both their specific functions and their wider roles in past cultural systems (Bahn 1987; Briuer 1976; Cattaneo *et al.* 1993; Copley et al 2001; Fullagar 1991; Haslam 2003; Kooyman *et al.* 1992; Kooyman *et al.* 2001; Loy 1983, 1987; Loy and Dixon 1998; Loy and Hardy 1992). Here we apply this approach to ceramic artefacts and report, for the first time, the discovery of blood residues on Maya ceramic vessels.

Determining the precise function of ceramic wares is a difficult task and Maya vessels are no exception. While a vessel's morphology may suggest its contents at a coarse-grained level and occasional glyphs on Maya pots provide clues to their contents (Stuart 1986, 1988), only microscopic residues can precisely inform us about what a vessel contained and perhaps what it was used for (Hurst *et al.* 1989; Hall *et al.* 1990; Henderson *et al.* 2007). Because prolonged or repeated contact of a particular substance on a ceramic surface will eventually trap and preserve remnants of the substance as a thin residue, modern laboratory techniques allow the identification of the chemical constituents of such residue. These techniques actually identify the molecules that characterize the residue. For example, calcium tartrate, tartaric acid and resins (e.g. terebinth) identify fermentation products and wine residues (McGovern 1996; McGovern *et al.* 1996), theobromine and caffeine identify cocoa residues (Hall *et al.* 1990; Hurst *et al.* 1989) and heme and globin molecules serve to identify blood residue (Loy 1992; Loy and Dixon 1998). Blood and other biological fluids can be further identified using techniques such as histological staining and chemical and immunological testing (Bahn 1987; Gurfinkel and Franklin 1988; Kooyman *et al.* 2001; Loy 1983; Petraglia *et al.* 1996). Once the nature of the residue is established, further testing can provide information pertinent to the function of ceramic vessels.

Under the umbrella of the University of Queensland's research project at Copán, Honduras, Classic Maya vessels of the Ventaron Ceramic Group were targeted for a pilot residue study aimed at assessing the survival of blood residues on ceramics. The entire known sample at the time of the study comprised five complete vessels, one semi-complete vessel and some 30 small fragments. These vessels, produced during the Acbi Period (AD 400-600) at Copán, are essentially shallow flat-bottomed dishes with low rims (one of them with short supports) that are generally decorated with animal effigies (Viel 1993: Figs 47 & 48; Viel and Cheek 1983: 573,580). All exhibit a large pouring spout that is level with the bottom of the vessel on one side (Figure 1). Four of them are decorated with bat heads and five with stylized bat wings. This vessel type was selected in order to test Viel's suggestion (1993:89) that it was used in sacrificial contexts for the collection and pouring of blood. Indeed, one of the four specimens available for this study clearly exhibited a reddish pigment stain on a broad section of its inside surface and was thus considered a particularly likely candidate for the test.

Although all four specimens came from the same period in Copán's cultural history, they were excavated at different times in different contexts over past decades (e.g. three were associated with burials [see Viel and Cheek 1983]). Nevertheless, they all underwent similar treatment before being stored for later analysis. Following discovery and extraction by hand, they were removed to the laboratory where they were washed with water and a soft brush to remove adhering soil. For whole vessels such as these, a second cleaning with a dilute alcohol solution was undertaken prior to storing them in a sealed glass cabinet.

METHODS AND RESULTS

The study involved a battery of six residue-identification techniques including microscopic analysis, staining, three biochemical tests and one immunological test. After gaining access to the vessels from the ceramic collection held at the Center of Investigations of the Honduran Institute of Anthropology and History in Copán Ruinas, Honduras, incident light microscopy

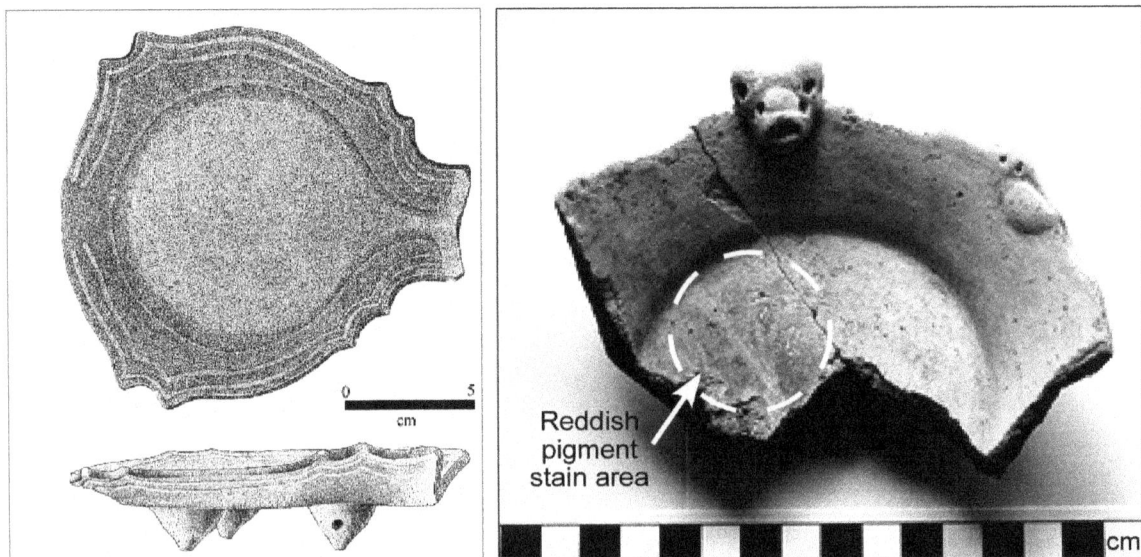

Reddish pigment stain area

Figure 1. Two of the Ventaron vessels used in this study (left, VV1; right, VV2).

was employed to determine the presence of residue and its extent. As shown in Table 1, this step revealed residue films that were visually similar to experimental blood films; that is, they appeared dark reddish brown and exhibited the cracking that is characteristic of a protein-rich residue (Loy 1993a, 1993b). Furthermore, anucleated red blood cells were observed within these films, indicating a mammalian origin ('RBC' column, Table 1). This identification was verified by transmitted light microscopy after some of the residue was transferred to a microscope slide and stained with a modified Wright's stain (Loy 1993b).

Table 1. Results of microscopic analysis of Ventaron vessels

Specimen No.	Microscopy result	RBC
VV 1	Dark globular protein-rich residue. Surface cracking present. Dark red-brown in colour.	Present
VV 2	Very little residue found & only in cracks in of the ceramic surface. Surface cracking present. Dark red- brown in colour.	Present
VV 3	Residue found in small dark red-brown globules. Surface cracking present in the larger globules.	None visible
VV 4	Large dark red-brown globules with surface cracking present in the globules.	Present

Biochemical testing was conducted following the microscopic analyses. Numerous colourimetric techniques have been developed for blood detection, all of which are based on the identification of the heme group of the hemoglobin and myoglobin molecules. This study employed the forensically accepted methods of a phenolphthalein technique (Gurfinkel and Franklin 1988) and a tetramethylbenzidine (TMB) technique (Cox 1991), along with a more easily applied method of a chemical reagent test strip (Hemastix) (Loy 1983; Loy and Wood 1989). The latter method of using chemical reagent test strips has been scrutinised for its accuracy and efficiency (Custer et al 1988; Manning 1994). We present below a modified version of the original method that uses a chelating agent, which has been demonstrated to remove all the cross-reactive possibilities presented as problematic for the method. This modified method works very effectively for the detection of heme-containing units. In this study a sample of residue was first removed from the vessel's surface by applying ultrapure water, allowing the residue to rehydrate with continual agitation with the pipette tip before removal to vials for application to each of the tests. Some samples were also stored for later testing. Samples for heme testing were taken from residue loci identified by microscopy, as well as from non-positive areas including vessel rims and undersides. Randomly selected local soil samples were tested to determine whether or not some constituents of the soil impinging on the vessel surface might produce a positive result.

Phenolphthalein solution was prepared and used following the method outlined by Gurfinkel and Franklin (1988; and Fiori 1962). Aliquots of the aqueous samples removed from the ceramic were placed into a small reaction tube with 2.5 volumes of prepared phenolphthalein solution, which develops a pink colour in the presence of heme. The reaction was scored after 30 seconds using a five-point colour-intensity scale, and then left to fade to ensure any positive reactions were not caused by non-heme substances. If there was no colour change in 30 seconds a volume of 3% hydrogen peroxide and absolute ethanol (v/v) equal to the volume of phenolphthalein was added to each tube to enhance the colour development.

Aliquots of the aqueous samples were also tested with a tetramethylbenzidine (TMB) solution, which was prepared according to Cox (1991). The samples were placed in separate 1.5 ml sterile tubes with two times the volume of TMB solution. If heme is present, the solution develops a green-blue colour, the intensity of which was recorded against a five-point scale. Because this solution-based test has been shown to react positively with many other substances (Cox 1991; Garner *et al.* 1976) the sample was also mixed with an equal volume of 500 mM ethylenediaminetetraacetic acid-sodium salt (Na-EDTA). Any non-heme positive results are eliminated by the chelating action of the Na-EDTA.

Lastly, the residues were analysed using the Hemastix test, which like the TMB test is based on 3,3',5,5'-tetramethylbenzidine (Bayer Diagnostics) but differs in that the sample is applied to a dry reagent test strip. The presence and intensity of a colour change (from yellow to dark green) on the Hemastix strip after 60 seconds was recorded on a scale of 0 (no reaction) to 5 (an instant colour saturation). Although the chemical reagent test strips are more sensitive and easier to use than the phenolphthalein test (Gurfinkel and Franklin 1988), caution must be taken because substances other than heme can produce positive reactions (Cox 1991; Custer *et al.* 1988; Garner *et al.* 1976; Gurfinkel and Franklin 1988; Loy 1983; Manning 1994). As in the TMB test above, any non-heme positive reactions were eliminated by repeating the test after mixing the sample with an equal volume of 500 mM Na-EDTA, thereby providing a specific reaction with hemoglobin/myoglobin.

Soil particles adhering to the sides and underside of each vessel as well as one approximately 500 mg bulk soil sample were also analysed using the three biochemical tests reported above. The soil was rehydrated with ultrapure water and processed in the same manner as the aqueous extracts from the vessel surfaces.

Results of the biochemical tests are shown in Table 2. Each test was performed several times on each vessel (see numbers in brackets, Table 2) and they returned a consistent result for the presence of heme in varying concentrations. The Hemastix test strip and the TMB-based solution testing of the microscopically-identified residues fell within the 2-5 range on the colour intensity scale. All control tests performed on the underside of the vessels produced negative results. The soil testing also produced a negative result. When repeated using Na-EDTA the TMB-based tests all produced a colour ranging from 2 to 5; nevertheless, the intensity following treatment was reduced because of the repeated resampling of the same location and/or the dilution of the aqueous sample by mixing with the Na-EDTA. All phenolphthalein tests yielded less intense positive results (range = 1-4; see Table 2) than the more sensitive TMB tests, and the negative control remained negative. The greater sensitivity of the TMB tests compared to the phenolphthalein test can be seen in the higher value recorded for Ventaron vessels 1, 2 and 3. However, Gurfinkel and Franklin (1988) state that these tests are less specific for the identification of hemoglobin. Repeating the tests using Na-EDTA provides a more specific reaction with hemoglobin and the results are still an intense reaction. The less sensitive phenolphthalein test and the TMB-based solution test are not as quick and easy to perform as the Hemastix chemical reagent strip test but they have much greater specificity to hemoglobin and their results confirm those of the test strips.

Table 2. Chemical test results

Specimens	Chemical reagent test strips[a]		Test strips with EDTA		TMB solution test[b]		Phenol-phthalein[b]	Dot–blot SpA test
	Residue	Control	Residue	Control	Residue	EDTA		
VV 1	3-4 (3)	0 (3)	3 (2)	0 (2)	4 (1)	2 (1)	2(1)	Not performed
VV 2	2-4 (2)	0 (3)	2-3 (2)	0 (2)	3 (1)	3 (1)	1 (1)	Not performed
VV 3	2-3 (3)	0 (3)	2 (2)	0 (2)	3 (1)	2 (1)	1 (1)	Not performed
VV 4	4-5 (6)	0 (3)	4-5 (3)	0 (3)	5 (1)	4 (1)	4 (1)	Positive
Soil sample	N/A	0 (3)	0 (3)	0 (3)	N/A	0 (1)	0 (3)	Negative

[a]The chemical reagent test strips were recorded on a 0-5 scale. 0 = no color change thus no reaction after 60 seconds. 1= a particulate trace reaction. 2= a trace but evenly reacting over the whole strip. 3-5 are grades of intensity above the trace reactions.
[b]The tetramethylbenzidine solution test and the phenolphthalein test were also recorded on a simple arbitrary 5-point intensity scale for rough comparison with the test strips, however the controls are not shown.
The 'Residue' column is samples removed from locations where microscopy indicated the presence of a potential blood residue. The 'Control' samples are removed from regions of the underside of the vessel in areas where microscopy has shown no residue. The number in brackets indicates the number of tests performed. Due to the small amount of residue on the first three vessels only residue of the fourth vessel was taken for a Staphylococcal protein A test. Again this test confirmed the results of the other tests with the presence of mammalian immunoglobulin G, which is found in blood.

Stored samples from one vessel (VV4) that produced a positive result for hemoglobin in the biochemical tests were then subjected to a Staphylococcal protein A (SpA) dot-blot test. SpA is a protein from *Staphylococcus aureus* that binds to the F_C region of most mammalian immunoglobulins, and with the highest affinity to immunoglobulin G (IgG) (Forsgren and Söquist 1967; Loy and Dixon 1998; Loy and Hardy 1992; Manning 1994; Tijssen 1985). In this test SpA (supplied by Sigma) is conjugated to colloidal-gold in order to enhance the test's sensitivity. The SpA dot-blot test is used as a multi-species mammalian detection system. Some researchers have shown that SpA can react with many mammalian immunoglobulins, not just IgG (Langone 1982; Lindmark *et al.* 1983). Nevertheless, there is only weak binding to other modern immunoglobulins (e.g. IgE) (Lindmark *et al.* 1983) and it would be expected that this affinity would be reduced further in ancient residues. In this test, 10 μl of residue is allowed to bind to PVDF membrane (Millipore), then washed and incubated in the colloidal gold SpA solution for IgG detection (following the manufacturer's protocol for detection without blocking). A positive reaction produces a pink colour on the dot blot. SpA dot blot results alone are not sufficient to prove the presence of blood given the weak reaction to other non-blood-borne immunoglobulins. However, given the predominance of IgG in the blood, and considering the concentration that would likely remain in a blood residue, the detection of predominantly IgG can be assumed (Loy and Hardy 1992). In any event, a positive reaction is indicative of the detection of immunoglobulins of mammalian origin.

Only one vessel (VV 4) had sufficient residue aliquots remaining to be analysed with the dot-blot. The residue reaction yielded a pink colour, indicating a positive result, while the associated soil sample remained colourless and was thus negative (Table 2).

In summary, the results of the microscopy, Wright's staining and biochemical and immunological analyses that were carried out in this pilot study demonstrate the presence of blood residues on the Ventaron ceramic vessels. First, microscopy highlighted a number of indicators for blood residues including red-brown pigmentation, a 'cracked-mud' appearance and some red blood cells. The latter were later confirmed by the modified Wright's stain test. Biochemical testing using three different methods, while employing Na-EDTA to eliminate non-hemoglobin reactions in the controversial chemical reagent test strips and the more sensitive TMB-based solution test. These indicated the presence of heme in the residues on the four vessels, but not in the soil or in samples taken from microscopically 'barren' locations. Although only one vessel (VV4) was tested with the SpA colloidal conjugated dot-blot test, the results indicated the detection of immunoglobulins of mammalian origin, which is consistent with the microscopic observations and staining of anucleate red blood cells in the residues.

DISCUSSION

The results of this pilot study are significant for both their methodological and archaeological implications. First, the results lead us to argue strongly for the possibility that blood residues are preserved after deposition on ceramic material in much the same way as they are on stone (Loy 1983, 1993b; Loy and Dixon 1998). These findings support the proposition by Craig and Collins (2002) that proteinaceous residues can be preserved on ceramic material. We propose that rapid drying and subsequent tertiary denaturation of some of the blood molecules (e.g. serum albumin) creates an unorganized polymer trapping and preserving more resistant molecules (e.g. hemoglobin) by forming a somewhat hydrophobic surface. The tertiary-denatured molecules also bond (presumably through a variety of chemical bonds) to highly charged, surface-reactive clay minerals in the pottery, and subsequent to burial, through bonds with impinging clay and silt particles (Cattaneo *et al.* 1993).

Craig and Collins (2002) present the best summary for the chemical bonds that may form between a protein residue and the mineral components of ceramic, demonstrated here by the highly resilient binding of the residue to the ceramic surface to survive the depositional environment, ground water and in this case the water wash and the mild alcohol wash. However this study's result is contrary to their finding that water is not a good solution for the removal of mineral bound protein. This contrast is most likely due to several differences between the Ventaron samples and those studied by Craig and Collins including: 1) method of preparation of their simulated protein residue; 2) the water removal protocol employed; and 3) differences in ceramic composition. Craig and Collins (2002) simulated a protein residue by heating to 85°C for seven days, which they stated may reflect residues associated with ceramic vessels that have been used for cooking. We suggest that the Ventaron vessels have not been used for cooking and have not been heated, and that this distinction may alter the predominant types of bonding between the residue and the mineral, allowing a water solution to extract residues more successfully than in the study by Craig and Collins (2002). Their study's water removal protocol involved submerging finely-ground protein-bound ceramic particles in water and rocking at 4°C for a range of different incubation times and is thus quite different from the protocol presented here, which was performed at room temperature with agitation. Finally, the finely-ground ceramic used by Craig and Collins (2002) was identified as quartz and illite which may not represent the mineral composition of the Ventaron ceramics.

The archaeological importance of these results significantly narrows the range of possible functions for the Ventaron vessels at Copán. As it is doubtful that blood appears on the flat internal surfaces of all four vessels by chance, its presence offers a limited number of possibilities including food preparation or meals involving butchered animals, medicinal practices and ritual involving butchered animals or human bloodletting. The fact that a large proportion of Ventaron vessels are decorated with effigy heads and wings of Copán's animal emblem, the bat, is a tantalising one that begs the question as to whether this animal's blood is represented, and if so, how it may have been incorporated into ritual practice at Copán. Whatever the case, the residues detected on the entire Ventaron sample supports the suggestion that they contained blood (Viel 1993:89) and thus do not refute the hypothesis that these vessels were used in a ritual sacrificial context. One additional implication of the preservation of blood proteins on ceramic wares in some quantity is the potential for direct radiocarbon assay using accelerator mass spectrometry (Loy 1993b; Loy *et al.* 1991). Such dating would assist the correlation of ceramic chronologies with corrected radiocarbon ages.

In closing, we reiterate that this was a pilot study carried out on a small sample of vessels of one ceramic group at Copán. While we claim to have identified mammalian blood on the vessels, we do not claim to have proven their function. We recognize the need for more extensive and thorough testing of these vessels and of a wider selection of ceramic material from the Ventaron Group, as well as perhaps other ceramic groups, before definitive statements about human behaviour can be asserted. We also need to discern the specific origin of the blood, a task that will be undertaken using genetic analysis.

ACKNOWLEDGMENTS

Our deepest gratitude to the late Tom Loy, without whose inspirational teaching and research input this project would neither have been conceived nor carried out. This research was supported by the Australian Research Council, The University of Queensland Institute for Molecular Biosciences and the Instituto Hondureño de Antropolgía e Historia.

REFERENCES

Bahn, P.G. 1987. Archaeology: getting blood from stone tools. *Nature* 330:14.

Briuer, F.L. 1976. New clues to stone tool function: plant and animal residues. *American Antiquity* 41(4):478-484.

Cattaneo, C., K. Gelsthorpe P. Phillips and R. J. Sokol 1993. Blood residues on stone tools: indoor and outdoor experiments. *World Archaeology* 25(1):29-43.

Copley, M.S., P.J. Rose, A. Clapham, D.N. Edwards, M.C. Horton and R.P. Evershed 2001. Detection of palm fruit lipids in archaeological pottery from Qasr Ibrim, Egyptian Nubia. *Proceedings of the Royal Society of London Series B Biological Science* 268(1467):593-7.

Cox, M. 1991. A study of the sensitivity and specificity of four presumptive tests for blood. *Journal of Forensic Sciences* 36:1503-1511.

Craig, O.E., and M.J. Collins 2002. The removal of protein from mineral surfaces: implications for residue analysis of archaeological materials. *Journal of Archaeological Science* 29:1077-1082.

Custer, J.F., J. Ilgenfritz, and K.R. Doms 1988. A cautionary note on the use of chemstrips for detection of blood residues on prehistoric stone tools. *Journal of Archaeological Science* 15:343-345.

Fiori, A 1962. Detection and identification of blood stains. In F. Lundquist (ed.) *Methods of Forensic Science*, pp. 243-290. New York: Interscience.

Forsgren, A., and J. Sjöquist 1967. Protein A from *Staphylococcus aureus*: reaction with rabbit gamma globulin. *Journal of Immunology* 99(1):19-24.

Fullagar, R.L.K 1991. The role of silica in polish formation. *Journal of Archaeological Science* 18:1-24.

Garner, D.D., K.M Cano, R.S. Peimer, and T.E. Yeshion 1976. An evaluation of tetramethylbenzidine as a presumptive test for blood. *Journal of Forensic Sciences* 21:816-821.

Gurfinkel, D.M., and U.M. Franklin 1988. A study of the feasibility of detecting blood residues on artifacts. *Journal of Archaeological Science* 15:83-97.

Hall, G.D., S.M. Tarka Jr., W.J. Hurst, D. Stuart and R.E.W. Adams 1990. Cacao residues in ancient Maya vessels from Rio Azul, Guatemala. *American Antiquity* 55(1):138-143.

Haslam, M. 2003. Evidence for maize processing on 2000-year-old obsidian artefacts from Copán, Honduras. In D. M. Hart and L. A. Wallis (eds) *Phytolith and Starch Research in the Australian-Pacific-Asian regions: The state of the art*, pp. 153–161. Canberra: Pandanus Books.

Henderson, J.S., R.A. Joyce, G.D. Hall, J. Hurst and P.E. McGovern 2007. Chemical and archaeological evidence for the earliest cacao beverages. *Proceedings of the National Academy of Science* 104(48):18937-18940.

Hurst, W.J., R.A. Martin, S.M. Tarka Jr. and G.D. Hall. 1989. Authentication of Cocoa in ancient Mayan vessels using HPLC techniques. *Journal of Chromatography* 466:279-289.

Kooyman, B., M.E. Newman and H. Ceri. 1992. Verifying the reliability of blood residue analysis on archaeological tools. *Journal of Archaeological Science* 19:265-269.

Kooyman, B., M.E. Newman, C. Cluney, M. Lobb, S. Tolman, P. McNeil and L. V. Hills 2001. Identification of horse exploitation by Clovis hunters based on protein analysis. *American Antiquity* 66(4):686-691.

Langone, J.J. 1982. Protein A of *Staphylococcus aureus* and related immunoglobulin receptors produced by *Streptococci* and *Pneumonococci*. *Advances in Immunology* 32:157-252.

Lindmark, R., K. Thoren-Tolling and J. Sjoquist. 1983. Binding of immunoglobulins to Protein A and immunoglobulin levels in mammalian sera. *Journal of Immunological Methods* 62:1-13.

Loy, T.H. 1983. Prehistoric blood residues: detection on tool surfaces and identification of species of origin. *Science* 220:1269-1271.

Loy, T. H. 1987. Recent advances in blood residue analysis. In W.R. Ambrose and J.M.J. Mummery (eds) *Archaeometry: Further Australasian Studies*, pp. 57-65. Canberra: Australian National University.

Loy, T.H. 1993a. Prehistoric organic residue analysis: the future meets the past. In W. Ambrose, A. Andrews, R. Jones, A. Thorne, M. Spriggs and D. Yen (eds) *A Community of Culture*, pp. 56-72. Department of Prehistory, Research School of PacificStudies. Canberra: The Australian National University.

Loy, T.H. 1993b. The artefact as site: an example of the biomolecular analysis of organic residues on prehistoric tools. *World Archaeology* 25(1):44-63.

Loy, T.H., and E.J. Dixon 1998. Blood residue on fluted points from Eastern Beringia. *American Antiquity* 63(1):21-46.

Loy, T.H., and B.L. Hardy 1992. Blood residues analysis of 90,000-year-old stone tools from Tabun Cave, Israel. *Antiquity* 66:24-35.

Loy, T.H., and A.R. Wood 1989. Blood residue analysis at Cayön Tepesi, Turkey. *Journal of Field Archaeology* 16:451-460.

McGovern, P.E. 1996. Vin Extraordinaire. *The Sciences* 36: 27-31.

McGovern, P.E., D.L Glusker, L.J Exner and M.M. Voigt 1996. Neolithic resinated wine. *Nature* 381:480-481.

Manning, A.P. 1994. A cautionary note on the use of hemastix and dot-blot assays for the detection and confirmation of archaeological blood residues. *Journal of Archaeological Science* 21:159-162.

Petraglia, M., D. Knepper, P. Glumac, M. Newman and C. Sussman 1996. Immunological and microwear analysis of chipped-stone artifacts from piedmont contexts. *American Antiquity* 61(1):127-135.

Stuart, D. 1986. The hieroglyphs on a vessel from Tomb 19, Rio Azul. In R.E.W. Adams (ed.) *Rio Azul Reports Number 2, the 1984 Season*, pp. 117-121. San Antonio: Centre for Archaeological Research University of Texas.

Stuart, D. 1988. The Rio Azul cacao pot: epigraphic observations on the function of a Maya ceramic vessel. *Antiquity* 62:153-157.

Tijssen, J.M. 1985. Practice and theory of enzyme immunoassays. In R.H. Burdon and P. H. van Knippenberg (eds) *Laboratory Techniques in Biochemistry and Molecular Biology*, Vol 15. Amsterdam: Elsevier.

Viel, R. 1993. *Evolucion de la Ceramica de Copán, Honduras*. Instituto Hondureño de Antropología e Historia, Tegucigalpa D.C.

Viel, R. and C.D. Cheek 1983. Sepulturas. In C. Baudez (ed.) *Introduccion a la Arqeuología de Copán,* pp551-610. Instituto Hondureño de Antropología e Historia y Secretaria de Cultura y Turismo. Tegucigalpa D.C.

15

A molecular study of a rare Maori cloak

Katie Hartnup[1], Leon Huynen[1], Rangi Te Kanawa[2], Lara Shepherd[1], Craig Millar[3] and David Lambert[3 & 4]

1. Allan Wilson Centre for Molecular Ecology and Evolution
Institute of Molecular Biosciences, Massey University
Private Bag 102 904 NSMC, Auckland, New Zealand.
Email: K.Hartnup@massey.ac.nz

2. Te Papa Tongarewa
PO Box 467, Wellington, New Zealand.

3. Allan Wilson Centre for Molecular Ecology and Evolution
School of Biological Science, University of Auckland
Private Bag 92019, Auckland, New Zealand.

4. Griffith School of Environment and School of Biomolecular
and Physical Sciences, Griffth University, 170 Kessels Road, Nathan
4111, Australia.

ABSTRACT

Kakahu or Maori cloaks are *taonga* (treasures) and are iconic expressions of Maori culture. Unfortunately much of the original information relating to the 'origins of the cloaks' has been lost. We present mitochondrial 12S sequence data from feathers sampled from a rare cloak that appeared to have been adorned with feathers from New Zealand moa. These species belonged to the ratite group of birds and have been extinct since soon after human arrival in New Zealand. Using microscopic amounts of feather tissues from this cloak, we have been able to show that this garment was actually adorned with Australian emu feathers. At the likely time of construction of the cloak, the then Governor of New Zealand, George Grey, kept emu on Kawau Island in the Hauraki Gulf. It seems probable that the remains of these individuals were the source of the feathers used, although we are not able to exclude the possibility that Maori obtained them as a result of early trading with Australia. To our knowledge this study is the first to use genetic techniques to identify the species of bird used in feather adorned Maori cloaks and illustrates the potential for molecular techniques to provide important information about these *taonga*.

KEYWORDS

Maori cloaks, species identification, ancient DNA

INTRODUCTION

Museum specimens and artefacts are now widely regarded as important genetic resources that can be utilised in a broad range of molecular studies (Wanderler *et al.* 2007). Such studies are aimed at many issues such as the taxonomic status of specimens, their provenance, past levels of genetic diversity and how changes in genetic diversity affect the population structure and the diversity of modern populations. The latter has implications for the conservation and management of fragile populations. Studies such as these often require museum specimens with detailed accompanying records, stating species, location, and time of collection. However, it is often the case that museum records are incomplete, particularly for historic samples. Alternatively, we can apply known ecological and genetic information from modern populations to uncover information about museum specimens that has either been lost or not originally collected. For example, this approach has been used to test the assumptions that a skeleton of a 19th century lion housed in a museum in Amsterdam belonged to the now extinct cape lion (*Panthera leo melanchaita*) (Barnett *et al.* 2007). Similar methods have been used to determine the specific status of kiwi remains that are indistinguishable using skeletal morphology alone (Shepherd and Lambert 2008). In the case of Maori feather cloaks, it is potentially possible to recover a wealth of information pertaining to the provenance, sex and species of the birds used in cloak construction and relate these findings to Maori culture and practices.

Maori cloaks (*kakahu*) are treasured items, or *taonga*, and are examples of one of the earliest forms of weaving (Best 1952), a process of hand knotting without the use of a loom. When Maori reached New Zealand from Polynesia in the 13th century their preferred material for clothing was the bark of the paper mulberry tree (*Broussonetia papyrifera*). This tree was brought to New Zealand with Maori but failed to flourish in the temperate climate (Best 1952). When searching for a viable alternative, the native flax (*Phormium* spp.) was discovered. From this flax a strong, pliable fibre called *muka* could be extracted and this was woven into cloaks. The earliest written records of Maori cloaks are from Captain Cook's first visit to New Zealand in 1769-1770 (Pendergrast 1997). At this time a crudely woven flax rain cape was the most common cloak type. Finely woven *muka* cloaks, often covered with strips of skin from the Polynesian dog (*kuri*), were worn only by chiefs.

Feathered garments are commonly mentioned in the oral histories of Maori and other Polynesian groups, but were not a common feature of Maori culture when the first European explorers reached New Zealand (Pendergrast 1997). Early flax cloaks sometimes had feathers, or skin with feathers attached, scattered across the cloak surface or woven into the cloak borders (Ling Roth 1923; Pendergrast 1987). Production of Maori cloaks completely covered in feathers began in the second half of the 19th century. The most prestigious of Maori cloaks were adorned with kiwi (*Apteryx* spp.) feathers and were known as *kahu kiwi* (Pendergrast 1987). Cloaks were held in high regard in Maori society, took a considerable amount of time to construct (up to eight months; Te Kanawa 1992), and were associated with high status. Prestigious cloaks such as *kahu kiwi* were empowered by a chief's *mana* - a Maori term signifying a combination of authority, integrity, power and prestige. There are cloaks in museums within New Zealand and overseas with good accompanying records, however a large number lack any information with respect to age, provenance and the species used to adorn cloaks. Our research programme is aimed at providing precise data regarding the origin and construction of cloaks.

During part of a larger study of Maori feathered textiles, our team encountered an unusual cloak at the Hawkes Bay Museum and Cultural Trust. This was the first cloak of this type to have been observed. The cloak's construction comprised a finely woven *muka* body or *kaupapa*, completely adorned with feathers that were extremely similar in morphology to moa feathers (Worthy and Holdaway 2002) (Figure 1). Moa (Aves: Diornithiformes) were foremost among the evolutionary novelties of New Zealand. Richard Owen first brought the presence of moa in New Zealand to the attention of Western scientists in 1842, when he described a femur shaft. Since that time the number and age of fossil specimens has grown considerably and suggests that moa inhabited New Zealand from over two million years ago (Worthy and Holdaway 2002:8-10). Although small groups of moa likely survived in remote locations for slightly longer, the main populations were probably extinct by AD 1400 (Worthy and Holdaway 2002).

Very few moa specimens have been excavated with their feathers intact. The few feathers recovered possessed a range of colours including white, reddish brown grading distally to black with a white tip, and purplish brown with a yellow stripe. Feathers were typically no longer than 18 cm in length, although feathers up to 23 cm in length have been recorded (Worthy and Holdaway 2002).

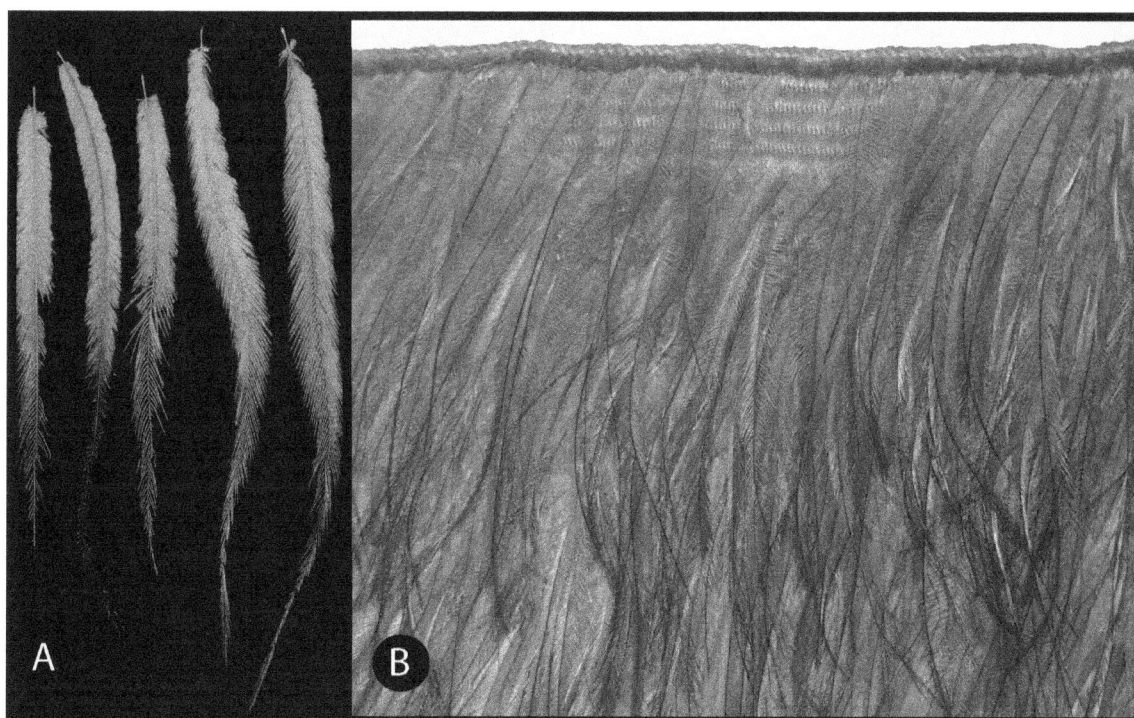

Figure 1. A comparison of known moa feathers (A) with those from the cloak under study (B).

It is possible to estimate the age of Maori cloaks due to variations in weaving techniques over time. The potential 'moa' cloak has been estimated by one of us (Rangi Te Kanawa, Maori Textile Conservator) to have been constructed in ~1850. Therefore, despite the similarity in feather morphology between the cloak and moa feathers, there is a large disparity between the time that moa became extinct and the estimated time of cloak construction. Despite this disparity, it is possible that moa feathers were stored for some time prior to their use in the construction of the Hawkes Bay cloak.

In order to test the possibility of a 'moa' cloak, ancient DNA techniques were employed to recover DNA sequences from cloak feathers and to compare these sequences with those obtained from moa bones and from a range of other ratite species.

MATERIALS AND METHODS

Seven feather shafts were kindly provided to us from the suggested moa cloak #45_264 from the Hawkes Bay Museum and Cultural Trust. Feathers are woven into Maori cloaks twice (Figure 2). This enables the removal of an approximately 2 mm section from the shaft of the feather with sterilised forceps and surgical scissors. This method minimises any detrimental effects on the integrity and the appearance of cloaks. DNA was extracted from each feather shaft by overnight incubation, with rotation, at 55°C in 300 µl of extraction buffer (10mM Tris-HCl pH 8.0, 50mM NaCl, and 1mM EDTA) supplemented with 30 µl of 10% SDS, 5µl of 1M DTT, and 5µl of 20mg/ml proteinase K. 200 µl of each mix was then purified using a QIAamp DNA Mini Kit (Qiagen) as outlined by the manufacturer.

Ratite-specific mitochondrial 12S primers, ratite12S1 (5'-CCTCAGAAGGCGGATTTAGCAGTAA) and ratite12S4 (5'-ATCTTTCAGGTGTA AGCTGAATGCTT), were designed using sequences retrieved from GenBank: rhea (*Rhea americana* – AJ002923), ostrich (*Struthio camelus* – AF069429), great spotted kiwi (*Apteryx haasti* – AF338708.2), cassowary (*Casuarius casuarius* – AF338713.2) and emu (*Dromaius novaehollandiae* – AF338711.1). DNAs extracted from cloak feathers and moa bone (*Emeus crassus*, Canterbury Museum CM_Av9132) were then amplified using the primers ratite12S1

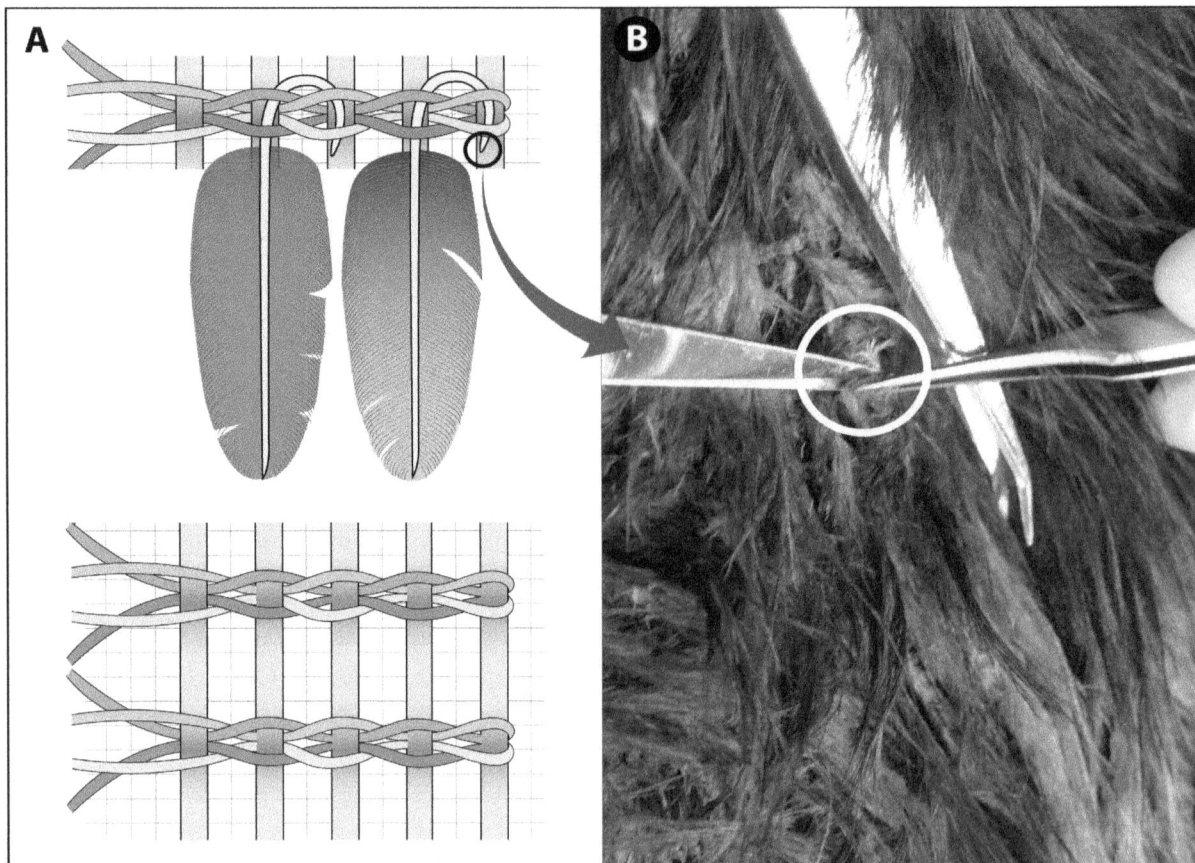

Figure 2. The DNA sampling method used for feather cloaks. (A) Details of the method used to weave feathers into the flax backing of a Maori cloak. The circle indicates the part of the feather shaft that was removed for DNA analysis. (B) Removal of a ~2 mm tip of a feather shaft, indicated by a circle, from a cloak.

and ratite12S4 as described in Huynen *et al.* (2003). Successfully amplified fragments of ~220 bp were sequenced in both directions using Applied Biosystems BigDye Terminator v3.1 chemistry and aligned to homologous sequences from other ratites using the Sequencher programme.

RESULTS

Two of the seven feather shafts sampled from cloak #45_264 amplified for a 220 bp sequence from the 12S region of the mitochondrial genome. These two sequences were aligned with homologous data from ratite species as shown in figure 3A. The two cloak sequences differed from each other at just two sites of the ~220 bp fragment (sites 5 and 16), suggesting the use of feathers from at least two different individuals in cloak construction. Cloak sample 2 was identical to the sequence of emu from Genbank. Both of the cloak sequences varied substantially from the moa sequence (11% average), and from other ratites for which sequences were available (rhea - 14.3%, kiwi - 8.6%, ostrich - 6.8% and cassowary - 5.9%). On average, the cloak sequences differed from the emu sequence by just 0.45%. Figure 3 presents an unrooted, neighbour-joining tree of 12S sequences created in PAUP* (Swofford 2002). The tree groups the sampled cloak feathers with emu. As the 12S mtDNA sequences targeted in this study are highly variable and difficult to align, they do not effectively resolve relationships amongst ratites. Haddrath and Baker (2001) used whole mitochondrial genomes to successfully investigate ratite phylogeny. However, the variability within the 12S region is ideal for species identification, making it a suitable choice for identifying the cloak samples.

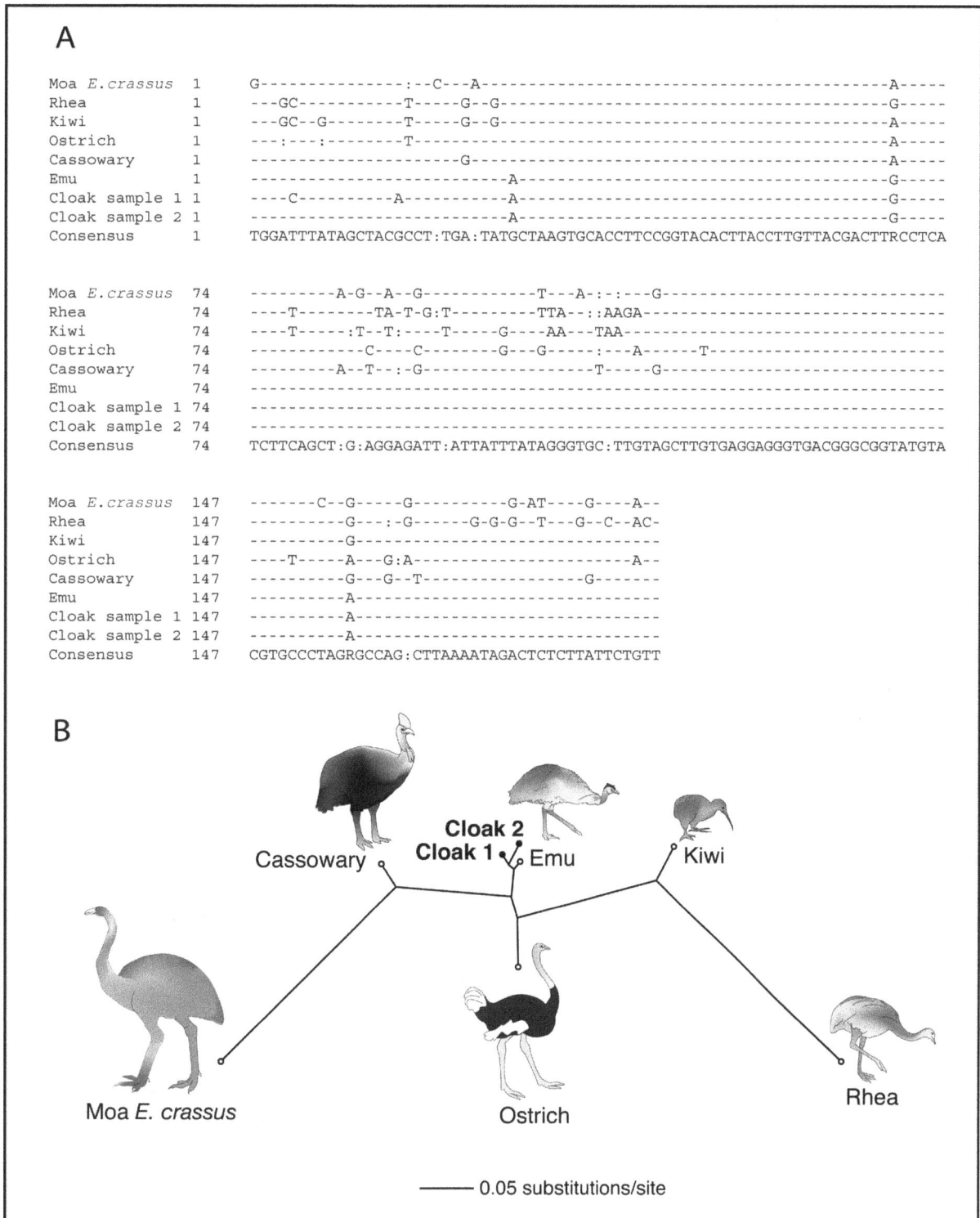

A

```
Moa E.crassus     1   G----------------:--C---A-----------------------------------------------A-----
Rhea              1   ---GC------------T-----G--G---------------------------------------------G-----
Kiwi              1   ---GC--G--------T-----G--G---------------------------------------------A-----
Ostrich           1   ---:---:--------T------------------------------------------------------A-----
Cassowary         1   ----------------------G-----------------------------------------------A-----
Emu               1   -------------------------A---------------------------------------------G-----
Cloak sample 1    1   ----C----------A------------------------------------------------------G-----
Cloak sample 2    1   -----------------------A----------------------------------------------G-----
Consensus         1   TGGATTTATAGCTACGCCT:TGA:TATGCTAAGTGCACCTTCCGGTACACTTACCTTGTTACGACTTRCCTCA


Moa E.crassus    74   ---------A-G--A--G-------------T---A-:-:---G--------------------------------
Rhea             74   ----T---------TA-T-G:T---------TTA--:AAGA---------------------------------
Kiwi             74   ----T------:T--T:----T-----G----AA---TAA---------------------------------
Ostrich          74   -----------C----C--------G---G----:---A------T---------------------------
Cassowary        74   ---------A--T--:-G----------------T-----G--------------------------------
Emu              74   ------------------------------------------------------------------------
Cloak sample 1   74   ------------------------------------------------------------------------
Cloak sample 2   74   ------------------------------------------------------------------------
Consensus        74   TCTTCAGCT:G:AGGAGATT:ATTATTTATAGGGTGC:TTGTAGCTTGTGAGGAGGGTGACGGGCGGTATGTA


Moa E.crassus   147   -------C--G-----G-----------G-AT----G----A--
Rhea            147   ----------G---:-G-------G-G-G--T---G--C--AC-
Kiwi            147   ----------G---------------------------------
Ostrich         147   ----T-----A---G:A---------------------A--
Cassowary       147   ----------G---G--T----------------G-------
Emu             147   ----------A---------------------------------
Cloak sample 1  147   ----------A---------------------------------
Cloak sample 2  147   ----------A---------------------------------
Consensus       147   CGTGCCCTAGRGCCAG:CTTAAAATAGACTCTCTTATTCTGTT
```

B

Figure 3. Aligned mitochondrial 12S DNA sequences of cloak feather samples, moa and other ratites (A). (-) indicates a base identical to the consensus sequences, (:) indicates a deletion in relation to the consensus sequence. An unrooted, neighbour-joining tree of 12S sequences from a range of ratite species (B), together with the two sequences recovered from the cloak samples.

DISCUSSION

We can conclude that this unique cloak from the Hawkes Bay Museum and Cultural Trust was not constructed using moa feathers. It was, however, adorned with feathers from emu, a ratite that originated in Australia and is not found in wild populations in New Zealand. Taking these findings into account, how did Maori in the second half of the 19th century obtain feathers from a native Australian bird species? There are two possible explanations for this. First, the emu feathers may have come from Sir George Grey's exotic flora and fauna collection on Kawau Island, north of Auckland. Second, the emu feathers may have been brought from Australia during a period of extensive timber and flax trading.

Sir George Grey

George Grey had a relationship with New Zealand spanning many years. He was appointed governor of New Zealand in 1845. Arguably his greatest success during this nine-year period was his management of Maori affairs. He scrupulously observed the terms of the Treaty of Waitangi and assured Maori that their rights to their land were fully recognised. He subsidised schools for Maori children, built several hospitals and encouraged Maori agriculture (Sinclair 2007). Grey enjoyed great *mana* among Maori, often travelling with chiefs. He was instrumental in efforts to record Maori traditions, legends and customs in written form. Te Rangikaheke, a Te Arawa tribal leader, taught Grey to speak Maori and lived with Grey and his wife in their house. Although Grey's second term as Governor from 1860-1868 was less successful because of extensive battles between Maori and settlers, he remained respected by Maori.

Grey purchased Kawau Island, located North of Auckland in the Hauraki Gulf, in 1862 (Figure 4). He poured a great deal of his energy, effort and fortune into the 2000 ha island. He turned the existing copper miners cottage into the formidable Mansion House and turned the land around the house into a botanical and zoological park. Grey imported seeds and cuttings from all over the world including redwood (*Sequoia sempervirens*) from Western USA, the Chilean wine palm (*Jubaea spectabilis*), the giant bird of paradise (*Strelitza nicolai*) from South Africa and the Japanese cedar (*Cryptomeria japonica*). Grey also imported various exotic fauna. Four species of wallaby (*Macropus eugenii, Macropus parma, Petrogale penicillata* and *Wallabia bicolour*) (Eldridge *et al.* 2001) were introduced and remain on Kawau today. Other animals, such as the zebra imported to pull his carriage (Eldridge *et al.* 2001), failed to acclimatise to their new home. Grey also imported birds such as peacocks (*Pavo* spp.), kookaburra (*Dacelo* spp.), and notably for this study, emu (Graham 1919).

It is highly likely that the feathers used to construct the emu feather cloak originated from Kawau, given the estimated construction of the cloak in around 1850, Grey's purchase of Kawau in 1862, the presence of emu on Kawau, and Grey's favorable association with Maori. It should be noted that emu were also found in the Hauraki Gulf on Motutapu Island, which was purchased in 1869 by the Reid brothers from Victorian entrepreneur Richard Graham. They introduced exotic fauna such as emu, deer, ostriches and wallabies (McClure 2007). It is known, however, that the flock of emu on Motutapu Island was provided from Governor Grey's flock on Kawau Island (Graham 1919).

Trade with Australia

European explorers visiting New Zealand in the 1700s quickly sought to make use of and to export resources, including timber and flax. Maori fashioned flax into ropes for visiting ships and bartered flax and weaving for European goods. Merchants in Sydney showed an interest in flax fibre and by the 1820s a trade began with Australia, peaking in the 1830s (Wigglesworth 1981). Trading stations were set up around the coast of New Zealand. Stations were present on the coasts of Northland, Waikato, Taranaki, the Coromandel, the Bay of Plenty, the East Cape, Southland, both sides of the Cook Straight, and the Banks Peninsula. Taking into account the extent of the trade between these two countries, and the timing of that trade, at present it is not possible to rule out these trade routes as the source of the emu feathers used to adorn the cloak. It is known that Maori flax producers were not paid in cash but in goods, usually muskets, although other goods such as feathers cannot be discounted (Swarbrick 2007).

Emu were known to be present on Kawau Island in about 1862. This coincides with the peak of the flax and timber trade with Australia (~1830). In addition, our estimate of the date

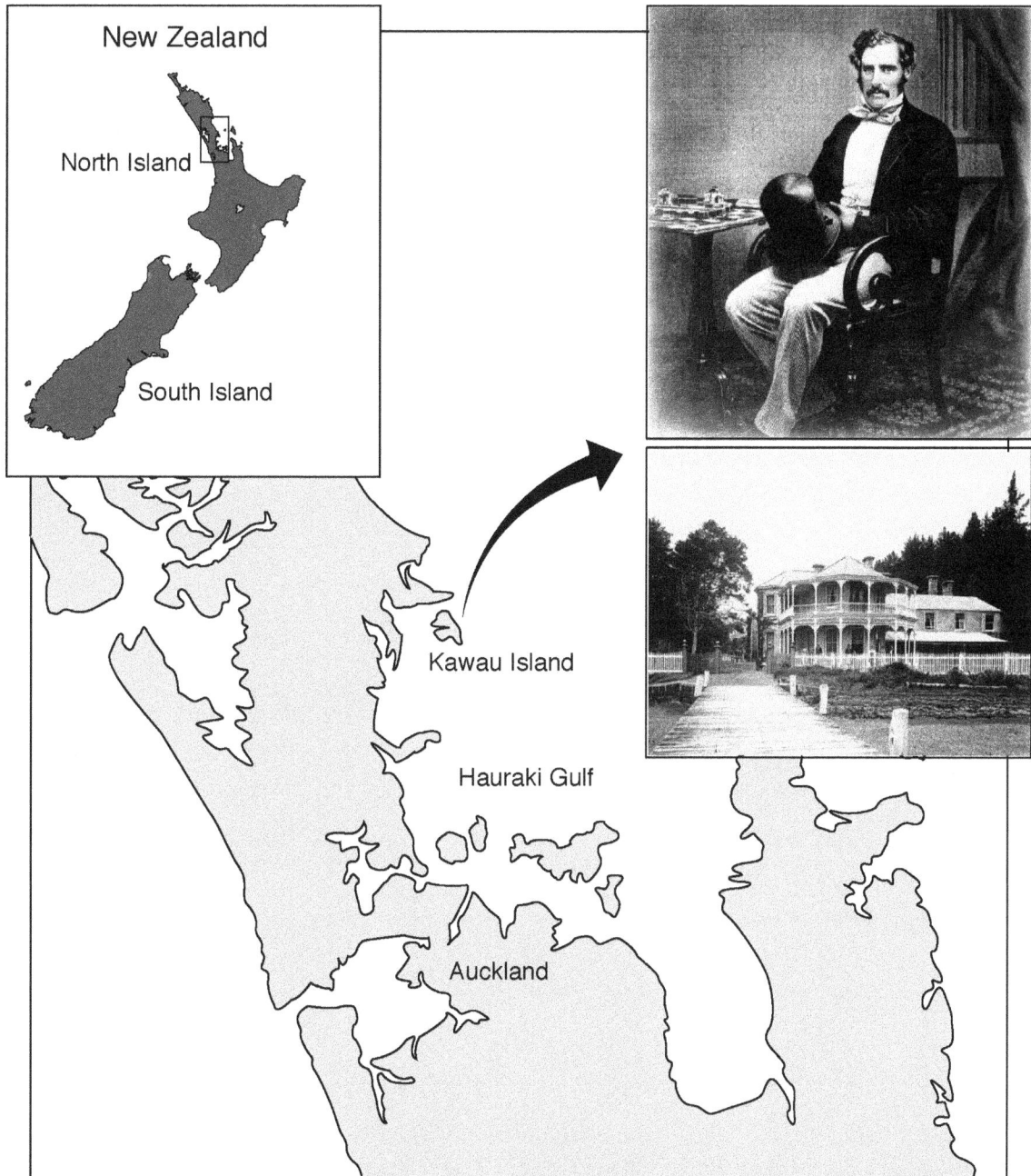

Figure 4. The location of Kawau Island off the coast of New Zealand north of Auckland, where Governor Grey kept his Zoological Park which included emu, together with a portrait of Governor Grey and an early photograph of Mansion House.

of construction of the emu cloak is approximately 1850. This makes it difficult to confirm with certainty that the emu feathers adorning the cloak came from George Grey's emu on Kawau Island. However, the definite presence of emu on Kawau versus only the potential for emu to be brought from Australia during the timber and flax trade makes the Kawau Island option more compelling. Further investigation could be conducted to test this idea. For instance, it is possible to look at more variable regions of the mitochondrial DNA genome to distinguish between different emu populations. If there were emu remains on Kawau, it would be possible to see if the feathers adorning the cloak match genetically to those remains and to compare these to Australian populations.

Recently, our team has come across two cloaks in the cloak collection at the Auckland War Memorial Museum. Both were constructed from feathers similar to those observed on the emu feather cloak from the Hawkes Bay. One of these is a *kahu hurhuru* which is a cloak adorned with feathers from many different species of birds and the other is completely adorned with what appears to be emu feathers. It is estimated that both cloaks were manufactured more recently than the Hawkes Bay example. The bodies of both cloaks are constructed from candlewick as opposed to *muka*. Cloak making using this material was typically observed from 1890 onwards. Future work will be conducted to determine if the feathers of the two newly observed cloaks are indeed emu, and if they are, how genetically similar these feathers are to those from the Hawkes Bay cloak. Generally, this study highlights the effectiveness of genetic analyses in recovering lost history from important ethnological artifacts.

ACKNOWLEDGEMENTS

We are grateful to the Hawkes Bay Museum and Cultural Trust for allowing us to sample from their cloak collection and to Paul Scofield from Canterbury Museum for the provision of the moa bone sample. We would also like to thank Lisa Matisoo-Smith and Mere Roberts for providing valuable comments on the manuscript. This project was supported by a grant from the Marsden Fund and the New Zealand Centres of Research Excellence Fund. We also thank Vivian Ward for graphics.

REFERENCES

Barnett, R., N. Yamaguchi, B. Shapiro and V. Nijman 2007. Using ancient DNA techniques to identify the origin of unprovenanced museum specimens as illustrated by the identification of a 19th century lion from Amsterdam. *Contributions to Zoology* 76(2):87-94.

Best, E., 1952. *The Maori As He Was*. Wellington, New Zealand: A.R. Shearer, Government Printer.

Eldridge, M.D.B., T.L. Browning and R.L. Close 2001. Provenance of a New Zealand bush-tailed rock-wallaby *(Petrogale pencillata)* population determined by mitochondrial DNA sequence analysis. *Molecular Ecology* 10:2561-2567.

Graham, G., 1919. Rangi-Hua-Moa, a legend of the moa in the Waitemata district, Auckland. *Journal of the Polynesian Society* 28(110):107-110.

Haddrath, O. and A.J.Baker 2001. Complete mitochondrial DNA genome sequences of extinct birds: ratite phylogenetics and the vicariance biogeography hypothesis. *Proceedings of the Royal Society B: Biological Sciences* 268:939-945.

Huynen, L., C.D. Millar, R.P. Scofield and D.M. Lambert 2003. Nuclear sequences detect species limits in ancient moa. *Nature* 425:175-178.

Ling Roth, H. 1923. *The Maori Mantle*. Halifax: Bankfield Museum.

McClure, M. 2007. Auckland places. *Te Ara - The Encylopedia of New Zealand* [URL:http://www.TeAra.govt.nz/Places/Auckland/AucklandPlaces/en].

Pendergrast, M. 1987. *Te Aho Tapu, The Sacred Thread*. Auckland: Reed Publisher Ltd,.

Pendergrast, M. 1997. *Kakahu, Maori Cloaks*. Auckland: Auckland Museum.

Shepherd, L.D., and D.M. Lambert 2008. Ancient DNA and conservation: lessons from the endangered kiwi of New Zealand. *Molecular Ecology* [doi:10.111/j.1365-294X.2008.03749.x].

Sinclair, K. 2007. Grey, George 1812-1898. *Dictionary of New Zealand Biography* [URL:http://www.dnzb.govt.nz].

Swarbrick, N. 2007. Flax and flax working. *Te Ara - the Encyclopedia of New Zealand* [URL:http://www.TeAra.govt.nz/TheBush/NativePlantsAndFungi/FlaxAndFlaxWorking/en].

Swofford, D.L. 2002. *PAUP* Phylogenetic Analysis Using Parsimony (*and other methods)*. Sinauer Associates, Sunderland.

Te Kanawa, D.R. 1992. *Weaving a Kakahu*. Wellington, New Zealand: Bridget Williams Books Limited.

Wanderler, P., P.E.A. Hoek and L.F. Keller 2007. Back to the future: museum specimens in population genetics. *Trends in Ecology and Evolution* 22(12):634-642.

Wigglesworth, R.P. 1981. The New Zealand timber and flax trade 1769-1840. Unpublished PhD thesis, Massey University.

Worthy, T.H. and R.N. Holdaway 2002. *The Lost World of the Moa*. Christchurch, New Zealand: Canterbury University Press.

16

Tools on the surface: residue and use-wear analyses of stone artefacts from Camooweal, northwest Queensland

Jane L. Cooper and Suzanne J. Nugent
School of Social Science
University of Queensland
St Lucia QLD 4072 Australia
Email: s.nugent@uq.edu.au

ABSTRACT

Although much of the Australian archaeological record lies on the surface, such assemblages are often seen as having reduced archaeological potential when compared with subsurface deposits. However, a microscopic residue and use-wear analysis of surface-collected Aboriginal stone tulas from Camooweal, Queensland, revealed use-related residues including blood, bone collagen, woody plant tissue and resin, along with use-wear indicative of adzing and scraping functions. The results suggest the tulas were employed primarily for butchery, bone-working and woodworking tasks and were hafted in various orientations to the handle. Residues and use-wear were also detected on blades, hand axes, points, and cores from the same site, demonstrating that a large range of residues survive on artefacts from both surface and subsurface sites, enabling the reconstruction of details of subsistence and tool use.

KEYWORDS

tulas, use-wear, residue analysis, surface collected artefacts, hafting, woodworking, bone-working

INTRODUCTION

The techniques of residue and use-wear analysis have commonly been applied to stone artefacts from excavated contexts, where organic components and wear patterns relating to their use are considered more likely to have been preserved (Loy and Nugent 2002:20). Taphonomic processes acting upon surface artefacts have been considered too substantial to warrant this type of analysis (Davis 1975:52; Barton this volume; however see Briuer 1976). Despite advances in understanding post-depositional disturbance processes (see Ebert 1992; Fanning and Holdaway 2001; Greenfield 2000; Rossignol and Wandsnider 1992; Sullivan 1998; Wilkinson 2001), excavated materials continue to provide the major source of evidence for inferring past lifeways, and this bias is particularly evident on the scale of the individual artefact. This study draws on documented observations of tula use as an integral component of the Aboriginal toolkits to explore the legitimacy of applying microscopic residue and use-wear analyses to surface collected examples of these stone artefacts. We are greatly indebted to Dr Tom Loy for instigating the analysis of artefacts in the field and inspiring us to continue with his research.

SITE BACKGROUND

The project area is situated along the construction corridor of a new 425 m two-lane bridge crossing the upper Georgina River at Camooweal, approximately 13 km from the Queensland / Northern Territory border (Figure 1). The Georgina River, banked mainly by box-eucalypt woodland, riparian woodland and Mitchell grasslands, consists of a chain of water holes during the dry winter months that are joined by floods during the wet season from January to February. Decomposition of the predominantly dolomite terrain has produced extensive residual deposits of nodular and tabular chert, known locally as 'ribbonstone', that is mantled by a thick, heavy clay soil. Many of the stone artefacts found in extensive surface scatters across the project area were knapped from this chert (Archaeo Cultural Heritage Services and Dugalunji Aboriginal Corporation 2002; Orr and Holmes 1990:243).

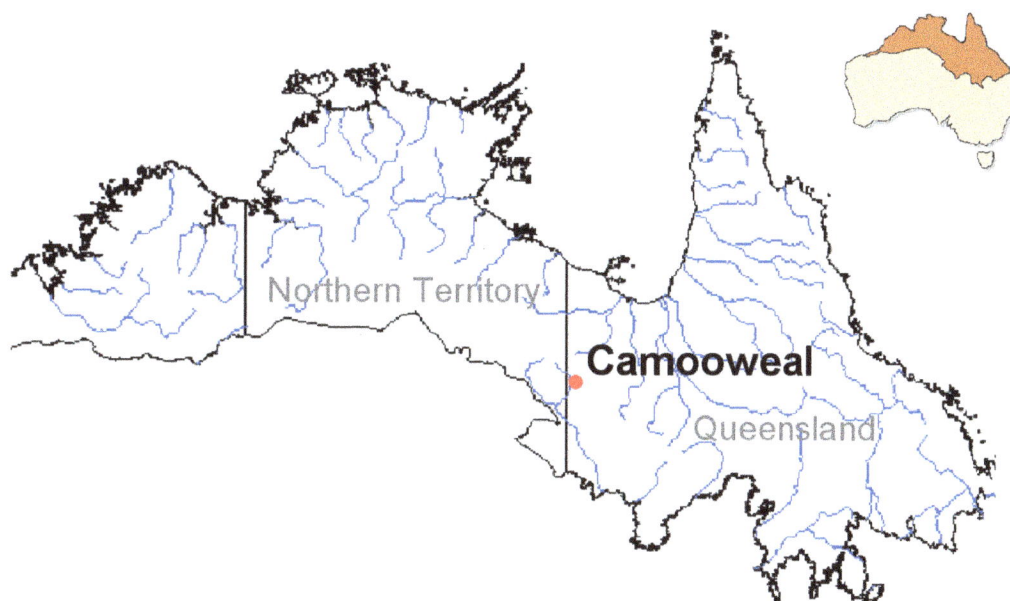

Figure 1. Map of northern Australia showing the location of Camooweal (after Tropical Savannas Cooperative Research Centre 2008).

Artefacts found on the surface were collected from a grid of 25 m x 50 m cells across the 3.825 ha site. In addition, 381 metre square pits were excavated, eight of which were randomly selected on the bridge pier footprints. Excavations usually terminated at 30 cm. The entire area of the footprints was subsequently excavated with shovels and bobcats to the same depth and the removed soil sieved to recover artefacts. In general, a mixture of Aboriginal stone artefacts and modern and historical materials were distributed vertically, although not evenly, through the excavated soil to similar depths. A total of 16,645 stone artefacts was recovered from the area, among which were 8 hand-axes, 7,789 retouched flakes, 6,765 blades, and 917 tulas (Archaeo Cultural Heritage Services and Dugalunji Aboriginal Corporation 2002:31-3,37,42).

When it rains, the predominantly black soil of the Georgina River area rapidly becomes saturated. The high clay content impedes percolation and run-off with water tending to pool on the surface. When the saturated soil swells, buried stones are forced upwards towards the surface. During the dry season, the moisture evaporates, resulting in soil shrinkage and cracking. Some of the cracks observed during fieldwork were 4 cm wide at the surface and extend as deep as 40 cm. These cracks provide an explanation for the migration of artefacts down the soil column. In addition, some mixing of artefacts would have occurred on the surface and approximately the top 10 cm of soil during grading of the road reserve (Archaeo Cultural Heritage Services and Dugalunji Aboriginal Corporation 2002:37-8). This post-depositional disturbance process has resulted in a site that is clearly disturbed and complex, as many excavated artefacts are likely to have previously seen multiple surface exposures, while those labelled 'surface collected' are just as likely to have been buried at some stage.

PRELIMINARY ANALYSIS

A total of 23 stone artefacts were examined first in the field (at Camooweal) and subsequently at the Archaeological Science Laboratory, University of Queensland in order to assess the potential for residue and use-wear analyses (Loy and Nugent 2002). The sample comprised four cores, 11 blades, four hand-axes and four tulas from surface and excavated contexts. The results are presented in Tables 1 and 2. In the field, an Olympus BHS microscope (50x, 100x, 200x, 500x and 800x nominal magnifications) was used to examine at least one surface of 10 of these tools. For the remainder of this sample, a Wild stereo-binocular low-magnification microscope (6-30.6x) was used to observe indications of use-wear and traces of residues. An Olympus BX60 microscope (50x, 100x, 200x, 500x, 1000x) with incident light was used to identify residues and observe very fine striations on those artefacts other than the hand-axe (3666-5). Photographs were taken at all stages of the analysis using microscope-mounted Olympus DP10 cameras. Slides were prepared of residues extracted from artefact surfaces, mounted with Aquamount (refractive index = 1.400), and examined using a BX60 microscope with transmitted light.

All 23 artefacts display possible use-wear marks and residues indicative of use. More importantly, when the results of the surface collected artefacts were compared to the excavated tools, some interesting trends became apparent. All four cores display indications of woodworking, and one surface and one excavated core also have evidence indicative of bone-working. The blades appear to have been used predominantly for bone- and woodworking, although two of the surface blades retain blood residues. It was inferred that residues including diatoms (naviculoid, cymbelloid and elongate types), sponge spicules, algae and plant tissue that were observed (in addition to bone collagen) on excavated blade 22666-2 are likely to result from regular inundation of the site by the Georgina River rather than use of the artefact for aquatic plant processing. Starch and plant tissue were identified on one of the surface hand-axes, suggesting the same task association as the two excavated hand axes, and all four tulas appear to have been employed in woodworking. Such consistent indications of use found on both the surface and excavated artefacts stimulated a detailed analysis of a further 16 surface collected tulas from the Camooweal site (Tables 1 and 2).

TULA ANALYSIS

Ethnographic background

The tula is one of the most common and distinctive chipped stone tools represented ethnographically and archaeologically in Australia's arid regions (Gould 1978:820). Roth (1897:101-2) recorded the manufacture and use of what he termed 'pot lid' flakes and 'native-gouge' composite adzes. In 1924, following their observations of the Wonkonguru people of the Lake Eyre region, Horne and Aiston (1924:80-9) recorded the use of a specific woodworking adze called 'koondi tuhla'. The 'tuhla', described as a broad semi-circular stone flake, was hafted to the end of a curved wooden handle or 'koondi' (the name meaning curved) with an organic fixative, such as spinifex (*Triodia* sp) or beefwood (*Grevillea* sp) resin, and functioned generally as a chisel, scraper, axe and/or adze. The tuhla flake was periodically resharpened with a lightweight hammerstone or boomerang, and later unhafted and replaced when considered to be no longer of any use (Horne and Aiston 1924:89). Subsequent to Horne and Aiston's (1924, see also Aiston 1928, 1929) observations, the name of this tool was abbreviated to 'tula' (Figure 2).

Tulas were observed to have a variety of uses dependent on the dimensions of the flake, the hafting arrangement, and the amount of resharpening the flake sustained. Hafted tulas were

Table 1. Results of residue and use-wear analyses of surface collected artefacts

Tool Type	Artefact Number	Traces of Use-Wear		Identified Residues		Suggested Task Association
		Dorsal View	Ventral View	Dorsal View	Ventral View	
Core	999-53	Lower RLE - retouch/ use-wear scars; RLE – fine 60° & 45° striae; OAR – abraded & rounded.	Lower LLE – retouch/ use-wear scars.	OAR - Bone collagen, opaque resin; DA red/ black resin; S - plant cellulose & starch 2-4 μm.	DE – vivianite, red ochre; DA – red/black resin; S - plant cellulose & starch 2-4 μm.	Bone- & resinous woodworking.
Blade	777-2(7)	ROAR – 45° curved striae & abraded.	Adjacent to PE - 45° drag marks through sediment.	Edges and OAR – opaque resin, granular bone collagen; RLE – collagen fibril; S - charcoal & starch 2-3 μm.	RLE – hair; S - charcoal & starch 2-3 μm.	Bone- & resinous woodworking.
	1600-4(3)	OAR – abraded; DE, LLE, RLE – retouch/ use-wear scars.	DE, LLE, RLE – retouch/use-wear scars; Upper LLE - 45° drag marks.	PE – plant cellulose; S - opaque, brown & red resin, starch 2-4 μm.	RLE – plant cellulose; LLE – green resin; S - opaque, brown, & red resin, starch 2-4 μm.	Resinous woodworking.
	7900-2(1) (TL*)	-	-	-	Surface centre – nucleated thin blood; Red ochre.	Ritual activity, male gender (interpreted by Ruby Saltmere, Indjilandji Elder).
	7900-2(2) (TL*)	RLE – retouch/ use-wear scars.	LLE – retouch/use-wear scars.	-	LLE & RLE – thin blood smear with anucleate red blood cells; S - downy and pennaceous *Chenonetta jubata* (wood duck) feather fragments	Ritual activity. (interpreted by Ruby Saltmere, Indjilandji Elder).
	20700-4	OAR – abraded.	None observed.	Adjacent to PE – granular bone collagen; PE, LLE & OAR – resin & plant tissue; S - starch 2-3 μm.	Adjacent to PE – granular bone collagen; S - starch 2-3 μm.	Bone- & plant working.
Hand Axe	999-67 (TL*)	None observed.	Heavy 25° drag marks away from LLE.	Lower LLE - plant tissue & ovate starch up to 6 μm.	Lower LLE - plant tissue & ovate starch up to 6 μm.	Starchy plant processing.
	40400-2 (TL*)	None observed.	None observed.	Distal LLE – hair & collagen fibril bundle.	-	Late stage butchery.
Tula	21500-3 (TL*)	Abraded retouched edges. Distal ridges – striae 20° & 45°.	Lower LLE – 45° & 65° striae.	-	1.2cm from DE – red/black & grey resin & plant debris; Left DE – woody tissue, abundant plant cellulose, plastids with starch & resin.	Once hafted; Woodworking & starchy plant processing.
	23500-5 (TL*)	Abraded retouched edges; LLE – vertical dragging.	Abraded retouched edges; DE - 20°, 45°, 60° & 90° striae, 60° dragging.	S - plant cellulose & starch 1–2 μm.	DE – red, yellow & black resin, DA – plant & grass fibres; S – starch 1–2 μm.	Woodworking & grass processing.

Key: TL* - Dr Tom Loy limited field examination;
DE – distal edge; DA – distal area; LLE – left lateral edge; RLE – right lateral edge; OAR – obtuse angle ridge; ROAR – right obtuse angle ridge; PE – platform edge; S – scattered on surface; μm – micron (starch granule diameter).

Table 2. Results of residue and use-wear analyses of excavated artefacts

Tool Type	Artefact Number	Traces of Use-Wear		Identified Residues		Suggested Task Association
		Dorsal View	Ventral View	Dorsal View	Ventral View	
Core	16666-4	OARs – abraded.	OARs – abraded; 45º, 60º and vertical dragging on negative scar.	S - starch 2-4 µm.	Bone collagen on negative scar edges; DA – wood tissue; PE – opaque & red/black resin; S - starch 2-4 µm.	Once hafted; Bone- & woodworking.
	17666-4	Left PE – pitting.	OAR – abraded; Upper RLE - 60º dragging.	S - opaque & red/black resin, charcoal & starch 2-4 µm.	S - opaque & red/black resin, charcoal & starch 2-4 µm.	Woodworking.
	23666-11	Retouch/use-wear scars; Abraded ridges; Right OAR – rough & pitted.	OAR – abraded; RLE – retouch/use-wear scars; PE - 45º dragging.	Opaque & red/black resin on ridges; Upper LOAR – charcoal; S - plant tissue & starch 2-4 µm.	PE – plant exudate & charcoal; OARs - opaque & red/black resin on ridges; S - starch 2-4 µm.	Possibly once hafted; Woodworking.
Blade	282-2	LLE - retouch/use-wear scars & 45º striae; RLE - 15º to 90º striae; OAR – abraded with 45º striae either side.	Lower LLE – vertical striae; RLE - retouch/use-wear scars; Lower RLE - 45º & 90º drag marks.	Edges – bone collagen; Lower RLE & DE – red ochre; Upper RLE – feather (Order – Galliformes); LLE – plant tissue; S - opaque & red/black resin & starch 2-4 µm.	Edges – bone collagen; RLE, LLE & DE – red ochre; RLE – plant tissue; S - opaque & red/black resin & starch 2-4 µm.	Bone- & woodworking & ceremonial use.
	4666-2(1)	OAR – abraded; LLE - retouch/use-wear scars & vertical striae; RLE –vertical striae; DE – almost vertical striae.	Upper LLE - 10º drag marks crossed by 45º striae; RLE - retouch/use-wear scars & curved 10º striae; DE – curved almost vertical striae.	OAR and all edges – opaque, red/black, orange & black resin; LLE – hair; S - starch 1-3 µm.	All edges - opaque, red/black, orange & black resin; RLE – plant tissue; S - starch 1-3 µm.	Possibly once hafted; Woodworking.
	5666-2(1)	OARs – abraded; Upper LLE - horizintal & vertical striae; Adjacent to centre PE - 15º drag marks; Lower RLE - 60º striae & retouch/use-wear scars.	Adjacent to upper & lower LLE – vertical dragging; PE – horizontal dragging; Upper RLE – 45º striae; Lower RLE & Distal edge – vertical striae.	LOAR – starch 6µm; S - opaque, red & black resin, charcoal & starch 2-4 µm.	S - opaque, red & black resin, charcoal & starch 2-4 µm.	Woodworking.
	14666-2(2) (TL*)	OAR – abraded with 45º drag marks to left; PE – drag marks.	None observed.	All edges – red ochre; LLE & PE – opaque & red/black resin & plant tissue; S - starch 1-3 µm.	PE – red ochre; PE & RLE - opaque & red/black resin; S - starch 1-3 µm.	Possibly once hafted; Decorative/ ceremonial use, bone- & woodworking.
	22666-2	OAE – abraded; Upper RLE & PE – 45º drag marks; PE - 90º drag marks.	LLE, RLE & PE – 45º crossed drag marks.	LLE & RLE – fibrous plant tissue & opaque resin; Lower LLE – bone collagen & algal tissue; S - diatoms (naviculoid, cymbelloid & elongate) & starch 1-2 µm.	S - diatoms (naviculoid, cymbelloid & elongate) & starch 1-2 µm.	Bone-working.
	52666-4(1) (TL*)	OAR – abraded.	Worn bulb of percussion.	S - fibrous plant tissue.	Right half – fibrous plant tissue; Upper RLE – anuclear red blood cells.	Butchery.
Hand Axe	3666-5	Some OARs abraded, Mid LLE - 45º scraping; LLE – abraded; DE - 90º striae & retouch/use-wear scars.	Some OARs abraded; Lower RLE - 90º curved drag marks; DE - retouch/use-wear scars.	LLE – plant tissue starch average 16µm; RLE – plant tissue, starch average 12µm; S - starch 2-3 µm.	LLE – plant tissue, starch average 18µm; RLE – plant tissue; DE – starch average 6µm; S - starch 2-3 µm.	Starchy plant processing.
	23666-12 (TL*)	OARs – abraded.	-	Lower half – plant tissue; DE – starch average 6 µm & raphide.	-	Starchy plant processing.
Tula	23666-10(1)	OARs – abraded; LLE - 45º striae & crossing curved rub marks; retouch/use-wear scars.	Lower LLE - 45º & 90º crossed striae; RLE - 60º & 90º striae.	All edges – bone collagen; S – wood & plant tissue, opaque & red/black resin patches & starch 2 µm.	PE – bone collagen; S - wood & plant tissue, opaque & red/black resin patches & starch 2 µm.	Bone- and woodworking.
	23666-10(2) (TL*)	-	Retouch/use-wear scars; Mid DE - 45º to 60º heavy striae.	-	All edges & centre – plant tissue; S - opaque & yellow resin & starch average 2 µm.	Woodworking.

Key:
TL* - Dr Tom Loy limited field examination;
DE – distal edge; DA – distal area; LLE – left lateral edge; RLE – right lateral edge; OAR – obtuse angle ridge; ROAR – right obtuse angle ridge; LOAR – left obtuse angle ridge; PE – platform edge;
S – scattered on surface; µm – micron (starch granule diameter).

generally employed as woodworking tools for the manufacture of weapons, ceremonial and sacred objects and various other wooden implements such as water carriers, digging bowls and spearthrowers (Horne and Aiston 1924:103; McCarthy 1967:17,28; Roth 1904:17; Spencer and Gillen 1969:637). McCarthy (1967:28; see also Thomson 1964:418; Tindale 1965-1968:154) observed hafted tulas employed for butchering animals, bone working and occasional digging. Similarly, Etheridge (1891:41) described the use of unhafted tula flakes in skinning kangaroos, and Mitchell (1949:9) mentioned a hafted disc-shaped flake used for making notches in the bark of trees for climbing footholds. Davidson (1935:160) observed Aborigines in the Northern Territory using hafted tulas as throwing weapons and clubs. Gould *et al.* (1971:155) recorded that

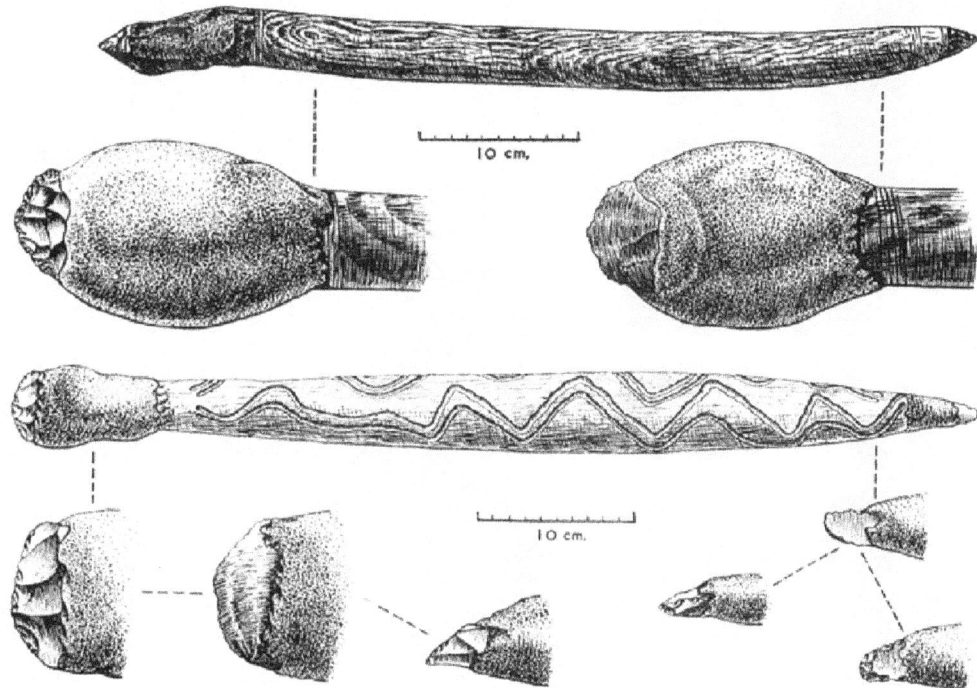

Figure 2. Illustration of hafted tula flakes employed by the Ngadadjara and Nakako people, South Australia (after Tindale 1965:134; courtesy Museum Board of South Australia).

the narrow edged flake, named 'pitjuru-pitjuru' by the Nyatunyatjara of the Western Desert, was employed exclusively for engraving decorations on sacred boards and spear-throwers.

During use, tula flakes were progressively resharpened along the distal/proximal extremity until they were too small to be hafted. At this point they were removed from the haft and either discarded, employed as hand scrapers and chisels, or rehafted to utilise other functional edges (McCarthy 1967:28-9). These 'worn-out' (Horne and Aiston 1924:89) flake remnants are commonly referred to as tula slugs (Campbell and Edwards 1966:161; Hiscock 1994:269; McCarthy 1967:28; McNiven 1993:23; Moore 2003:24).

Methods of analysis and data collection

Given the quantity of artefacts recovered, an analysis of the entire collected sample was beyond the scope of this study. Based on the notion that the individual artefact can be as informative as the larger site (Loy 1993:44), a sub-sample was examined with the aim of generating inferences about artefact use. Prior to the analysis, the tulas were divided into two morphologically defined categories of 'flake' and 'slug', based on size, shape and extent of retouch (see Table 3). From ethnographic accounts and current research into tula manufacture and use, it was expected that certain areas on the artefact's surface were more likely to contain residues than others. These areas guided the placement of the transect lines employed during the examination of each artefact, making a total of 512 analysed quadrants (Figure 3). The size of each quadrant depended on individual artefact morphology.

The data collection process was divided into macroscopic and microscopic examination, including low- and high-magnification (incident and transmitted light), and biochemical testing. During macroscopic examination, distinguishing features such as flake scars, fractures, and signs of retouch (as defined by Kamminga 1982:6) were noted. Low-magnification examination using the Wild microscope proceeded along the transect lines and the position, direction and quantity of residues were recorded, along with any signs of edge rounding (including bevelling), fractures, edge-chipping/scarring, striations, and polish. High-magnification incident microscopy was performed with an Olympus BX60. Residues were photographed and recorded in terms of colour,

Table 3. Morphological and macroscopic examination of tulas

Artefact Number	P L (mm)	P W (mm)	T L (mm)	T W (mm)	T T (mm)	Dist. Angle	Prox.D Angle	Prox.V Angle	Artefact Description
20100-7a (Flake)	35	7	27	33	7	60°	60°	120°	**DF** Flake scars along left lateral margin to point at distal edge, same for opposite margin: **VF** Eraillure scar on bulb of percussion, two feather fractures on right dorsal edge of platform, fine edge chipping along left lateral margin.
22300-4 (Flake)	42	11	30	48	11	45°	45°	120°	**DF** Flake scars from platform overhang removal, flake scars along left, right and distal edges: **VF** Small step and feather fractures on distal edge.
999.428 (Flake)	31	11	27.5	38	7	45°	45°	120°	**DF** Feather fractures along platform edge, flake removal scars along left, right and distal edges: **VF** Large feather fracture originating from left dorsal edge on platform, edge chipping on left, right and distal edges.
34700-3 (Flake)	41.5	8.5	28.5	49	13.5	45°	85°	127.5°	**DF** Large flake removal scar on right platform edge, step fractures along right lateral margin, large step fractures along distal edge with edge chipping: **VF** Large feather fracture on left lateral margin and distal edge, edge chipping along dorsal edge of platform.
66800-3 (Flake)	34	9.5	34	35	9	60°	60°	120°	**DF** Overhang removal along platform edge with two large step fractures on left and right sides, flake removal scars along left, right and distal edges, moderate fracturing along right lateral margin: **VF** "6680-3" written in pencil along bulb of percussion.
60400-2 (Flake)	42	18	31	43	9.5	45°	60°	127.5°	**DF** Large obtuse angle ridge upper left lateral margin to lower right lateral margin, large step fracture left platform edge, chipping along left and distal edges, concave edge lower right lateral margin: **VF** Step fractures along right lateral margin, eraillure scar on bulb of percussion, edge chipping along distal edge.
20100-7b (Flake)	31	11	40	47.5	10.5	60°	60°	120°	**DF** Small step fractures middle of platform edge, concave edge lower right lateral margin to distal edge: **VF** Edge chipping upper left lateral margin, step fractures along distal edge, two feather fractures on dorsal edge of platform.
1100-4a (Flake)	36	11	27	43	9	60°	45°	127.5°	**DF** Edge chipping and step fractures along platform edge, flake removal scars along left, right and distal edges: **VF** Feather fracture on distal edge, compression rings along bulb of percussion, feather fractures on upper left lateral edge, two large feather fractures on dorsal edge of platform.
777-7 (Flake)	40	11.5	26.5	42	10	60°	45°	120°	**DF** Edge chipping along platform edge, large flake removal scars along left and right lateral margins and along distal edge, obtuse angle ridge from upper right lateral margin to lower left lateral margin: **VF** Slight edge chipping on dorsal edge of platform, slight edge chipping along lower left lateral margin and distal edge.
1600-3 (Flake)	24	6.5	22.5	28.5	7	52.5°	60°	105°	**DF** Slight chipping along platform edge, large feather and step fractures upper right lateral margin, flake removal scars along left and right lateral margins and along distal edge: **VF** Large feather fracture from distal edge across face, flake removal scars below right platform edge, edge chipping along distal edge.
1100-5 (Flake)	35	9	27.5	38	9	60°	67.5°	112.5°	**DF** Feather fractures along platform edge, large feather fracture on right platform edge, flake removal scars along distal edge and left lateral margin: **VF** Slight edge chipping along left lateral margin and distal edge, small step fracture upper left lateral margin.
62600-1 (Flake)	34	9.5	29.5	38	9.5	45°	60°	120°	**DF** Large step fracture along platform edge, large concave flake removal on right lateral edge, flake removal scars along left and distal edges, obtuse angle ridge above distal edge: **VF** Edge chipping along right and distal edges, compression rings along bulb of percussion, feather and step fractures along platform.
30900-2 (Slug)	40	9	21.5	47	12	75°	60°	125.5°	**DF** Multiple retroflexed hinge flake scars along distal edge, overhang removal and large feather fractures along platform edge: **VF** Feather fracture on dorsal edge of platform, feather fracture on left lateral margin, slight edge chipping on right lateral margin.
31800-1 (Slug)	41	11.5	17.5	45	11	75°	60°	97.5°	**DF** Edge chipping and feather fracture along platform edge, multiple retroflexed hinge flake scars along distal edge, slight edge chipping along left and right lateral margins: **VF** Slight edge chipping and edge fracture along distal edge, slight edge chipping along left lateral margin.
1100-4b (Slug)	35	10	17	40	10	75°	75°	97.5°	**DF** Multiple retroflexed hinge flake scars along distal edge, overhang removal and step fractures along platform edge, flake scar along right lateral edge: **VF** Edge chipping along distal edge.
9800-4 (Slug)	30	8	15.5	37	8	82.5°	67.5°	120°	**DF** Step fractures and edge chipping along platform edge, multiple retroflexed hinge flake scars along distal edge: **VF** Edge chipping along distal edge.

Key:
PL = Platform Length, PW = Platform Width, TL = Tool Length (proximal to distal), TW = Tool Width, TT = Tool Thickness (at bulb of percussion), Prox. D = Proximal Distal Face, Prox. V = Proximal Angle Ventral Face, DF = Dorsal face. VF = Ventral face

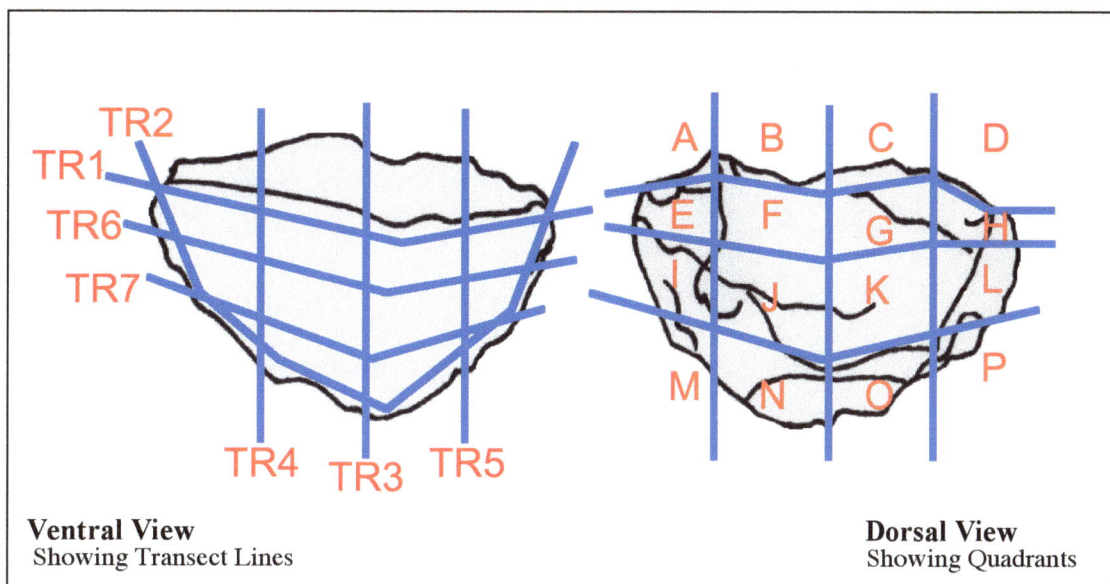

Figure 3. Detail of transect lines and quadrants used in analysis.

form, texture, identifiable structures, and amount (*small*, *moderate* and *large*) on the artefact surface. Striations, edge rounding, fractures and polish were also recorded and photographed. Residues requiring further examination were removed and placed on microslides and viewed using an Olympus BX50 polarising microscope. A reference collection of plant and animal residues specific to Australian flora and fauna, the Laboratory Microscopy guide (Archaeological Sciences Laboratory, University of Queensland), illustrated literature of plant and animal tissues, the Hemastix colourimetric test (Loy 1983) and samples collected from the site assisted in the identification of observed residues.

RESULTS

Macroscopic examination
With the exception of size and curvature of the distal edges, all tulas in the sample have distinct morphological similarities (Table 3). Within the 'flake' category, moderate to extensive flake scarring was identified along all edges (excluding the ventral platform edge), predominantly of step and hinge termination types. Other observations included: fine edge retouch along the dorsal side of the proximal edge and some degree of fine lateral and distal retouch; a smooth convex ventral surface (excluding artefacts 20100-7 and 60400-2, which have errailure scars along the bulb of percussion); a maximum width greater than the maximum length; an acute distal edge and dorsal platform edge angle; and an obtuse ventral platform edge angle. The tula 'slugs', although possessing the same morphological attributes, have additional characteristics dissimilar to the tula 'flakes'. These were observed as: a marked reduction in the length of the flake; a straight or slightly convex distal edge; large feather and step flake scars along dorsal platform edges (excluding artefact 31800-1, which has minimal retouch along this edge); and multiple flake scarring along the distal edge on the dorsal face, particularly of hinge and retroflexed hinge types.

Low-magnification examination
All of the tulas display some degree of microfracture scarring and rounding along the distal, lateral and dorsal platform edges. The most frequently occurring wear types observed were micro step, feather and hinge termination scars, edge rounding and micro edge-chipping (Figure 4). The term edge-chipping was applied to consecutive indiscriminate microfractures, occasionally of a rounded or angular morphology, as the individual flake scars were non-diagnostic of any of the above mentioned termination types. Edge-chipping constitutes the most prominent wear attribute, occurring in varying degrees on all artefacts in the sample. Step fracturing was also observed

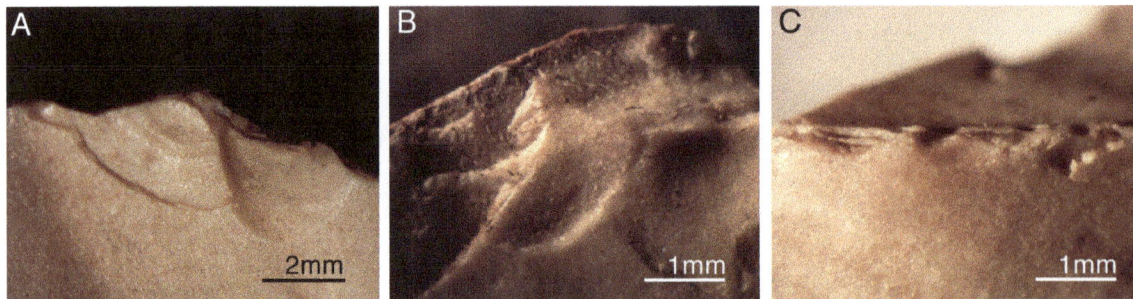

Figure 4. Examples of step, feather and hinge termination scars, and micro edge chipping. (A) Step and feather fractures. Artefact 1100-5, ventral right lateral margin, 8x (B) Multiple step and hinge fractures. Artefact 20100-7b, ventral distal edge, 16x (C) Edge chipping. Artefact 20100-7a, ventral left lateral margin, 16x.

on all artefacts (excluding 1100-5), predominantly on the distal edges, and the dorsal left lateral margin, although 20100-7a, 1600-3 and 30900-2 displayed step fracturing proximal to the bulb of percussion.

Although edge rounding was common it was usually restricted to the smoothing of microfractures along the perimeter edges. Some degree of edge rounding is also present on the obtuse angle ridges on artefacts 20100-7a, 60400-2, 1100-4a and 30900-2. Striations, surface polish, bending fractures, and compression ring fractures were observed relatively infrequently on only a few artefacts at low magnifications. All tulas in the sample have varying amounts of sediment and other residue adhering to their surfaces. Seven tula flakes and three slugs have plant fibres attached to their surfaces, which are probably an environmental additive rather than use-related. Various other residues were observed at low magnifications such as red resinous exudate and bone collagen (later identified during high-magnification microscopy) (Figure 5). Black fungal spores were also identified on some of the artefacts.

High-magnification examination

During high-magnification analysis, four types of use-wear were observed and a total of 37 individual residue types were identified across the sample. Suspected blood residues were located on the ventral face of six artefacts (22300-4, 34700-3, 60400-2, 777-7 and 31800-1 and on both faces of artefact 1100-4a). These deposits occur as either thin, colourless plaques with raised cracking or as dark reddish/brown cracked masses (Figure 6A). Blood films were occasionally observed in association with deposits of bone collagen (Figure 6B). Bone collagen was the most frequently observed animal residue, occurring in varying amounts on all artefacts in the sample (with the exception of 1100-4b). Artefact 31800-1 displays the largest occurrence of collagen on both the ventral and dorsal faces. Collagen, identified primarily as white, granular smears or fragments (Figure 6C) but also as translucent/grey amorphous sheets with a distinctive 'basket-weave' appearance (Figure 7A), was occasionally found in the presence of vivianite (see below). Feather fragments are present on five of the tulas in small and moderate quantities, although no diagnostic downy barbules were identified (Figure 7B). Degraded hair fragments, often embedded in resin, were also observed (Figure 7C).

Plant residues consist of tissue, cells, cellulose, fibres, exudate, phytoliths, raphides, and pollen. These residues were observed predominantly in association with each other, occurring in moderate to large quantities on many of the tulas. Cellulose, of varying amounts, was the most common plant residue type observed across the sample and when found in association with other plant tissue components, such as vessel elements, bordered pits, wall thickenings and tracheids, was assumed a derivative of the associated plant material. Large deposits of plant tissue frequently occur as inclusions in resin or are caught beneath micro fractures and along edges. Occasionally, plant tissue was observed with the dermal and ground cellular structure intact. These tissue cells, located on artefacts 20100-7a, 22300-4, 66800-3, 62600-1, 30900-2, 31800-1 and 9800-4 are similar in structure to the epidermis and parenchyma of woody plants (Raven *et al.* 1999:586)

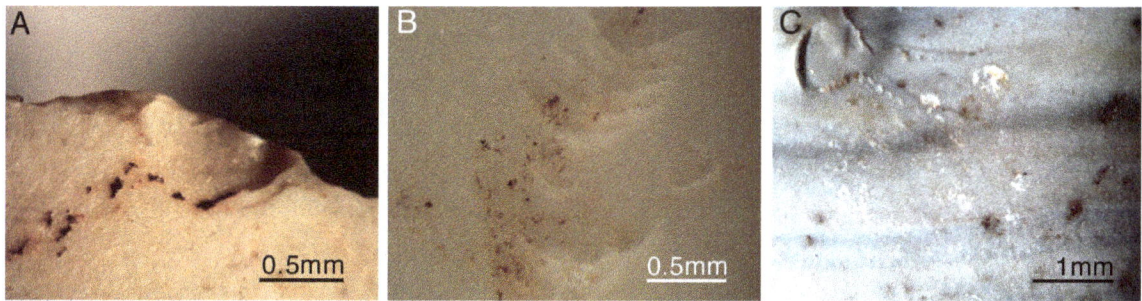

Figure 5. Residue in association with signs of use-wear. (A) Step fracture with resin. Artefact 60400-2, ventral distal edge, 25x. (B) Resin and compression ring fractures. Artefact 1100-4a, ventral mid section, 25x. (C) Step fracture, resin and bone collagen. Artefact 34700-3, ventral mid section, 16x.

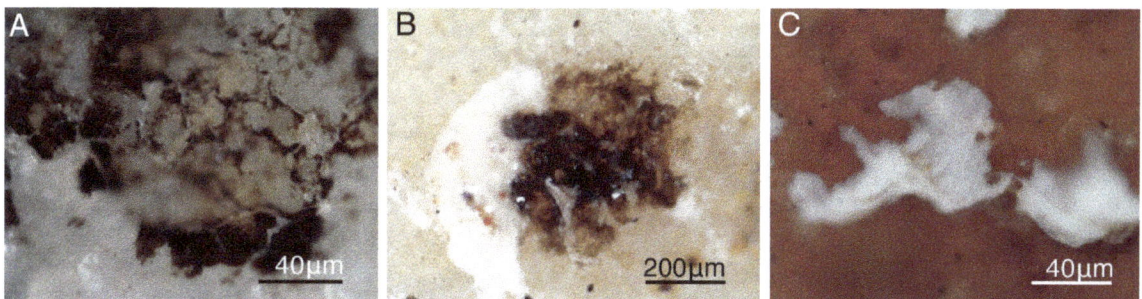

Figure 6. (A) Blood residue. Artefact 60400-2, right mid section ventral face, 500x bright field illumination, cross-polarised (BF xpol.). (B) Blood and collagen residue. Artefact 22300-4, distal edge ventral face, 100x BF xpol. (C) Granular bone collagen. Artefact 999.428, left platform ventral face, 500x BF xpol.

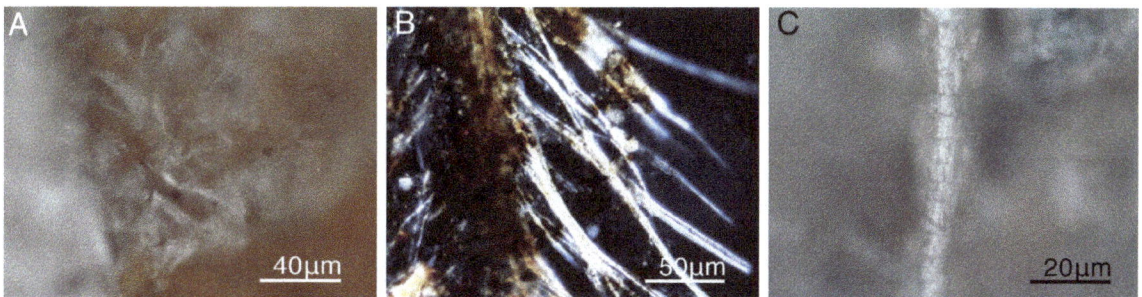

Figure 7. (A) Sheet collagen. Artefact 30900-2, middle left lateral margin dorsal face, 500x brightfield illumination plane-polarised (BF ppol.). (B) Feather barb. Artefact 62600-1, middle left lateral margin ventral face, 400x BF ppol (transmitted light). (C) Degraded hair. Artefact 62600-1, upper right lateral margin ventral face, 1000x BF ppol.

(Figure 8A). Plant exudate or sap (produced in vacuoles) was also observed in conjunction with plant tissue, occurring as colourless, glossy films, although generally in small quantities (Raven, *et al.* 1999:173). Phytoliths, primarily of festucoid and panicoid types, were observed on many of the tulas either individually or in rows within plant tissue (Figure 8B). Dumbbell phytoliths range in size from 15 μm to 25 μm in length, while ovate phytoliths vary in diameter from 10 μm to 15 μm. Bundles of whisker raphides were observed on the edges of artefact 34700-3 (Figure 8C) and on the ventral right lateral margin of artefact 66800-3. Clusters of pollen were identified on the upper left lateral margin of artefact 30900-2, and a few scattered pollen grains were observed on the dorsal face of artefacts 20100-7a and 20100-7b. Starch grains (1-4 μm diameter), some of which are damaged or gelatinised, were observed on the majority of tulas.

Resin was the most frequently observed residue type, occurring in varying quantities on all tulas. The types of resin exhibited are red, orange/amber, clear/grey and charred. The most

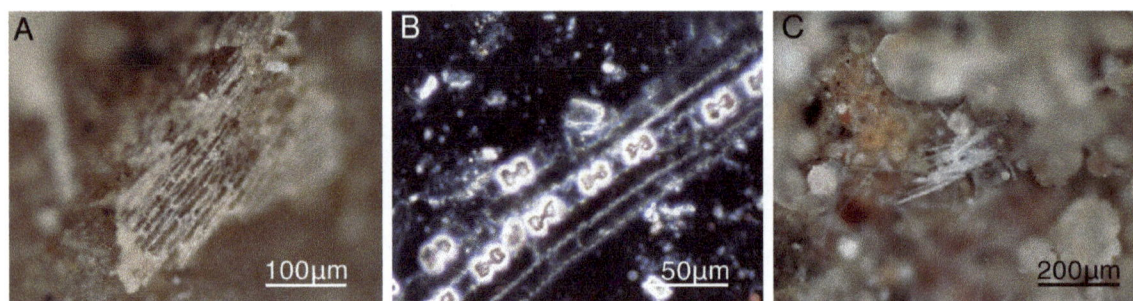

Figure 8. (A) Woody plant tissue. Artefact 22300-4, left platform edge ventral face, 200x dark field illumniatoin (DF). (B) Festucoid grass phytoliths. Artefact 22300-4, distal edge ventral face, 400x DF (transmitted light). (C) Bundle of raphides. Artefact 34700-3, distal edge dorsal face, 100x BF xpol.

consistently occurring type is red resin, predominantly observed in moderate and large quantities on both the dorsal and ventral faces of all artefacts except 9800-4. Red resin appeared as globular masses or smears and is generally located perpendicular to an edge, caught in microfractures, or along ridges (Figure 9A). Charred resin, regularly observed in association with red resin, is similar to the latter in appearance but is birefringent in plane-polarised light due to a high charcoal content. The physical similarities and location suggest charred resin is probably a carbonised form of red resin. Clear/grey resin, observed in similar frequency to red resin, appears either as thin grey smears or thick deposits containing a high proportion of grass phytoliths in addition to other plant material such as fibres and trichomes (plant hairs) (Figure 9B). Orange/amber resin with its distinct translucent, cracked morphology (Figure 9C) was observed in small and moderate quantities on the majority of the tulas.

Charcoal was observed primarily in association with resin (particularly charred resin) and various plant residues on all tulas. Charcoal appeared either scraped onto the surface in linear deposits or as individual fragments, with the largest quantities occurring on the dorsal faces of artefacts 20100-7a, 999.428 and 62600-1, and on the ventral face of artefacts 66800-3 and 1600-3. Both red and yellow ochre (Figure 10A) were also found on the majority of the tulas, with the largest deposits occurring predominantly toward the lower mid section and distal edges of artefact 22300-4 and 1600-3.

Possible post-use residues, or environmental additives were identified in varying quantities on both the ventral and dorsal faces of the majority of tulas. These include plant tissue, algae, fungal spores and hyphae, pennate (freshwater) diatoms, insect remains, including spidermite webs, eggs, wings and legs, silt, mineral crystals (possibly calcium oxalate), spherulites (calcium carbonate crystals often formed in avian and reptile uric acid) (Canti 1998:442), and graphite (pencil).

Use-wear identified at high-magnifications included striations, edge rounding, fractures, and surface polish. Striations were commonly observed either through residue, such as resin and bone collagen (Figure 10B), or directly on the stone surface. The latter striations were generally noted as fine, multidirectional and intersecting and probably formed as a result of soil abrasion (Kamminga 1982:14). Conversely, striations through residue are generally unidirectional and indicate either how the residue was applied to the surface or the subsequent direction of use. Edge rounding or smoothing occur largely along the distal edges of most tulas and along the dorsal obtuse angle ridge of 34700-3 and 777-7. Edge rounding was often observed with striations perpendicular or sub-perpendicular to the edge (between 65° and 80°), and with some degree of surface polish. Surface polish or abrasion was identified as a marked increase in surface shine and occur in pitted and interlocking patches along surface protrusions and/or parallel to an edge (Figure 10C). All forms of use-wear increase in intensity towards the distal edges, with some degree of wear observable along the platform edges and the upper lateral margins. A summary of the residues and use-wear observed per artefact is presented in Table 4.

Figure 9. (A) Red resin along ridge. Artefact 66800-3, right lateral margin dorsal face, 100x BF xpol. (B) Mass of clear/grey resin. Artefact 22300-4, upper left lateral margin dorsal face, 100x DF. (C) Orange/amber resin. Artefact 34700-3, upper platform edge dorsal face, 500x BF xpol.

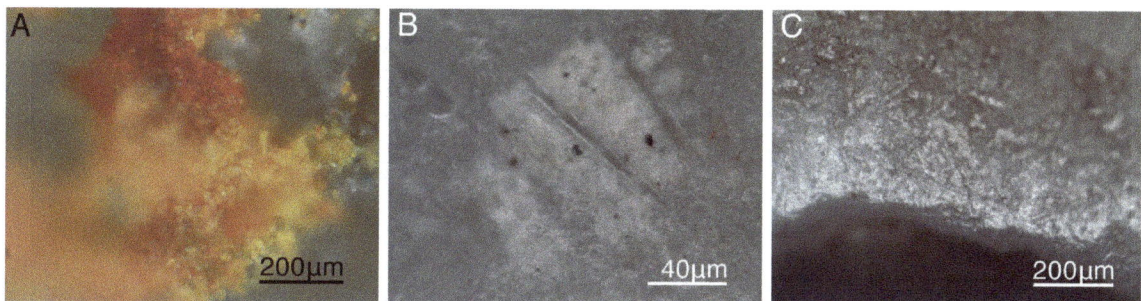

Figure 10. (A) Red and yellow ochre. Artefact 22300-4, distal edge ventral face, 100x BF xpol. (B) Striations through bone collagen. Artefact 31800-1, right lateral margin dorsal face, 500x DF. (C) Edge polish and striations. Artefact 66800-3, distal edge dorsal face, 200x BF ppol.

Subsequent to high-magnification microscopy, six samples of suspected blood residue were removed and tested using the Hemastix colourimetric test for the presence of hemoglobin or myoglobin. Three of the six returned a positive reaction prior to the addition of sodium EDTA (a chelating agent that removes chlorophyll and/or heavy metals that may also react positively). Artefact 22300-4 scored the highest concentration reaction of five (equivalent to three nanograms of blood per microlitre), and with the addition of the chelating agent returned a result of three (0.5 nanograms per microlitre). Artefact 60400-2 scored a reaction of three and with sodium EDTA, scored a trace reaction. The sample from 1100-4a scored an initial trace reaction but failed to produce a reaction with the addition of sodium EDTA.

ANALYSIS AND DISCUSSION
Taking into consideration the unknown conditions of the open site and the variety and extent of environmental additives, the presence of a particular residue such as plant tissue alone was not considered sufficient to suggest task association. Based on the various combinations of residues observed on the tulas and in further consideration of previous residue analyses of archaeological, ethnographic and replicated tools, a list of common residue types and the associated tasks from which they were produced was formulated (Table 5). A particular task was inferred if the combination of identified residues on the artefacts was consistent with the residue types presented in this table.

Residues
The most commonly occurring residues (excluding clear/grey resin) identified on both the dorsal and ventral sides of the tulas were bone collagen, charcoal, red resin, and woody plant tissue. The combination of the latter two residues suggests that working resinous wood was a consistent task

Table 4. Summary of the quantities of use-related residue and wear observed at high magnification

Identified Residues	20100-7a		22300-4		999.428		34700-3		66800-3		60400-2		20100-7b		1100-4a	
	D	V	D	V	D	V	D	V	D	V	D	V	D	V	D	V
Blood	-	-	-	L	-	-	-	M	-	-	-	S	-	-	S	S
Collagen	M	-	S	L	M	L	S	L	S	L	M	M	M	L	M	M
Vivianite	-	-	-	-	-	S	-	S	-	S	-	-	-	S	-	-
Feather	-	-	-	-	M	-	-	-	-	-	-	-	-	S	-	-
Hair	S	-	-	-	-	S	S	-	-	S	S	-	-	-	-	-
Plant Tissue	L	-	L	M	L	M	L	S	-	L	M	-	-	S	-	-
Cellulose	L	M	M	S	M	S	M	M	S	M	S	S	S	S	S	-
Phytoliths	-	-	S	S	S	-	-	-	-	M	S	-	-	-	S	-
Raphides	-	-	-	-	-	-	L	S	-	S	-	-	-	-	-	-
Starch Granules	SG	-	SG	ST	-	-	MTG	ST	-	LTG	SG	-	SG	SG	-	-
Red Resin	L	M	L	L	M	L	L	L	L	L	M	L	M	M	L	L
Orange/Amber Resin	-	S	S	S	S	M	M	S	S	S	M	S	S	M	M	S
Clear/Grey Resin	-	-	M	S	M	L	M	L	S	L	L	S	S	S	S	L
Charred Resin	L	S	L	M	S	S	M	S	S	M	-	L	M	-	S	S
Charcoal	L	M	S	S	L	S	S	M	S	L	S	M	M	S	S	-
Ochre	M	-	S	S	S	-	-	-	-	S	-	S	S	-	S	-
Diatoms	S	-	-	-	-	-	-	-	-	-	-	S	S	-	-	-
Identified Use-Wear																
Striations	M	L	-	L	L	L	S	M	M	S	-	S	-	-	M	S
Edge Rounding	M	-	L	-	L	L	L	M	M	M	S	S	S	S	M	L
Fractures	-	-	-	S	-	-	-	-	-	S	S	S	S	S	S	S
Surface Polish	-	-	-	-	S	-	-	M	M	-	-	-	-	-	M	M

Identified Residues	777-7		1600-3		1100-5		62600-1		30900-2		31800-1		1100-4b		9800-4	
	D	V	D	V	D	V	D	V	D	V	D	V	D	V	D	V
Blood	-	S	-	-	-	-	-	-	-	-	-	M	-	-	-	-
Collagen	M	S	M	L	M	M	S	L	M	M	L	L	-	-	M	M
Vivianite	-	-	-	-	-	-	-	-	-	-	-	M	-	-	-	-
Feather	S	-	-	-	-	-	-	M	-	-	-	S	-	-	-	-
Hair	S	-	-	S	-	-	-	S	-	-	-	S	-	S	-	-
Plant Tissue	-	M	S	M	L	-	M	L	-	L	-	L	L	-	L	-
Cellulose	S	S	-	S	S	-	S	M	S	S	-	M	S	-	S	-
Phytoliths	-	-	S	-	-	-	-	M	-	S	-	M	S	-	S	-
Raphides	-	-	-	-	-	-	-	-	-	-	-	-	-	-	-	-
Starch Granules	-	-	-	-	SG	SG	SG	MTG	ST	SG	-	ST	ST	-	-	-
Red Resin	M	L	M	L	L	L	L	M	M	L	M	S	M	L	M	-
Orange/Amber Resin	S	-	S	M	S	S	S	S	S	S	-	-	-	S	S	-
Clear/Grey Resin	L	L	S	L	M	M	M	L	L	L	M	L	L	S	L	-
Charred Resin	S	S	-	-	S	L	-	S	S	L	M	S	-	-	S	S
Charcoal	M	M	S	L	M	M	L	M	M	M	S	S	S	S	M	S
Ochre	-	-	L	-	S	M	-	-	S	-	-	S	S	-	-	-
Diatoms	-	-	-	-	-	-	-	-	-	-	-	-	-	-	M	S
Identified Use-Wear																
Striations	S	S	-	S	S	S	M	M	S	S	M	L	S	S	-	-
Edge Rounding	S	S	S	S	S	S	S	M	S	S	M	S	L	M	M	L
Fractures	L	S	-	S	-	-	S	-	M	M	M	S	M	-	M	-
Surface Polish	-	L	-	M	-	-	-	-	S	S	S	M	S	L	L	S

Key: D = Dorsal, V = Ventral, S = Small quantity, M = Moderate quantity, L = Large quantity, T = transient granules, G = gelatinized or damaged granules also present

Table 5. Residue co-occurrence and inferred task association

Possible Task Association	Residue Types Identified	References
Woodworking	Woody plant tissue, cells, cellulose, bordered pits, exudate	Hardy and Garufi (1998)
Plant Processing	Plant tissue, cellulose, phytoliths, transient starch	Briuer (1976)
Boneworking	Bone collagen	Loy (1993)
Charred Woodworking	Charred resin, plant fibres, cells, cellulose, charcoal	Anderson (1980)
Root/Tuber Processing	Starch, plant tissue, cellulose, phytoliths	Fullagar et al. (1992)
Butchery	Blood, collagen, vivianite, feather*, hair*	Loy (2000)
Secondary Bone-working	Collagen, vivianite, feather*, hair*	Loy (2000)
Tertiary Bone-working	Bone collagen, ochre*	Loy and Nugent (2002)
Ceremony / Decoration	Ochre, feathers, blood, hair*	Akerman et al. (2002)
* Non-essential residue type for associated task		

performed with the tulas in the sample. These results concur with the ethnographic documentation cited above. Charcoal was also consistently observed in association with red and charred resin, which, as suggested above, indicates the wood may have been charred prior to being worked.

The presence of gelatinised starch and plant cellulose on the surface of an artefact suggests contact with heated or cooked starchy plants, such as corms, tubers, and rhizomes (Crowther 2001). The processing of cooked starch-rich foods using a tula has not been documented ethnographically, although a hafted tula may have served to remove food from a fire or similarly to remove the fleshy parts of the plant from its casing or skin. It is also plausible that gelatinised starch was deposited on the surface of the artefacts by touch-transfer.

Plant material was also identified in conjunction with small and moderates amounts of ochre. As suggested by Horne and Aiston (1924:109), replacement tula flakes were stored in hair, fur or plant fibre bags decorated with bands of ochre. Tindale (1965-1968:147) observed flakes and their wooden handles being decorated in ochre before they were traded, and Spencer and Gillen (1969:668) documented tula hafted woomeras being used as receptacles for holding material, such as feathers, ochre and blood for use as decoration or in ceremony. The transfer of ochre onto the tula surfaces may have been caused by any or all of these activities. Alternatively, this residue may have originated as an environmental additive, but this cannot be determined without conducting soil analyses. Blood, feather and hair, found in association with bone collagen, may also indicate butchery or fresh (secondary stage) bone-working.

As vivianite is formed in the presence of lipids such as animal fats (McGowan and Prangnell 2006) its presence with bone collagen further suggests that butchery and/or secondary stage bone-working was conducted with the tulas, tasks that are both reported in the ethnographic literature. The presence of granular bone collagen and bone pieces, occasionally with ochre, are indicative of dry (or tertiary stage) bone-working and/or decorative engraving. Although the working of dried bone with hafted tulas has not been observed ethnographically, the concentration of bone collagen on the artefact surfaces in association with relevant residues strongly suggests that bone-working at each stage was undertaken using the tulas.

Use-wear

Although the sample of tulas was small, obvious variation in the distal edge morphology was observed. This variation resonates with the ethnographies that suggested the curvature of the working edge determined the flake's primary application and, accordingly, the depth and width of the incision made. The extent of retouch exhibited on the tulas is also in direct accordance with ethnographic observation and archaeological research into tula reduction (Hiscock 1988; Kamminga 1982; Roth 1904). However, apart from the distal edge retouch evident on the slugs, it was difficult to conclude with any certainty which scarring was produced through manufacture or as a result of use. Similarly, the fine edge-chipping evident on the distal, lateral and proximal edges of the majority of tulas in the sample was more likely to have been produced during the final stages of reduction, as observed by Aiston (1928:127). Alternatively, and especially in view of the disturbed nature of the site, this chipping with its irregular morphology may have resulted from post-use damage (McBrearty *et al.* 1998; Moss 1983). The angles of the distal and dorsal platform edges on the tulas, however, are synonymous with the range of angles applied by Kamminga (1982:73,76), and thus, are suggested as suitable for efficient adzing and scraping.

The observation of particular types of scarring present on the tulas is also in accord with Kamminga's (1982) results concerning the use-wear produced from hafted adzing and scraping. The frequent occurrence of large step, feather and, to a lesser degree, hinge termination fractures along the distal edge and lateral margins on the majority of analysed artefacts indicates that these edges were used. This combination of termination fractures has been suggested by Kamminga (1982:69-78) to be a product of medium-light wood scraping and/or dense wood adzing. Moreover, artefacts displaying bending fractures may also have been employed for dense wood scraping. Edge rounding, striations, and surface polish are also indicative of use (Hayden and Kamminga 1979; Semenov 1964). Some edge and surface damage was observed in association with inferred use-related residue, such as resin and bone collagen, further suggesting the tulas' application.

Edge utilisation

By examining the distribution of use-related residues and wear patterns along the four 'working' edges, it was possible to infer particular functions performed by the tulas. Bone collagen was the most frequently occurring residue type observed along the proximal edge and red resin was the highest occurring residue type for the distal edge and lateral margins. Given that a predominance of plant material, cellulose, fibres, and mircofractures were observed along the distal edge in association with red resin, it may be inferred that this edge was primarily used for woodworking. Surface polish resembling Binneman's (1983:94) results for dried woodworking was also observed along the distal edge of most tulas. The severity of wear, the concentration of polish and the occurrence of red resin lessened towards the platform edge where the highest occurrence of bone collagen was observed.

It has been documented ethnographically that the degree of curvature of the working edge influenced the flake's application to a particular stage of woodworking (Aiston 1928:125; McCarthy 1967:29; Spencer and Gillen 1969:637-9). Furthermore, Kamminga (1982:77) reported that a curved working edge is necessary to lessen the impact of dynamically loaded functions such as hafted adzing. These observations support the suggestion that woodworking activities conducted with the tulas in the sample were confined to the curved perimeter edges of the distal edge and upper lateral margins. The presence along the platform edge of bone collagen with fine striations and compression ring fractures, which are generally associated with scraping (Kamminga 1982:70), suggests that bone-working activities may not require a curved edge. Such evidence also implies predetermined tula manufacture.

Evidence of hafting

As recorded in the ethnographic literature, tulas functioned primarily as hafted composite tools. Therefore, the presence of resinous material identified on all of the tulas may not only be associated with woodworking, but also hafting. Inferences concerning hafting, including the orientation of the flake in the haft and multiple hafting arrangements, were determined by analysing the nature, location and frequency of the types of resins observed. It is unlikely that all five resins types identified represent separate hafting mediums. The resin types more likely attributable to woodworking are red, charred and purple resin, and those suggested here as hafting mediums, are clear/grey and orange/amber resin.

According to ethnographic observations, both the distal and platform edges were primary working edges for adzing and scraping, having both acute edge angles adjacent to the convex ventral surface. This patterning is particularly evident on the slugs, supporting the idea that continued use of the flake necessitated the utilisation of the proximal end. Furthermore, both red and charred resin occur predominantly in linear smears perpendicular or sub-perpendicular to these edges and occasionally adjacent to or caught within edge microfractures. Linear smears, as opposed to a scattered distribution of residue, are generally indicative of the artefact's movement along the work surface and similarly through the residue. Accordingly, the specific resin patterning and location on the tula's surface may be the products of adzing or scraping red resinous wood. The presence of charred resin may indicate prior charring of the wood to facilitate smoothing and shaping, as recorded in the ethnographic literature (Thomson 1975:98; Tindale 1972). Tree species common to Camooweal and the surrounding area that produce a red resinous exudate are desert or western bloodwood (*Corymbia terminalis*), red bloodwood (*Eucalyptus gummifera*), and beefwood (*Grevillea striata*).

Clear/grey resin was observed in similar quantities to red resin, although differences in appearance and location on the tulas' surfaces suggest this resin is more likely to be the remnants of hafting rather than woodworking. Unlike the red and charred resins, clear/grey resin did not display evidence of direction of application in the form of striations or linear smearing but generally occurs in dense, cracked patches and scattered deposits across the artefact surface. This pattern suggests the resin was applied to the surface of the artefact directly such as a hafting medium rather than resulting from artefact use. Comparison with samples of spinifex resin (*Triodia pungens*) obtained from Camooweal revealed similarities in appearance, suggesting clear/grey resin may have originated from this species. In comparison, orange/amber resin was observed in similar deposits and patterning along the lateral margins and percussion bulb but in much smaller quantities and less frequently across the sample. Orange/amber resin is similar in appearance to grass tree resin (*Xanthorrhoea* sp.), which has also been ethnographically observed as a hafting medium (Gott and Conran 1991). Although its presence or use has not been documented at Camooweal, grass tree resin may have been traded into the area.

The distribution of clear/grey resin across the tula surfaces, in association with use-related residues and wear patterns on all perimeter edges, strongly suggests that these artefacts were hafted in more than one orientation to the handle. This distribution concurs with the ethnographic literature where it was noted that a combination of hafting arrangements were employed at different stages of tula manufacture, (re)sharpening, and use (Tindale 1965-1968:154).

The resin patterning on 999.428, 66800-3, 62600-1, 60400-2 and 9800-4 indicates that these tulas were hafted at some point to utilise the proximal end, given the concentration of clear/grey resin along the lower mid section and distal edge. Artefacts 1100-4a and 1600-3 display evidence of a distal hafting arrangement, and artefacts 22300-4, 34700-3, 1100-4b were once hafted to utilise the lateral margins. The remaining tulas in the sample display either a fairly even distribution of clear/grey resin across the surface, or insufficient quantities to infer the orientation of the flake in the haft. An increase in frequency of hafting resin toward a particular edge may indicate the most recent hafting arrangement prior to discard, as subsequent use of a rehafted flake may remove some, if not the majority, of resin from prior hafts. Prior hafting is indicated by the presence of use-related residues and wear patterns on more than one edge in association with the proposed hafting resin. A scattered distribution of resin across the artefact surface in association with use-related residues and wear patterns along all edges suggests that a combination of hafting arrangements were employed. In the absence of hafting resin, it may be inferred that artefacts were either used unhafted or that exposure to post-depositional processes removed this evidence.

Inferred functions and task associations

Integration and comparison of results of the macroscopic, low-magnification and high-magnification examinations of the tulas enabled their possible hafting arrangements, functions and task associations to be determined (see Table 6). These results suggest that all 16 tulas were employed not only for woodworking but also for butchery and secondary and tertiary bone-working. Possible woodworking tasks include the manufacture of water carriers, digging bowls and other

Table 6. Inferred hafting arrangement, functions and task associations

Artefact number	Inferred hafting arrange.	Inferred function	Wood-work.	Plant process.	Bone-work.	Charred wood-work.	Tuber process.	Butchery	Secondary bone-working	Tertiary bone-working	Ceremony/ decoration
20100-7a	Inconclusive	Scraping	Yes	Yes	Yes	Yes	Yes			Yes	Yes
22300-4	Lateral	Adze / Scrape	Yes	Yes	Yes	Yes		Yes	Yes	Yes	
999.428	Proximal	Scraping	Yes	Yes	Yes	Yes			Yes	Yes	Yes
34700-3	Lateral	Adze / Scrape	Yes	Yes	Yes	Yes	Yes	Yes	Yes		
66800-3	Proximal	Adze / Scrape	Yes	Yes	Yes	Yes	Yes		Yes	Yes	
60400-2	Proximal	Adze / Scrape	Yes		Yes	Yes		Possible		Yes	
20100-7b	Inconclusive	Scraping	Yes	Yes	Yes		Yes	Possible	Possible	Yes	
1100-4a	Distal	Scraping	Yes		Yes	Yes		Possible			Yes
777-7	Prox /Dist	Adze / Scrape	Yes		Yes	Yes		Possible			Yes
1600-3	Distal	Scraping	Yes	Yes	Yes	Yes				Yes	
1100-5	Inconclusive	Scraping	Yes		Yes	Yes				Yes	
62600-1	Proximal	Scraping	Yes	Yes	Yes	Yes	Yes			Yes	
30900-2	Lat / Prox	Adze / Scrape	Yes		Yes	Yes				Yes	
31800-1	Prox / Dist	Adze / Scrape	Yes	Yes	Yes	Yes		Yes	Yes	Yes	
1100-4b	Lateral	Scraping	Yes		Possible						
9800-4	Proximal	Scraping	Yes		Yes	Yes				Yes	

items from local tree species such as *Corymbia terminalis*, *Eucalyptus gummifera*, and *Grevillea striata*, The tulas were also hafted to the handle in several orientations, possibly with spinifex (*Triodia* sp.) resin or grass tree resin (*Xanthorrhoea* sp.). Alternating hafting arrangements allowed for the use of all 'working' edges, thereby increasing the inherent functionality, adaptability and utilitarian 'use-life' of the tula.

Why analyse surface collected artefacts?

Some of the most abundant and diverse archaeological information in Australia lies on the surface (Fanning and Holdaway 2001:668; Hiscock and Hughes 1983:87). Nevertheless, a large proportion of this information remains hidden beneath the labels of 'disturbed', 'complex' and 'exposed'. Traditionally, these views have fuelled the perception that surface assemblages lack the integrity of subsurface archaeological deposits. Consequently, their full potential as an important component of the archaeological record has not always been realised (Fanning and Holdaway 2001:668; Lewarch and O'Brien 1981:297). Much of the rationale behind these statements was based on the assumption that subsurface materials are 'fossilised' records of the past, and that post-depositional disturbance processes acting upon surface material are apparently unique to contemporary ground surfaces and do not affect subsurface deposits (Fanning and Holdaway 2001; Lewarch and O'Brien 1981:312). However, it has been more than two decades since Dunnell and Dancy (1983) made the observation that all archaeological deposits now buried were once 'exposed' on the surface, albeit for a lesser period of time. In addition to factors that affect artefact use, curation, reuse and discard, post-depositional natural and cultural processes continually modify the content, condition and patterning of archaeological deposits (Lewarch and O'Brien 1981:312; Schiffer 1987). As a result few artefacts, surface or subsurface, are *in situ* in the conventional sense of the term (Foley 1981:170). Based on the consistent indications of tool use on both the labelled excavated and surface collected artefacts presented above, we argue traces of use are as likely to survive post-depositional disturbance processes in either context.

CONCLUSION

The results of the initial residue and use-wear analyses of 23 stone artefacts from the Georgina River Bridge site, Camooweal, suggest similar uses for certain tool types, whether they had been excavated or surface collected. The subsequent detailed examination of 16 surface collected tulas identified traces of use and a wide variety of residues indicative of multiple hafting episodes and specific tasks performed, included woodworking, plant processing, bone-working and butchery. Examination of these results in light of ethnographic documentation and previous archaeological research suggests the manufacture of the tula was predetermined to achieve certain tasks, and that these artefacts may have been used for several tasks before being discarded. The consistent indications of tool use found in both studies strongly supports the application of use-wear and residue analyses as suitable methods for examining stone artefacts recovered from complex and disturbed surface sites.

ACKNOWLEDGEMENTS

We are sincerely grateful for the inspiration and guidance always generously provided by Tom Loy. Thank you to Colin Saltmere and the Indjilandji People from the Camooweal area for access to the artefacts and permission to publish the results of the preliminary research. In addition, we acknowledge Ann Wallin for permission to cite from the Archaeo Cultural Heritage Services' report 'Georgina River Bridge Cultural Heritage Project'. We also thank members of Tom's Group for encouragement and support, and especially Dr Gail Robertson for comments on drafts of this paper. Funding for the research was provided by the School of Social Science at the University of Queensland, and Archaeo Cultural Heritage Services.

REFERENCES

Aiston, G. 1928. Chipped stone tools of the Aboriginal tribes east and north-east of Lake Eyre, South Australia. *Papers and Proceedings of the Royal Society of Tasmania* 123-131.

Aiston, G. 1929. Method of mounting stone tools on Koondi: tribes east and north-east of Lake Eyre. *Papers and Proceedings of the Royal Society of Tasmania* 44-46.

Akerman, K., R. Fullagar and A. van Gijn 2002. Weapons and *wunan*: production, function and exchange of Kimberley points. *Australian Aboriginal Studies* 1:13-42.

Anderson, P. 1980. A testimony of prehistoric tasks: diagnostic residues on stone tool edges. *World Archaeology* 12:181-194

Archaeo Cultural Heritage Services and Dugalunji Aboriginal Corporation. 2002. *Georgina River Bridge Cultural Heritage Project*. Report to the Department of Main Roads, Queensland.

Binneman, J. 1983. Microscopic examination of a hafted tool. *South African Archaeological Bulletin* 38:93-95.

Briuer, F. 1976. New clues to stone tool function: plant and animal residues. *American Antiquity* 41:478-484.

Campbell, T. and R. Edwards 1966. Stone implements. In B.C. Cotton (ed.) *Aboriginal Man in south and central Australia, Part 1*, pp. 159-220. Adelaide: South Australian Government Printer.

Canti, M. 1998. The micromorphological identification of faecal spherulites from archaeological and modern materials. *Journal of Archaeological Science* 25:435-444.

Crowther, A. 2001. Pots, Plants and Pacific Prehistory. Unpublished B.A. Honours thesis. St Lucia: School of Social Science, The University of Queensland.

Davidson, D.S. 1935. Archaeological problems of northern Australia. *The Journal of the Royal Anthropological Institute of Great Britain and Ireland* 65:145-186.

Davis, E. 1975. The "exposed archaeology" of China Lake, California. *American Antiquity* 40:39-53.

Dunnell, R.C. and W. Dancey 1983. The siteless survey: a regional scale data collection strategy. *Advances in Archaeological Method and Theory* 6:267-287.

Ebert, J. 1992. *Distributional Archaeology*. University of New Mexico Press, Albuquerque.

Etheridge, R. 1891. Notes of Australian Aboriginal stone weapons and implements. *Proceedings of the Linnaean Society of New South Wales* 6:31-43.

Fanning, P. and S. Holdaway 2001. Stone artefact scatters in western NSW, Australia: geomorphic controls on artefact size and distribution. *Geoarchaeology* 16:667-686.

Foley, R. 1981 . Off-site archaeology: an alternative approach for the short-sited. In G. Hodder, G. Isaac and N. Hammond (eds) *Patterns of the Past, Studies in Honour of David Clarke*, pp. 157-183. Cambridge University Press, Cambridge.

Fullagar, R., B. Meehan and R. Jones 1992. Residue analysis of ethnographic plant-working and other tools from Northern Australia. *Préhistoire de l'Agriculture : Nouvelle Approches Expérimentales et Ethnographiques*. Monographie du CRA No. 6, pp.40-53.

Gott, B. and J. Conran 1991. *Victorian Koorie Plants: Some Plants Used by Victorian Koories for Food, Fibre, Medicines and Implements*. Hamilton: Yangennanock Womens Group.

Gould, R. 1978. The anthropology of human residue. *American Anthropologist* 80:815-835.

Gould, R., K. Koster and A. Sontz 1971. The lithic assemblage of the Western Desert Aborigines of Australia. *American Antiquity* 36:149-169.

Greenfield, H.J. 2000. Integrating surface and subsurface reconnaissance data in the study of stratigraphically complex sites: Blagotin, Serbia. *Geoarchaeology* 15:167-201.

Hardy, B.L. and G.T. Garufi 1998. Identification of woodworking on stone tools through residue and use-wear analysis: experimental results. *Journal of Archaeological Science* 25:177-184.

Hayden, B. and J. Kamminga 1979. An introduction to use-wear: the first CLUW. In B. Hayden (ed.) *Lithic Use-wear Analysis*, pp. 1-13. New York: Academic Press.

Hiscock, P. 1988. A cache of tulas from the Boulia District, Western Queensland. *Archaeology in Oceania* 23:60-70.

Hiscock, P. 1994. Technological responses to risk in Holocene Australia. *Journal of World Prehistory* 8:267-292.

Hiscock, P. and P.J. Hughes 1983. One method of recording scatters of stone artefacts during site surveys. *Australian Archaeology* 17:87-98.

Horne, G. and G. Aiston 1924. *Savage Life in Central Australia*. London: Macmillan.

Kamminga, J. 1982. *Over the edge: Functional Analysis of Australian Stone Tools*. Occasional Papers in Anthropology No 12. St. Lucia: The University of Queensland Printery.

Lewarch, D. and M. O'Brien 1981. The expanding role of surface assemblages in archaeological research. *Advances in Archaeological Method and Theory* 4:297-342.

Loy, T.H. 1983. Prehistoric blood residues: detection on tool surfaces and identification of species of origin. *Science* 220:1269-1271.

Loy, T.H. 1993. The artefact as a site: an example of the biomolecular analysis of organic residues on prehistoric tools. *World Archaeology* 25(No 1):44-63.

Loy, T.H. 2000. *Analysis of six pottery sherds from the west coast of Guam.* Report to Micronesian Archaeological Research Services, Guam.

Loy, T.H. and S.J. Nugent 2002. *Residue analysis of a portion of the Georgina River Bridge collections, Camooweal, Queensland.* Report to Archaeo Cultural Heritage Services Pty. Ltd. Brisbane.

McBrearty, S., L. Bishop, T. Plummer, R. Dewar and N. Conard 1998. Tools underfoot: human trampling as an agent of lithic artefact edge modification. *American Antiquity* 63:108-129.

McCarthy, F.D. 1967. *Australian Aboriginal Stone Implements.* Sydney: Museum Trust.

McGowan, G. and J. Prangnell. 2006. The significance of vivianite in archaeological settings. *Geoarchaeology* 21:93-111.

McNiven, I. 1993. Tula adzes and bifacial points on the east coast of Australia. *Australian Archaeology* 36:22-31.

Mitchell, S.R. 1949. *Stone-age Craftsmen: Stone Tools and Camping Places of the Australian Aborigines.* Melbourne: Tait Book Co. Pty. Ltd.

Moore, M.W. 2003. Flexibility of stone tool manufacturing methods on the Georgina River, Camooweal, Queensland. *Archaeology in Oceania* 38:23-36.

Moss, E.H. 1983. Some comments on edge damage as a factor in functional analysis of stone tools. *Journal of Archaeological Science* 10:231-242.

Orr, D.M. and W.E. Holmes 1990. Mitchell grasslands. In G.N. Harrington, A.D. Wilson and M.D. Young (eds) *Management of Australia's Rangelands*, pp. 241-254. Melbourne: Division of Wildlife and Ecology, CSIRO.

Raven, P.H., R.F. Evert and S.E. Eichorn 1999. *Biology of Plants.* New York: W. H. Freeman and Company.

Rossignol, J. and L. Wandsnider 1992. *Space, Time and Archaeological Landscapes.* New York: Plenum Press.

Roth, W.E. 1897. *Ethnological Studies Among the North-west-central Queensland Aborigines.* Brisbane: Government Printer.

Roth, W.E. 1904. *Domestic Implements, Arts and Manufactures. North Queensland Ethnography: Bulletin No. 7.* Brisbane: George Arthur Vaughan, Government Printer.

Schiffer, M.B. 1987. *Formation Processes of the Archaeological Record.* Albuquerque: University of New Mexico Press.

Semenov, S. 1964. *Prehistoric Technology.* Cory, Adams and Mackay, London.

Spencer, B. and F.J. Gillen 1969. *Northern Tribes of Central Australia.* Oosterhout: Anthropological Publications.

Sullivan, A.P. 1998. *Surface Archaeology.* Albuquerque: University of New Mexico Press.

Thomson, D.F. 1964. Some wood and stone implements of the Bindibu tribe of central Western Australia. *Proceedings of the Prehistoric Society* 30:400-422.

Thomson, D.F. 1975. *Bindibu Country*. Melbourne: Thomas Nelson.

Tindale, N.B. 1965-1968. Stone implement making among the Nakako, Ngadadjara and Pitjandjara of the Great Western Desert. *Records of the South Australian Museum* Vol 15:131-164.

Tindale, N.B. 1972. The Pitjandjara. In M.G. Bicchieri (ed.) *Hunters and Gatherers Today*, pp. 217-268. New York: Holt Rinehart and Winston Inc.

Wilkinson, T.J. 2001. Surface collection techniques in field archaeology. In D.R. Brothwell and A.M. Pollard (eds) *Handbook of Archaeological Sciences*, Chichester: Wiley and Sons Ltd.

17

Starch residues on grinding stones in private collections: a study of morahs from the tropical rainforests of NE Queensland

Judith Field[1], Richard Cosgrove[2], Richard Fullagar[3] and Braddon Lance[4]

1. Australian Key Centre for Microscopy and Microanalysis, F09 and
School of Philosophical and Historical Inquiry
The University of Sydney
N.S.W. 2006 Australia
Email: judith.field@emu.usyd.edu.au

2. Archaeology Program, School of Historical and European Studies
La Trobe University
Victoria 3086 Australia

3. Scarp Archaeology
25 Balfour Rd
Austinmer, NSW 2515 Australia

4. Department of Statistics
Macquarie University
North Ryde NSW 2109 Australia

ABSTRACT

Morahs are incised grinding stones from the tropical rainforests of Far North Queensland. They are made from grey slate, are roughly ovate to rectangular in shape, and have distinctive incised parallel grooves running transversely across the body of the stone. The region in which they are found is also known for the processing of toxic starchy plants by Aborigines. The process involves a relatively complex processing schedule, including cooking, pounding and leaching before consumption. Ethnographic studies have documented the processing of a number of rainforest species with starchy kernels in which morahs may have been used for pounding these kernels before leaching. A selection of morahs from private collections were analysed to determine their potential for starch residue studies. The results show that incised grooves act as residue traps for starch. In some cases the starch recovered from these grindstones enabled starch identifications of economically important endemic rainforest species, particularly *Beilschmiedia bancroftii* (Yellow walnut) and *Endiandra insignis* (Hairy Walnut). The uneven surface created by the incised grooves may facilitate the breakup of the starchy kernels, and this proposal is supported by use-wear studies on similar artefacts where soft plant processing is indicated.

KEYWORDS

morah, grindstone, slate, starch residues, North Queensland rainforests, archaeology

INTRODUCTION

Grindstones from Australia are known to be used for a variety of functions including plant processing, preparation of ochre and also for the maceration of foods such as lizards and cats

(Gould 1980; Mitchell 1848). Grindstones are ubiquitous in the more arid parts of the Australian continent where they are generally manufactured from sandstone and are associated with grass seed grinding (Fullagar and Field 1997; Fullagar *et al.* 2008; Smith 1985; Tindale 1977). Grindstones are also common in the tropical rainforests of far North Queensland and have often been found by farmers ploughing paddocks (Woolston & Colliver 1973). One grindstone fragment has also been recovered from an occupation horizon excavated at Urumbal Pocket at Koombooloomba Dam in the Wet Tropics World Heritage Area (Cosgrove *et al.* 2007). One particular type of grindstone, referred to as the morah stone, comes from a well defined area from Tully in the south, to Cairns in the north and west to the Ravenshoe area on the Atherton Tablelands (Woolston and Colliver 1973:117; see Figure 1).

Morah stones (sometimes referred to as graters by locals) are made from grey slate, a soft and brittle stone that is available locally. They are distinguished from other grindstones by a series of incised grooves running perpendicular to the axis of the artefact (Figure 2). Maisie Barlow, a Jirrbal elder from Ravenshoe, relates that the incised grooves were made with quartz pieces, though bone points have also been suggested as tools for this purpose.

Morah stones occur in a region where the processing of toxic starchy plants is common (see Pedley 1993), and their use may be tied to the extended processing of these economically important foods. Morah stones are described by an unidentified Nutjen (Ngatjin) woman to be used

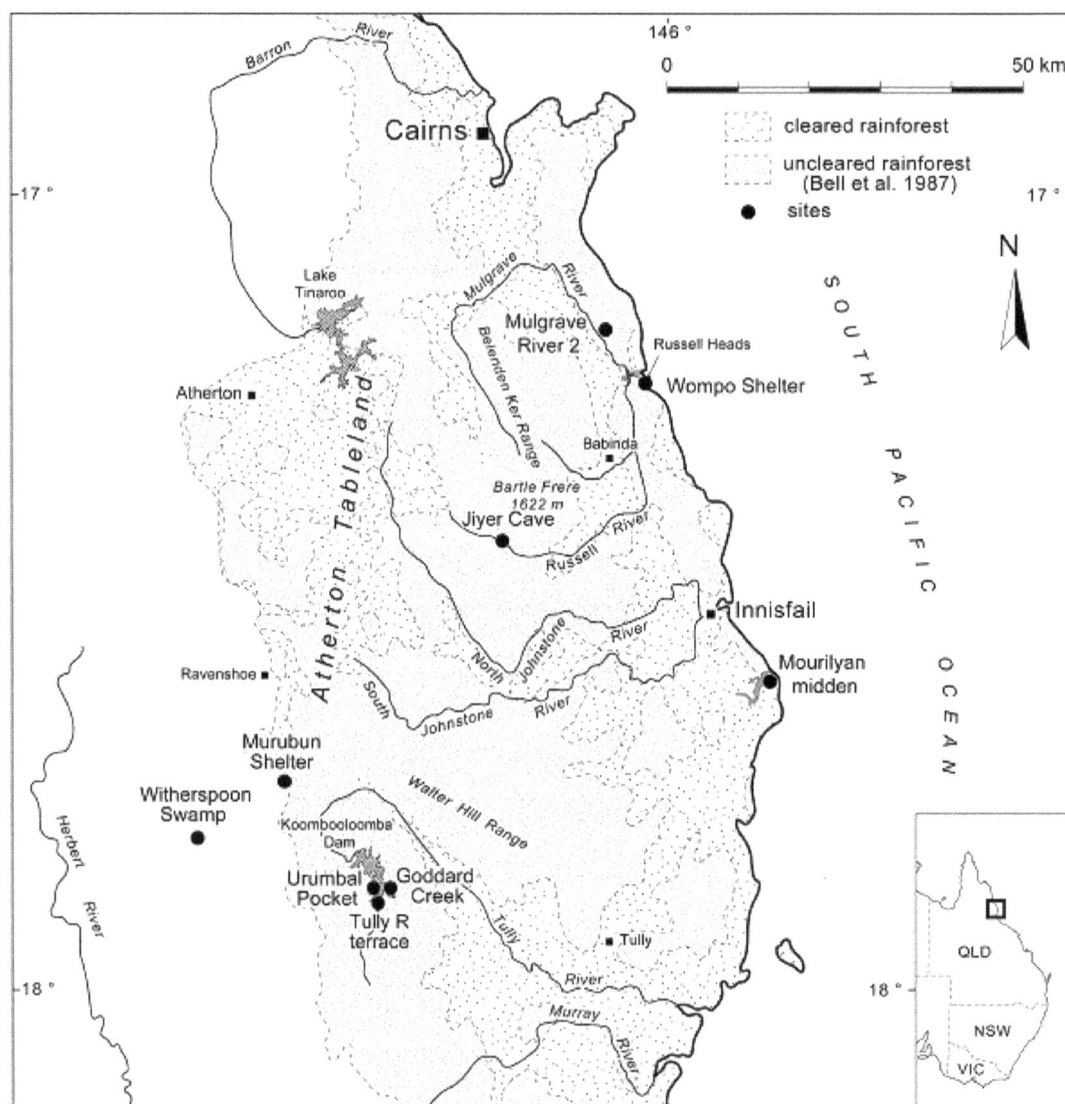

Figure 1. The study area in far North Queensland showing places mentioned in text. (Illustration: R. Frank, after Cosgrove *et al.*, 2007).

Figure 2. Incised slate grinding stone fragment from North Queensland. Darker patch in centre right of the grindstone is the location where the residue sample was collected. Note the parallel grooves running across the stone surface. Damage to the far left end of the stone is likely to be the result of damage from tractor blades prior to recovery. (Photo: J. Field, Courtesy of Pat and Alverio Croatto).

in a 'rolling crushing' motion rather than grinding (Woolston & Colliver 1973). These were said to be used to crush zamia and walnuts. Apart from these two references, the use of morahs has not been documented elsewhere, either in the early explorer journals or in ethnographic records about the area (e.g. Ferrier 1999:80). Harris (2006:S69) has claimed that their use for pounding toxic plants is supported by 'testimony of Aboriginal people who retained knowledge of their former use'. The upper stones are known as 'moogi', the name given by the same informant describing the use of morahs referred to by Woolston and Colliver (1973:118) and are documented in the Australian and Queensland Museum collections. Moogi are usually a granite raw material, which are much harder than the slate of the morah stones.

Horsfall (1987:209-211) measured over 58 morahs (22 whole and 36 fragments) and found that none of them were more than 65 mm thick, though in this study the thickness of the grindstones did not exceed 30 mm. The thickness of the grinding stones is argued to be a function of the slate raw material, which tends to cleave into relatively thin plates. Some of the morahs examined by Horsfall had incised grooves on both surfaces but most had incisions on only one surface. Used surfaces tend to be flat to concave, consistent with documented grindstone wear patterns (see Field & Fullagar 1998; Smith 1985). Very few complete morahs are known and the largest morah we have observed was nearly 50 cm in length, and while broken, was mostly complete and in a private collection near Innisfail. Morahs are also known from the tablelands and recent surveys identified numerous examples during a survey of the foreshores of Koombooloomba Dam at low water levels, which is adjacent to the now flooded channels of the Tully River (Cosgrove *et al.* 2007; see Figure 1).

This paper presents the results of a study examining residue preservation on incised slate grinding stones (morahs) from the Far North Queensland rainforest that are held in private collections. Combined with overall morphology and related use-wear studies, the function of morah stones will be discussed.

METHODS

The artefacts examined in this study were in private collections held at Innisfail and Babinda - on the coastal strip south of Cairns. The artefacts were generally stowed under houses or in sheds. These artefacts were collected by farmers when they were turned up during ploughing of fields in preparation for planting sugar cane. As such, some have been damaged and some had been refitted and glued together. Residue samples were generally collected in the field at the site where the collections were held.

Unused surfaces were smooth and featureless and not incised. Ethnographic observations of the processing of starchy foods indicate that all surfaces of the grindstone will generally become covered in starchy material (e.g. Figure 3) and any flaws, or incised surfaces are likely to act as residue traps. Combined with residue studies and morphological analyses, use-wear analysis is an informative adjunct for identifying patterns of wear and probable tasks (Fullagar *et al.* 1996). As use-wear analyses were not possible for these particular artefacts, the outcome of the analysis of similar artefacts by Richard Fullagar is used for comparison.

Eleven artefacts were sampled using a variable volume pipette to dispense c. 200 μl aliquots of distilled water to the used surface of the grindstone. A nylon pipette tip was used to scrape along the incised grooves to dislodge residues. The water plus residue was then transferred to a 1.5 ml microcentrifuge tube for transport back to the laboratory. As the size of the samples was

Figure 3. Biddy Simon processing water lily root in the Kimberley Region of north western Australia. She is using a sandstone grindstone with a small upper stone, or muller (to the right). The use of water in the milling of these seeds ensures that the starch paste is distributed over most surfaces and onto the surrounding ground surface. (Photo: Richard Fullagar/Lesley Head collection).

very small, they were not submitted for further processing (i.e. heavy liquid separation) in order to conserve the sample and reduce the loss of material. The water plus residue was centrifuged to concentrate the sample which was then were mounted on slides using either Permount or Karo mounting media. Slides were scanned for starch using a Zeiss Axioskop2 transmitted brightfield microscope with polarizing filters and Nomarksi optics. Images were captured using a Zeiss HRc digital camera and archived with Zeiss Axiovision software.

The maximum length of all starch grains was measured (using the Axiovision software) and the results presented in a boxplot as counts per sample against the modern comparative reference collection. Maximum length has been determined as the best attribute for the initial filtering of results in order to exclude those species which have not contributed to the assemblage (Field and Lance, unpublished results). This is especially important when large comparative reference databases are being used and allows the analyst to quickly sort the material into a manageable form for further investigation. The North Queensland comparative starch reference collection is comprised of both field collected material and Herbarium voucher specimens, the latter courtesy of the Atherton Tropical Herbarium. Modern comparative reference materials were collected from a range of known economically important species from the Atherton Tablelands and the coastal strip south of Cairns (see Pedley 1993). In order to provide a representative sample of starch grains, the comparative sample consists, in most cases, of measurements (from ≥100 granules) from three different specimens in order to document the range and variety of starches produced by each species.

RESULTS

Of the eleven morahs that were sampled for this analysis, one grindstone yielded no starch at all, while the remainder produced variable amounts of starch (Table 1; Figure 4). Five grindstones yielded over 30 grains each, allowing a very good estimate of the variability in the original population, assuming all extracted starch derives from a single taxon. Assuming an homogeneous sample, the estimate of the mean and variance is improved incrementally less as the sample size increases, and this commonly plateaus most noticeably around a sample size of 25-30.

Table 1. Starch grain counts from the eleven sampled grindstones

CROATTO	n	MAXWELL	n	STAGER	n	STAGER	n	STAGER	n
Sample	n	Sample	n	Sample	n	Sample	n	Sample	n
CU4/1	103	WC G1/1	0	ST1/1	0	RS05031-1	11	RS05035-1	0
CU4/2	70	WC G1/2	0	ST1/2	1	RS05031-2	15	RS05035-2	12
CU4/3	123					RS05031-3	16	RS05035-3	3
CR2/1	131			ST2/1	0	RS05032-1	3	RS05034-1	0
CR2/2	97			ST2/2	1	RS05032-2	13	RS05034-2	3
CR2/3	79			ST2/3	0	RS05032-3	31	RS05034-3	0

Of the three collections sampled, the Stager and Croatto material was productive, with the former providing variable counts. Of note are the two Croatto grindstones (CR2 and CU4) which yielded significant numbers of starch grains. Three of the Stager collection also produced significant numbers of starch grains (as a combined sample) and these were compared to the comparative reference collection for maximum size measurements. The maximum length of all starch grains was recorded and the results presented in a boxplot (Figures 5 and 6). Starch counts are shown above each grindstone sample number and for the individual species of reference material and are also presented in Table 1. The Croatto samples are consistent with *B. bancroftii* and *E. insignis*. While the whisker plots overlap with three other species, these are more likely to contribute only a small percentage of the sample, providing an indication of other species being processed. Some starch grains are very distinctive and require very few grains in order to estimate the likely species of origin - for example tubers and some grasses - however for the species likely to be represented here such as *Cycas media* and *B. bancroftii* the overlap in size and similarities in morphology demand a larger sample for examination (see Figure 7).

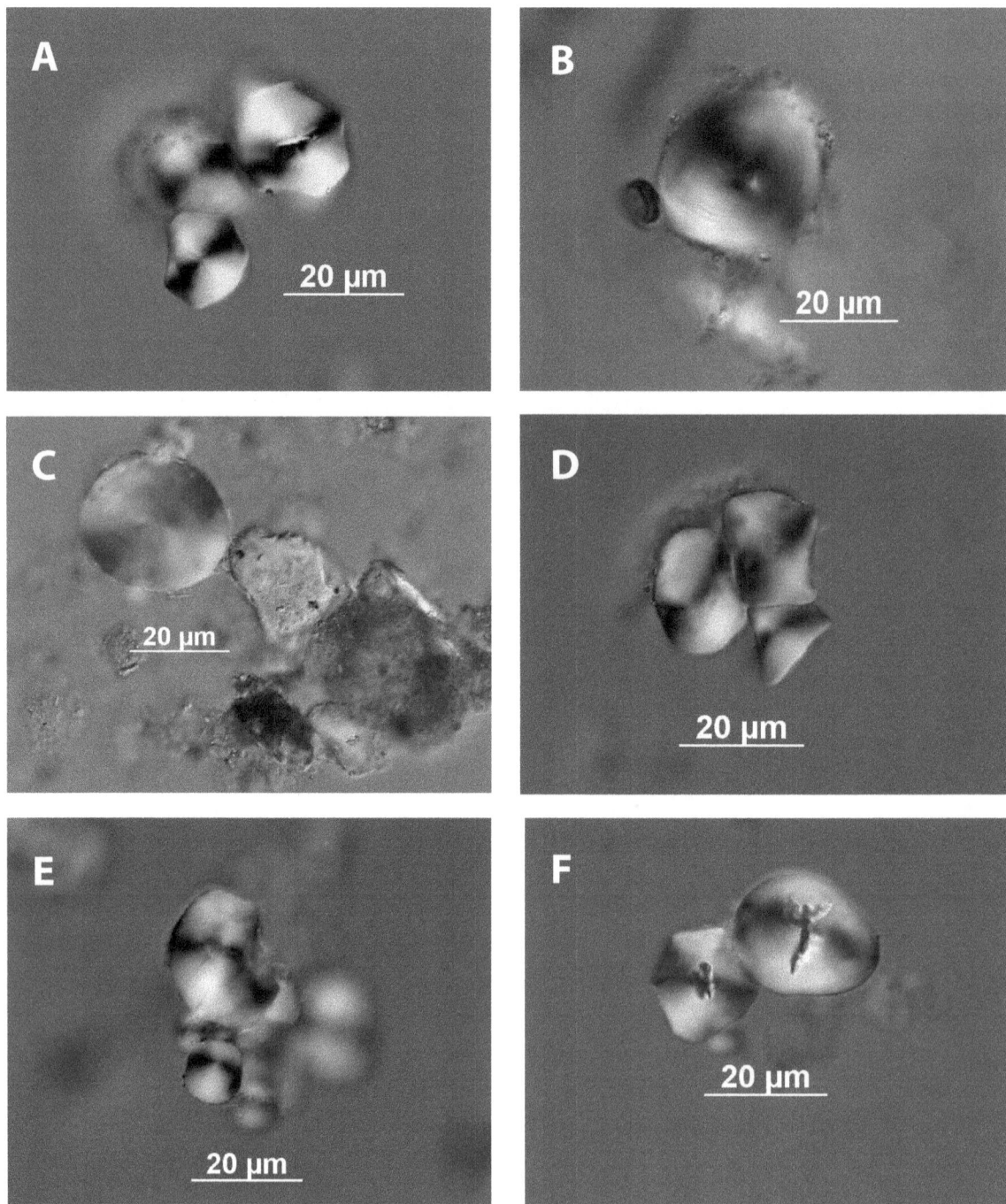

Figure 4. Starch grains identified on the surface of Morah grindstones which were sampled in private collections from North Queensland. A-D Stager Collection; E-F Croatto Grindstones. Many of the starch grains are facetted and some appear to exhibit damage from grinding. Fissures are also a common feature of many of the starch grains, e.g. F. Faceting and fissures at the hilum are common features of the Yellow Walnut and the Hairy Walnut. The processing of both species on these morahs is indicated by the maximum length measurements. All images collected on a Zeiss Axioskop II brightfield microscope with Nomarksi optics and a Zeiss HRc digital camera.

DISCUSSION

Grinding stones are durable formal tools that are commonly found as surface finds across the Australian continent. They are made of varying raw materials and, at least for sandstone grindstones, may be associated with restricted access quarries and traded for some considerable distances (McBryde 1997; Mulvaney 1998). Grinding stones are less common in the temperate margins of the Australian continent, except perhaps for far North Queensland rainforests where

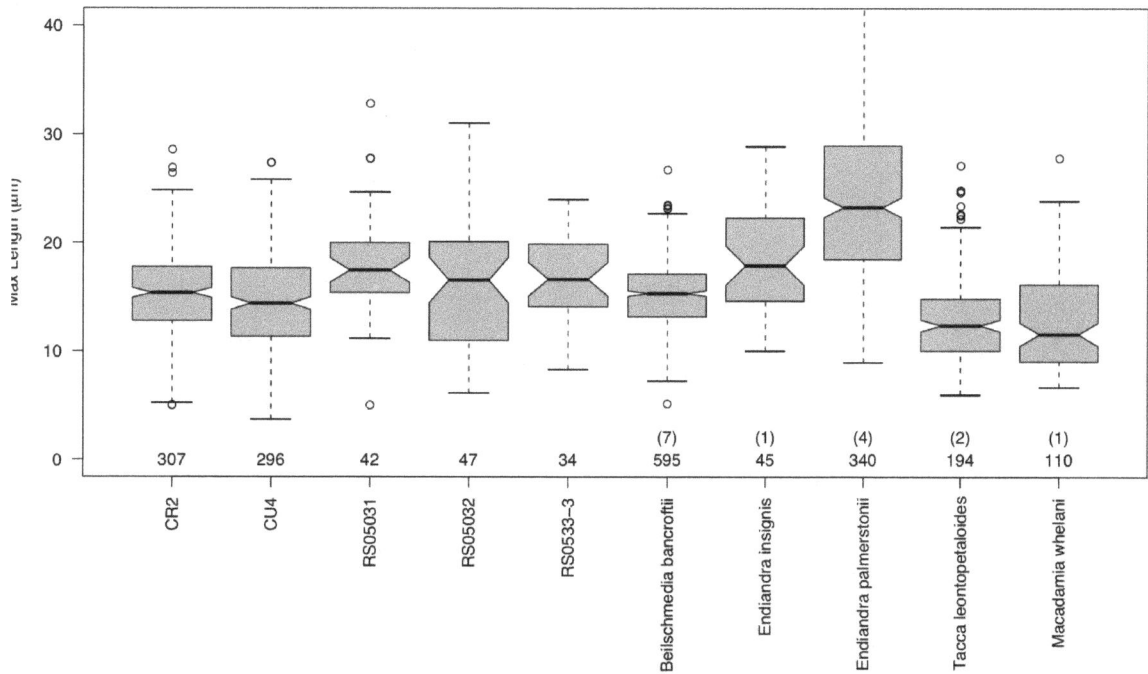

Figure 5. Boxplot of maximum lengths of starch for five morahs compared to modern reference collections. Note the two grindstones from the Croatto collection yielded significant starch counts, which are consistent with a predominant use of the Yellow Walnut, or *Beilschmiedia bancroftii*, but also falls within the parameters of the Hairy Walnut, *Endiandra insignis*.

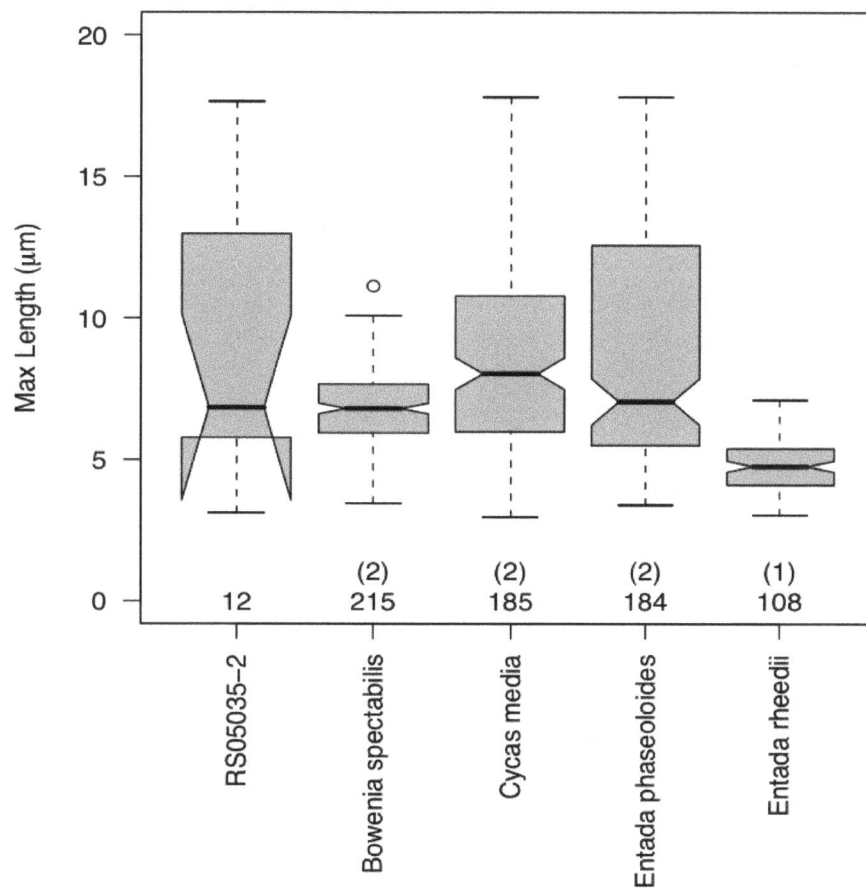

Figure 6. Boxplot of counts for a morah stone (RS05035-2) from the Stager collection. While the starch counts are low, the distinct difference in average size between this and those presented in Figure 5 indicated a different set of plants may have been processed with this implement, perhaps including one or all of those presented here.

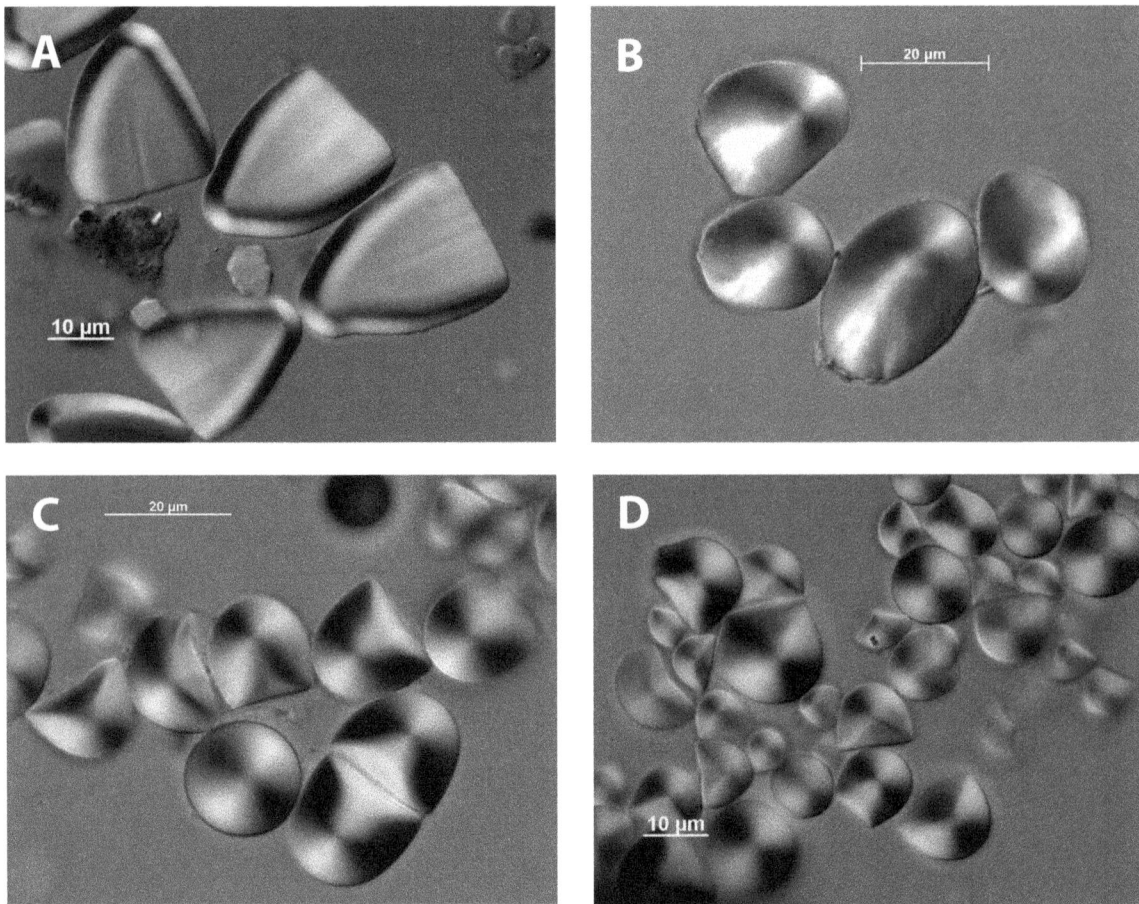

Figure 7. Examples of starch types from economically important plants of Far North Queensland rainforest and surrounds. A. *Dioscorea bulbifera*, or Hairy Yam, a tuber. B. *Endiandra palmerstonii*, Black Walnut, C. *Beilschmiedia bancroftii*, Yellow Walnut; D. *Cycas media*, Cycad. Note the morphological similarities between *Cycas media* and the Yellow walnut, though these can be separated on the basis of size measurements. (Photos: Nomarski optics with Axioskop 2 microscope and Zeiss HRc digital camera). (Photos: J. Field).

the morah or incised grinding stone is a common feature of the archaeological record (Cosgrove 1996; Cosgrove *et al.* 2007). Demonstrating their use for processing toxic starchy plants relies on the preservation of residue associated with use on the surfaces of these implements (see Cosgrove 2005; Harris 2006). Barton (2007) has reported the survival of starch on museum collections from tropical New Guinea, providing supporting data not only for the inferred use of particular artefacts, but also the survival of these microfossils in collections that are sometimes 'cleaned' and curated, not necessarily with the preservation of use-related residues in mind. In this study, the preservation of starch residues on morah stones held in private collections was investigated. These artefacts are generally stored on or adjacent to the locations where they were found and have not been curated or cleaned, as would perhaps occur in the museum collections described by Barton. The morahs examined here were found stacked together, covered in the sediment from which there were recovered and stored in 'dry' conditions.

This preliminary study was focussed on determining whether starch was preserved on the used surfaces of morah stones. The results are variable, but indicate that the residue traps provided by the incisions on the surface of morahs hold good potential for these types of studies. Combined with the morphology and associated use-wear patterns described below, there is compelling evidence to suggest that morah stones were used as grinding stones. On similar artefacts examined by Richard Fullagar, fine striations were found to occur at right angles to the incisions on the surface; they exhibit abrasive smoothing from grinding; and develop limited polish (polishes appear to develop on hard quartz grains within the stone matrix) - this is consistent with plant working. It has also been noted that undertaking use-wear studies on this type of stone is problematic as the raw material is very soft and residue films obscure the surface. Nonetheless, taken together, the morphology, residues and information about use-wear on similar implements are consistent with plant processing, in this case toxic starchy plants, most probably dominated by the Yellow Walnut (*B. bancroftii*) and also Hairy Walnut (*E. insignis*).

Establishing the use of morahs for the processing of toxic starchy plants has implications for our understanding of the timing and nature of permanent settlement of rainforest environments. Long term use of rainforest environments by people may hinge on access to a suite of toxic starchy plants (see Cosgrove *et al.* 2007). The development of the technologies for processing plants such as the Black Walnut, Yellow Walnut and Black Pine may have been transferred from the known methods for processing cycads, the antiquity of which has been reported as 13,000 years in Western Australia (Smith 1982, 1996). *Cycas media* is a common feature of the vegetation in the dry country on the margins of the rainforest, and the methods for processing these and the rainforest species are essentially the same – baking the starchy kernels for c. 6 hours followed by pounding to a paste, then leaching in running water for several days. The second step in this process, pounding to a paste, can also be achieved by grating with a shell (Pedley 1993).

In these studies, three separate lines of evidence are used to determine function (see Fullagar *et al.* 1996). Firstly, a technological study indicating that the morphology of morahs is typical of grinding stones – they are flat, with one or both surfaces showing evidence of use. The used surface is identified by the presence of concave areas where the surface has been worn down by continuous grinding. Peculiar to morahs are the incised parallel lines that run perpendicular to the maximum length of the stones.

Use-wear studies, the second stage in the functional analysis, indicate that these may have been made by quartz flakes as evidenced, by the sharp angular cuts and cross sectional characteristics. The raw material, slate, is a very soft stone and as such the incisions would have been relatively easy to produce. The incisions in the slate grindstones may serve the same purpose as the pecked surfaces observed on sandstone grindstones. In the latter case it is a method used to rejuvenate the grindstone surface to facilitate the breaking up of grass seed husks. The incised surfaces are always the used surfaces on the morahs.

The residue study which was the focus of this research has shown that the incised surfaces provided ideal locations for preservation/recovery of starch. Most of the artefacts sampled produced starch grains, though in greatly varying quantities. The variations may relate to the storage conditions of the morahs which were different in each case. Nonetheless the starch recovered has provided a clear indication of target species on the basis of maximum dimension measurements (see Lance *et al.* in press). While these stones appear to have been used for the processing of Yellow Walnuts and Hairy Walnuts, it is also clear that they were used to process other toxic nuts, perhaps as they became seasonally available. It is likely that these residues represent a record of the last use of the stones.

An interesting find is that the Stager grindstone, RS05035, was used to process a different set of plants to the other grindstones examined here, as indicated by the small sample of starch granules present on the stone surface. One of the species that may be present on this grindstone is *Macadamia whelani*. Pedley (1993:139) reports that no information was found in relation to the processing of *M. whelani* by Jirrbal-Girramay informants, nor from Murray Upper, near Tully, and it may not have been processed uniformly across the region where the morah stones are found. The identification of this plant as part of the assemblage on the Stager grindstone may be an indicator of detoxification closer to Babinda. A second species possibly identified on this stone, the Polynesian Arrowroot (*Tacca leontopetaloides*) is found on the coast in 'open forests and extends into rainforest behind sandy beaches' (Pedley 1993:117) and as such may act as an indicator of the location from which the stones are derived (i.e. coast as opposed to tablelands).

The identifications to plant species that are indicated in this study will need further corroboration by detailed analysis of the morphology of the starch, the degree of faceting and other surface features.

CONCLUSION

Morah stones are used in the processing of toxic starchy nuts. This study has established that residues documented for these incised grinding stones are consistent with plant food processing, especially when the morphology and comparative data on use-wear patterns are taken into account. Comparison with modern reference materials indicates that these plants are likely to include *B. bancroftii* (Yellow Walnut) and Hairy Walnut (*E. insignis*). Other economic species may have contributed to the residue assemblage as it appears that these artefacts are unlikely to be single use and are of a size that is easily portable.

ACKNOWLEDGEMENTS

We would like to thank Maisie Barlow (Jirrbal), Pat and Alverio Croatto, Ned Maxwell, Deanna and Ron Stager for access to their private collections of morah stones. The Atherton Tropical Herbarium staff, especially Rebel Elick and Bruce Gray, helped us compile the plant/starch reference collections. Thanks to Don Page, Bernadette McCall and Anna Charlton for technical assistance. We are also grateful to Ernie Raymont (Ngatjin) for advice and support. The project was funded by an Australian Research Council Discovery Project grant, La Trobe University and the University of Sydney. We appreciated the constructive comments of Alison Crowther and Catherine Westcott. The authors acknowledge the facilities as well as scientific and technical assistance from the staff in the Australian Microscopy and Microanalysis Research Facility (AMMRF) and at the Australian Key Centre for Microscopy and Microanalysis at the University of Sydney.

REFERENCES

Barton, H. 2007. Starch residues on museum artefacts: implications for determining tool use. *Journal of Archaeological Science* 34:1752-1762.

Cosgrove, R. 1996. Past human use of rainforests: an Australasian perspective. *Antiquity* 70:900-912.

Cosgrove, R. 2005. Coping with noxious nuts. *Nature Australia* 28(6):46-53.

Cosgrove, R., J. Field, and A. Ferrier 2007. The archaeology of Australia's tropical rainforests. *Palaeogeography, Palaeoclimatology, Palaeoecology* 251:150-173.

Ferrier, Å. 1999. A study of contact period Aboriginal material culture from the rainforest region of Northeast Queensland. Unpublished BA (Honours) thesis. Bundoora: Department of Archaeology, La Trobe University.

Field, J. and R. Fullagar 1998. Grinding and pounding stones from Cuddie Springs and Jinmium. In R. Fullagar (ed.) *A Closer Look: Recent Australian Studies of Stone Tools*, pp. 95-108. Sydney: Sydney University Archaeological Methods Series 6.

Fullagar, R. and J. Field 1997. Pleistocene seed grinding implements from the Australian arid zone. *Antiquity* 71:300-307.

Fullagar, R., J. Furby, and B. Hardy 1996. Residues on stone artefacts: state of a scientific art. *Antiquity* 70:740-745.

Fullagar, R., J. Field, and L. Kealhofer 2008. Grinding stones and seeds of change: starch and phytoliths as evidence of plant food processing. In Y. M. Rowan and J. R. Ebeling (eds) *New Approaches to Old Stones: Recent Studies of Ground Stone Artifacts*, pp. 159-172. London: Equinox Publishing P/L.

Gould, R.A. 1980. *Living Archaeology*. Cambridge: Cambridge University Press.

Harris, D.R. 2006 The interplay of ethnographic and archaeological knowledge in the study of past human subsistence in the tropics. *Journal of the Royal Anthropological Institute* Special Issue. S63-S78.

Horsfall, N. 1987. Living in the rainforest: the prehistoric occupation of North Queensland's humid tropics. Unpublished PhD thesis. Townsville: James Cook University.

Lance, B., J. Field and R. Cosgrove In press. Intra-taxonomic variability in starch reference collections and the implications for ancient starch studies. In A. Fairbairn and S. O'Connor (eds) *Proceedings of the 2005 Australasian Archaeometry Conference*. Canberra: ANU E Press.

McBryde, I. 1997. 'The Landscape is a series of Stories'. Grindstones, quarries and exchange in Aboriginal Australia: a Lake Eyre case study. In A.Ramos-Millan and M.A. Bustillo (eds) *Siliceous Rocks and Culture,* pp. 587-607. Granada: University of Granada.

Mitchell, T.L. 1848. *Journal of an Expedition into the Interior of Tropical Australia in Search of a Route from Sydney to the Gulf of Carpentaria*. London: Longmans.

Mulvaney, K. 1998. The technology and Aboriginal association of a sandstone quarry near Helen Springs, N. Territory. In R. Fullagar (ed.) *A Closer Look: Recent Australian Studies of Stone Tools*, pp. 73-94. Sydney: Sydney University Archaeological Methods Series 6.

Pedley, H. 1993. Plant detoxification in the rainforest: The processing of poisonous plant foods by the Jirrbal-Girramay people. Unpublished M.A. thesis. Townsville: Material Culture Unit, James Cook University.

Smith, M. 1982. Late Pleistocene zamia exploitation in southern Western Australia. *Archaeology in Oceania* 17:109-116.

Smith, M. 1996. Revisiting Pleistocene Macrozamia. *Australian Archaeology* 42:52-53.

Smith, M.A. 1985. A morphological comparison of Central Australian seed-grinding implements and Australian Pleistocene-age grindstones. *The Beagle, Occasional Papers of the Northern Territory Museum of Arts and Sciences* 2(1): 23-38.

Tindale, N.B. 1977. Adaptive significance of the Panara or grass seed culture of Australia. In R.V.S. Wright (ed.) *Stone tools as cultural markers: Change, evolution and complexity* pp. 345-349. Canberra: Australian Institute of Aboriginal Studies.

Woolston, F.P. and F.S. Colliver 1973. Some stone artefacts from North Queensland rainforests. *Occasional Papers in Anthropology* 1:104-125.

18

Aboriginal craft and subsistence activities at Native Well I and Native Well II, Central Western Highlands, Queensland: results of a residue and use-wear analysis of backed artefacts

Gail Robertson

School of Social Science,
The University of Queensland,
St Lucia QLD 4072 Australia
Email: g.robertson@uq.edu.au

ABSTRACT

This research provides insight into activities at two adjoining Aboriginal rockshelters in the Central Highlands in western Queensland, Native Well I and Native Well II. The study involved a residue and use-wear analysis of the backed artefact component of the stone assemblage. Prior to this, interpretation of the sites essentially relied on evidence of changes in stone technology over time, sequential and spatial patterning of artefacts and ethnographic analogy. This analysis revealed a range of activities occurring during the mid-to-late Holocene. Backed artefacts were used as knives, scrapers and/or incisors for wood-working and bone-working, as well as knives and scrapers for plant processing, including cooked starchy plants. Artefacts with ochre and feather residues may have been used for ceremonial purposes, while distribution of resin indicates more than half the artefacts had been hafted.

KEYWORDS

residue analysis, use-wear-analysis, microscopy, Australian backed artefacts, Aboriginal rockshelters

INTRODUCTION

This paper discusses the results of an integrated residue and use-wear analysis of the backed artefact component of stone assemblages from two Aboriginal occupation sites in the Central Highlands in western Queensland, Native Well I and Native Well II. The study was undertaken as part of a larger research project that addressed the question of backed artefact use in eastern Australia during the mid-to-late Holocene through an analysis of artefacts from six different sites (Robertson 2005). The two Central Highland sites were excavated by Morwood (1979) as part of his doctoral research, in which he employed a multi-attribute approach to the region's prehistory by documenting both the artwork and the artefact assemblages at a number of different sites.

THE SITES

Native Well I and Native Well II are adjacent rock shelters located in a quartzite outcrop overlying sandstone on a tributary of Sandy Creek, which is itself a tributary of the upper Warrego River (Figure 1). Native Well I is a relatively large low shelter and art gallery with a floor area of approximately 30m² situated on the western edge of the outcrop. Native Well II, which has a floor area of only 18m², is an adjacent but separate shelter directly to the south (Morwood 1979:168, 206). The coarse sandstone of the rear walls of the shelters is heavily engraved, mostly by an abrasion technique but with some peckings. There are also numerous stencils and paintings (Morwood 1981). The two sites are discussed sequentially below.

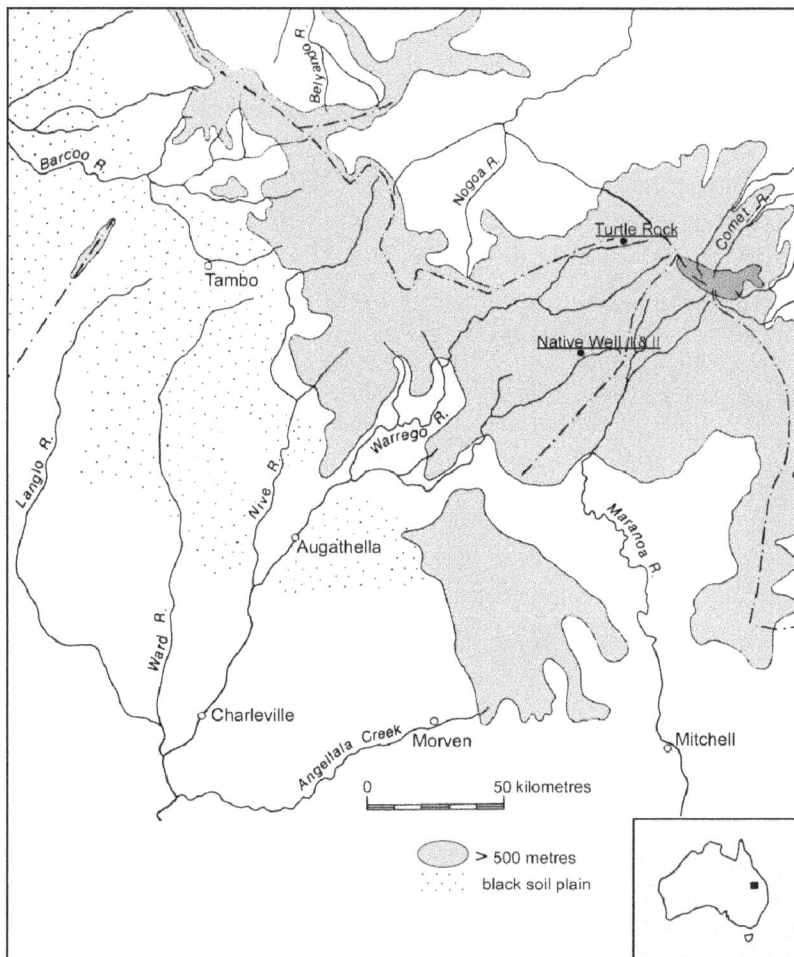

Figure 1. Map of Central Highlands, western Queensland illustrating the archaeological sites of Turtle Rock, Native Well I and II (from Morwood 1979, Fig. 1.1).

Native Well I

Almost 10,000 stone artefacts were recovered from excavations at Native Well I. Of these, 566 were identified as tools on the basis of either use-wear or retouch and 41 artefacts were recorded as being backed (Morwood 1979:181). As well, there were 675 pieces of ochre and pipe-clay of a wide range of colours, although red (60%) predominated (Morwood 1979:179). Organic material was poorly preserved with recovered bone mostly from the top two excavation units, possibly natural rather than cultural given that very little was burnt (Morwood 1979:178-179; 1981:27). Two emu feathers, an emu egg-shell fragment and two echidna quills were the only other faunal remains identified. There are no recorded plant remains in the site. Occupational deposits reached a depth of approximately 90 cm throughout most of the shelter, although squares outside the shelter contained artefacts to a greater depth. The deposits were sieved at the site through 2 mm, 5 mm and 9 mm sieves, with material remaining in the sieves bagged and eventually wet sieved in the laboratory.

Five radiocarbon dates on charcoal samples were obtained for the site (Table 1). Backed artefacts in the deposits are bracketed by the dates 1270±70 BP and 4320±90 BP. Morwood (1979:178) noted that artefacts were found below the earliest dated deposits and by extrapolation suggested that the site was initially occupied from ca. 13,000 BP, although there was no evidence of Aboriginal use inside the shelter until 6190±100 BP.

The criteria used to classify the diagnostic artefacts for the Central Highlands, including backed artefacts, were based on variants of those of McCarthy (1946, 1976) with regional

Table 1. Native Well I conventional ^{14}C dates on charcoal (Morwood 1979:176-7)

Excavation Unit	Depth below surface (cm)	Lab code	C14 dates
Charcoal from a hearth in Stratigraphic Layer 3, Square C4	20±2cm	ANU 2002	1270±70 BP
Charcoal from the top 10cm of Stratigraphic Layer 4A, Square B5	49±5cm	ANU 2003	4320±90 BP
Charcoal from the top 5cm of Excavation Unit 6, Square C2	56±2.5cm	ANU 2171	4230±90 BP
Charcoal from Stratigraphic Layer 5, Square B5	95±5cm	ANU 2001	6190±100 BP
Charcoal from Excavation Unit 14 at the base of Stratigraphic Layer E4, Squares C0 and C1	140±5cm	ANU 2035	10,910±140 BP

characteristics taken into account (Morwood 1981:2). Figure 2 illustrates the defined categories. According to this method of classification, 41 artefacts were recorded as being backed. However, for the purpose of uniformity across the sites in this study, the definition of Australian backed artefacts provided by Hiscock (1993, 2002:163) and Hiscock and Attenbrow (1996) was adopted for this study. A technological analysis was undertaken by Hiscock (Australian National University, pers. comm., 4. May 2001) and 26 artefacts were identified as backed. These are catalogued as geometric microliths (n=14), backed microliths (n=4), and backed pieces (n=8). All artefacts catalogued as backed microliths are asymmetric in morphology and are referred to in the text as asymmetric or Bondi points for consistency. All but one of the backed artefacts were manufactured on quartzite, with NWI#1 produced on silcrete. They range in size from 1–4 cm, with 10 artefacts between 2–3 cm, 12 artefacts between 1–2 cm, two artefacts 3–4 cm and two exactly 2 cm in length.

Native Well II
The Native Well II Aboriginal rockshelter also contains a gallery of rock art including engravings and red and white (and one black) stencils. The site was excavated by Morwood (1979:206) in 1977, and 3849 pieces of stone were recovered, with 178 identified as tools on the basis of morphology and either retouch or use-wear. Excavated faunal material comprised five pieces of bone, only

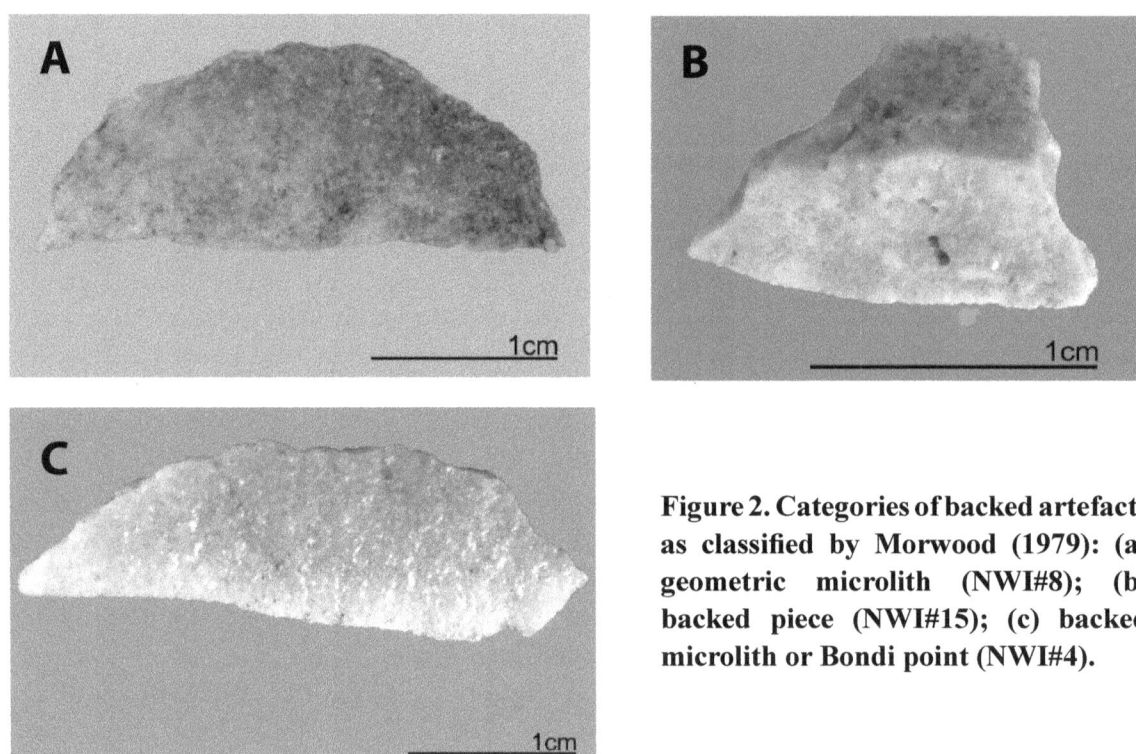

Figure 2. Categories of backed artefacts as classified by Morwood (1979): (a) geometric microlith (NWI#8); (b) backed piece (NWI#15); (c) backed microlith or Bondi point (NWI#4).

one of which was unburnt and identifiable, all recovered from the three uppermost excavation units (Morwood 1981:38). Abundant ochre and pipe clay fragments were also excavated with a distribution similar to that of the stone tools. However, because of the range and quantity of material, Morwood (1979:213) concluded that the pigments were applied to a variety of artistic endeavours not all represented by the rock art at the site.

Three radiocarbon dates on charcoal samples were obtained for the site (Table 2). Morwood (1979:210) suggests the most recent backed artefacts recovered may have been 'scuffed up' from lower levels. The corresponding radiocarbon date for the most recent backed artefacts would therefore be younger than 2170±80 BP. A date of 3700±500 BP for the earliest appearance of backed artefacts at the site was based on an age/depth curve and differs from that of 4320±90 BP for Native Well I.

Table 2. Native Well II conventional ¹⁴C dates on charcoal (Morwood 1979:210)

Excavation Unit	Depth below surface (cm)	Lab code	C14 dates
Charcoal from Excavation Unit 5 in Square L7	46±5cm	ANU 2117	2170±80 BP
Charcoal from Excavation Unit 10 Square L7	92±5cm	ANU 2091	2470±80 BP
Charcoal from Excavation Unit 18, near the base of Stratigraphic Layer 5 in Squares L7 and M7	173±5cm	ANU 2035	10770±135 BP

The same classification criteria have been applied to the Native Well II backed artefacts as described for Native Well I. Fourteen backed artefacts in the Native Well II assemblage were catalogued as geometric microliths (n=6), backed microliths, also referred to as asymmetric or Bondi points, (n=2), and backed pieces (n=6). All of the backed artefacts at Native Well II were manufactured on quartzite. This is particularly interesting as other artefacts recovered from the site are produced on chert, sandstone, fine-grained silcrete, volcanics, and a duricrust silcrete, with changes in raw material use over time (Morwood 1981:30). In fact, other tools in the 'small tool tradition' were produced on fine-grained silcrete (Morwood 1979:215). All artefacts were less than 4 cm in size with nine artefacts between 1–2 cm, two exactly 2 cm, two 2–3 cm, and one 3–4 cm.

METHODS

The techniques of residue and use-wear analysis have demonstrated their combined effectiveness as a means of inferring task association and function in archaeological material (for examples, see Fullagar 1986, 1994, 1998; Haslam 1999; Robertson 2002, 2005, 2006; Rots and Williamson 2004; Wadley and Lombard 2007). In combination with the use of powerful optical microscopes for observation, the analysis relied on access to a comprehensive comparative reference collection assembled during the course of the project by the author and colleagues in the Archaeological Science Laboratory, School of Social Science, University of Queensland, and also on published data. As a component of this research, tables were created to reflect anticipated residues and use-wear associated with various hypothesised tasks such as butchery, hunting, bone-working, skin-working, wood-working, general plant processing and also ceremonial or decorative activities (see Robertson 2005:Table 3.2 and Robertson 2005:Table 3.4). These tables were employed to make inferences on the use of backed artefacts at Native Well I and Native Well II.

Residue analysis
Residue analysis is the study of organic and inorganic materials left adhering to artefacts as a result of Locard's 'exchange principle' (Briuer 1976:478). In this discussion, the term residue analysis refers to the methodology associated with the identification and interpretation of archaeological residues. Although residues on stone artefacts are the focus of this study, residue

analysis is employed in a growing number of areas of archaeological research including the study of pottery, glass, coprolites and archaeological soils. Residues on stone tools may be from several different sources. They may be culturally derived, for example as the result of association with a task or as part of the manufacturing process (i.e. hafting and/or tool decoration). They may also arise incidentally from some other activity, or they may be due to taphonomic factors and result from the post-depositional environment or post-excavation processes. An essential part of residue analysis is the exclusion of the non-cultural elements present on an artefact.

The types of diagnostic organic residues potentially associated with excavated stone artefacts are generally subdivided into plant and animal residues, while inorganic residues such as ochre, aragonite, vivianite and some other minerals may also relate to stone tool use. Typical plant residues are cellulose (amorphous, tissues and fibres), sap, resin, starch grains, raphides and druses, each with specific characteristics that generally allow their microscopic identification (for identification characteristics see Franceschi and Horner 1980:381; Fullagar 1986:176; Gunning and Steer 1975:117; Haslam 2004; Horner and Wagner 1995:56; Langenheim 2003:46; Lombard 2005; Raven *et al.* 1999; Robertson 2005:54-85). Among the animal residues identified on stone artefacts are blood (including proteinaceous films and red blood cells), bone, animal tissue and fibres (including collagen and muscle), lipids, feathers and hair (Akerman *et al.* 2002; Balme *et al.* 2001; Cooper 2003; David 1993; Francis 2000, 2002; Fullagar 1986; Fullagar and Jones 2004; Lombard 2005; Lombard and Wadley 2005; Loy 1983, 1985, 1990, 1993, 1994; Loy and Dixon 1998; Loy and Hardy 1992; Loy and Wood 1989; Robertson 2002, 2005, 2006; Sobolik 1996; Tomlinson 2001; Wadley *et al.* 2004a; Wallis and O'Connor 1998; Williamson 2000). Microscopic identification to a specific taxonomic level is only possible with a limited number of these residue types, namely, hair and feather, and then only if the residue is relatively undamaged and there is access to a reference collection of local specimens (Loy 1985, 1990, 1993; Loy and Nelson 1986; Robertson 2002). For more specific information from the residues, further testing is usually necessary.

Despite the fact that the mechanisms of preservation are not yet fully understood (but see Jones this volume), there is still substantial evidence that organic residues on stone tools, unlike macro-remains, are capable of surviving for long periods of time under a variety of conditions, including caves and open air sites, swamps and deserts (Cattaneo *et al.* 1993:41; Loy 1990). Whether blood proteins survive in a biologically active form is still a matter for debate, although there appears to be a consensus on the relative stability of haemoglobin (or at least the haem portion). This suggests that the modified Hemastix test (see Matheson *et al.* this volume) is still the most useful indirect screening test for the presence of blood residue, although it cannot be used as the only indicator.

Use-wear analysis

Use-wear analysis comprises a series of techniques for obtaining functional information from stone tools to augment that available from conventional morphological and technological approaches. Information is obtained by studying 'the effects of the utilisation process on the tool itself' (Odell 2004:135). It is defined as 'the study of tool functions by examining modifications to the edges and surfaces of stone tools' (Fullagar 1986:9). Since such modifications may be cultural or taphonomic, use-wear analysis has two components: it is a method of defining or describing wear features attributable to cultural factors, that is, tool-use; and a means of interpreting function. The major forms of use-wear observed were edge rounding, edge-fracturing, striations, lineation and abrasive smoothing and polish (see Robertson 2005 for definitions of these terms with reference to Fullagar 1986 and Kamminga 1982). Use-wear analysis, in this study, was employed primarily to locate used edges and to determine the mode of action of a tool. The more complex wear patterns associated with sustained use were not consistently identified.

The potential for confusion of use-wear with non-use-related wear features is a major methodological issue. All tools will have been subjected to non-use wear usually from a number of different sources during their life histories, and unless this is identified as such or at least recognised as a possibility, any functional interpretation based on use-wear analysis alone is likely to be questionable. A number of researchers have attempted to address the problem through experimentation (see Burroni *et al.* 2002 for detail), but Hurcombe (1992:71) has adopted an interesting methodological approach to the issue by employing Schiffer's (1972) separation of archaeological and systemic contexts as a basis for constructing a table of phases in the life history of a tool which might produce non-use wear (see Robertson 2005:Table 2.1). This framework allows consideration of the various types of potential wear patterns and also the possibility of identifying at least some of them (Hurcombe 1992:71). Hurcombe (1992:71-78) provides an excellent discussion of numerous sources of non-use modifications and accidental damage to lithic artefacts within this framework. Those considered relevant to this research included some manufacturing techniques such as abrasion and retouch, 'bag-wear', and accidental damage due to trampling by either human or animal agency (see Kamminga 1982:7-8; McNiven 1993; Vaughan 1985:23).

Other taphonomic factors to be considered are soil processes, including patination caused by soil chemicals, soil movement causing friction, and the detrimental effects of exposure to wind, heat and water (Burroni *et al.* 2002). The site context of the artefacts should indicate which, if any, of these issues need to be accounted for during analysis. Sieving and cleaning of artefacts is known to cause significant modifications to an artefact surface, particularly abrasion and striations, and also of course the removal of residues (Hurcombe 1992:77; Kooyman 2000:154). Use-wear attributes were observed in less than ideal conditions in this study. Totally accurate use-wear analysis requires artefacts to be thoroughly cleaned, often with harsh chemicals, prior to microscopic examination. However, this research was an integrated study involving both use-wear and residue analysis and cleaning of the artefacts was not a viable option.

Hafting also has a significant influence on the mode of action of a tool and the presence of a haft may occasionally be inferred from wear traces. However, according to Rots (2003:812) the 'use of resin often hinders trace production'. This is is an interesting finding in relation to interpretations of Australian artefact use where the presence of resin traces is the most distinctive hafting evidence recorded in most previous research. Rots (2003:812) determined that 'absence of scarring and polish in a well-delimited area' usually signifies the use of resin, not necessarily the absence of hafting. Lombard (2005) used a multi-analytical approach that included both use-wear and the presence of resin to infer hafting on artefacts from Sibudu Cave in South Africa.

Microscopy

Both low and high magnification microscopes were employed in the analysis because each provided a different image of the artefact surface. This difference relies not only on their differing range of magnifications and degree of resolution, but, more significantly, on the angle of lighting (Fullagar 1986:27).

Initial examination of an artefact at low magnification allowed assessment and identification of traces of wear, including use-wear and wear due to taphonomic factors, location of potential use-related residues, hafting evidence, and contaminants. This analysis involved the use of a Wild stereo-binocular microscope with variable magnifications from 6x to 30.6x diameters mounted with an Olympus DP10 digital camera set at highest resolution (3.2 million pixels). The light source employed was a Microlight 150 fibre-optic light with adjustable arms. The latter allow observation of artefact surfaces with oblique lighting, which is essential for the identification of a number of use-wear attributes.

An Olympus BX60 metallographic microscope fitted with 10x eyepiece lenses and 5x, 10x, 20x, 50x and 100x objective lenses was used for high-power microscopy, providing nominal magnifications of 50x, 100x, 200x, 500x and 1000x diameters. The microscope employs vertical incident brightfield and darkfield illumination for observation of residues and artefact surfaces, but also has the capacity for use as a transmitted light microscope for observation of residue samples removed to slide. The Olympus microscope was also fitted with an Olympus DP10 digital camera.

RESULTS

Native Well I

Of the 26 artefacts analysed from Native Well I, 20 show evidence of having been used for an identifiable task or tasks, two exhibit only a slight indication of use with one of these being allocated a task association of possible ceremonial activities on the basis of residues, and a further four artefacts were unable to be assigned a specific task although they exhibited some evidence of use (Table 3). Half of the artefacts exhibited evidence for hafting and this is discussed below in conjunction with resins. Figure 3 provides a broad overview of the analysis results. Inferred task associations for each artefact are categorised separately, but several artefacts were used for more than one task, including a backed piece (NWI#9) used for both wood-working and processing of starchy plant food, and two symmetric artefacts (NWI#10, NWI#20) used for both general plant processing and processing of starchy plant food. One artefact (NWI#3) with numerous feather residues was also used for general plant processing. Ochre was present in varying quantities on 19 artefacts, one of which may have been used for pigment preparation or in a ceremonial context.

Plant processing

General plant processing includes cutting or shredding non-woody material and scraping or cutting to remove bark from woody stems, with the latter activity often adding a dark resinous plant sap, occasionally charred. Eleven artefacts were allocated the task association of general plant working on the basis of the presence and distribution of plant sap or exudate, amorphous and fibrous cellulose and small starch grains and, occasionally, resin. Associated use-wear features included small bending flake scars along a slightly rounded edge with occasional lineation in the residues. A further two artefacts (NWI#4, NWI#23) exhibited some of the above features, but the evidence was insufficient to confidently infer plant processing. Five artefacts showed evidence of hafting.

On five artefacts, cooked or heated starch was a significant residue, generally co-occurring with plant fibres, cellulose fragments and in one instance (NWI#19), with raphides. Most uncooked grains were approximately 2 µm, while cooked grains were generally larger (4-8 µm on these artefacts) and with a diffuse extinction cross (Figure 4). The source of starch is likely to be tubers or rhizomes. Two artefacts (NWI#10, NWI#20) were also used for processing resinous plant material at some stage, and the tip of one (NWI#9) was used to incise wood. None of these artefacts appeared to have been hafted.

Three artefacts (NWI#1, NWI#19, NWI#23) exhibited raphides, although not in any significant quantity. In each case they were associated with starch grains and other plant residues and their presence substantiates the inferred use of the artefacts for plant processing. The co-occurring starch grains on one artefact (NW#19) appeared cooked, signifying possible food preparation. All raphides observed were of similar morphology with lengths of 25 µm, widths of 2-3 µm and slightly uneven edges possibly due to variations in their crystalline structure producing banding (see Robertson 2005:61). This similarity in size and shape is a possible indicator of their source from the same plant family (see Crowther this volume), particularly as they were all found in conjunction with almost identical starch grains. Raphide identification on NWI#23 is confirmed by the presence of part of a raphide idioblast. An almost complete raphide idioblast was also observed on another artefact (NWI#3) and, although no raphides were identified, this artefact also had an inferred task association of general plant processing.

Wood and bone-working

Seven artefacts were allocated the task association of wood-working on the basis of residues and use-wear features such as moderate to pronounced rounding of edges (scraping) or tips (incising), the presence of striations, and edge-scarring in the form of bending and step microfractures (Figure 5). Residues consisted of smeared resin and/or plant exudate that was often charred,

Table 3. Task association/s and hafting assessment for Native Well I artefacts

NW I #	Artefact Type	Animal Task Association	Animal Function	Plant Task Association	Plant Function	Other Task Association	Hafting Assessment
1	Bondi point			general plant processing / fibrous	cutting and scraping		insufficient evidence
2	Geometric microlith			general plant processing / non-woody	cutting and scraping		probably hafted
3	Geometric microlith	feathers	uncertain	general plant processing /non-woody	scraping and shredding		hafted
4	Bondi point			uncertain - possible plant processing	uncertain		insufficient evidence
5	Bondi point				uncertain		hafted
6	Bondi point			wood-working / resinous	cutting and incising		probably hafted
7	Backed piece			wood-working / resinous	incising and/or scraping		insufficient evidence
8	Geometric microlith			probable wood -working / resinous	uncertain - probable scraping		probably hafted
9	Backed piece			starchy plant processing (cooked and fibrous); wood-working / resinous	cutting and scraping (starchy plant); incising (wood)		insufficient evidence
10	Geometric microlith			general plant processing; starchy plant processing	cutting and scraping		insufficient evidence
11	Geometric microlith	secondary or tertiary bone-working	cutting				insufficient evidence
12	Geometric microlith			wood-working	scraping and engraving		probably hafted
13	Backed piece			general plant processing / resinous	cutting		insufficient evidence
14	Backed piece			general plant processing / resinous	cutting		insufficient evidence
15	Backed piece			general plant processing / resinous	scraping and cutting		probably hafted
16	Geometric microlith			general plant processing / woody stems or sappy plant	scraping		hafted
17	Geometric microlith			wood-working / charred	scraping and burnishing		insufficient evidence
18	Geometric microlith			general plant processing/ soft, fleshy stems	cutting and/or scraping		hafted
19	Geometric microlith			starchy plant processing	cutting and scraping		insufficient evidence
20	Geometric microlith			general plant processing / resinous; starchy plant processing	cutting and/or scraping		insufficient evidence
21	Geometric microlith			starchy plant processing	cutting		insufficient evidence
22	Backed piece			wood-working / resinous	incising		probably hafted

23	Backed piece			uncertain - possible plant processing / resinous	cutting		hafted
24	Geometric microlith					possible ritual or pigment preparation	hafted
25	Geometric microlith			unknown	unknown		hafted
26	Backed piece			general plant processing / charred, resinous	cutting		insufficient evidence

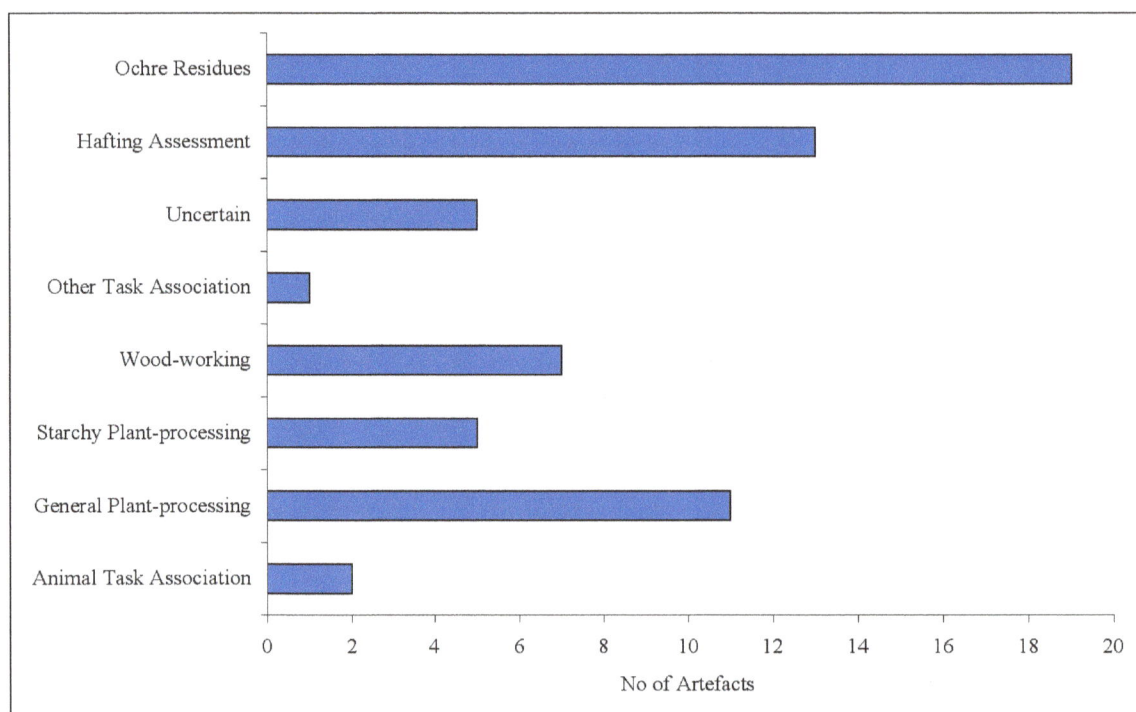

Figure 3. Inferred task associations, hafting assessment and ochre residues for Native Well I.

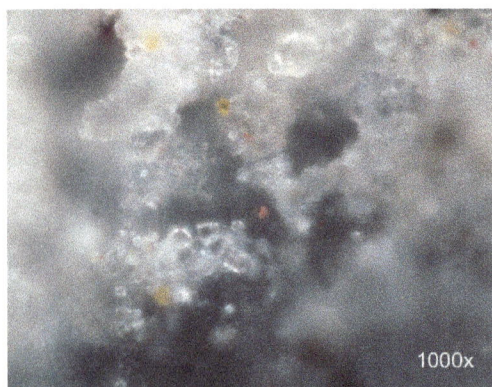

Figure 4. Cooked starch grains (2-4 μm) and cellulose, on the chord of NWI#9 (BFxp).

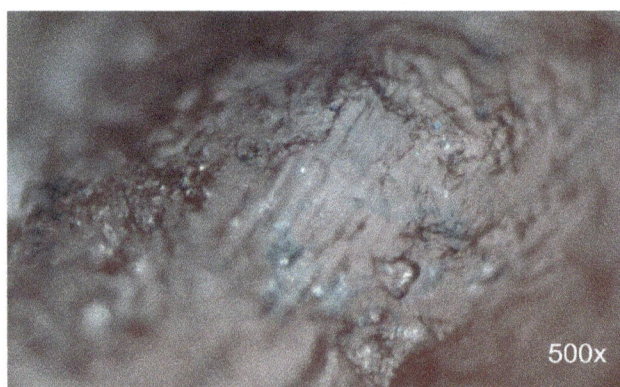

Figure 5. Striations sub-parallel to the flattened tip in a smoothed and polished section of the tip of NWI#6 (BF+a).

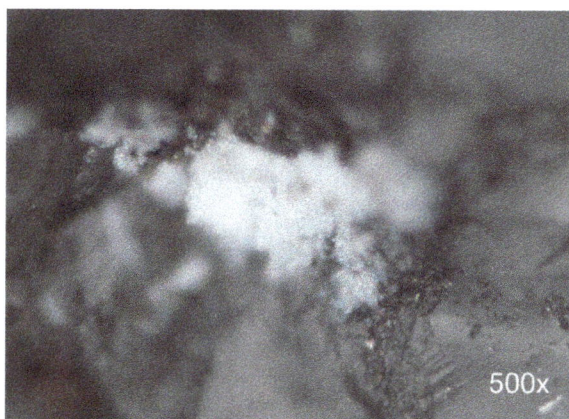

Figure 6. Granular bone collagen in a flake scar on one end of the chord (ventral surface) of NWI#11 (BFplp).

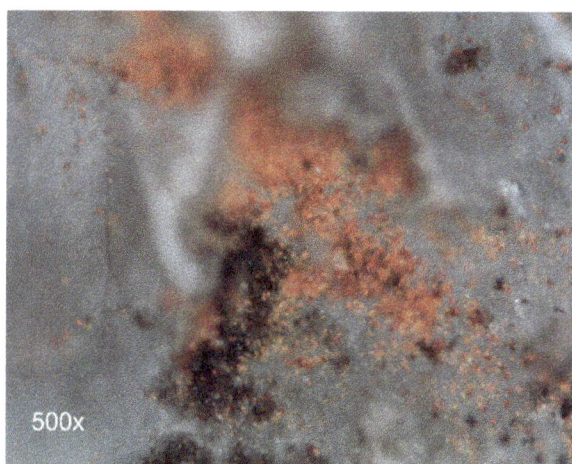

Figure 7. Red ochre and resin on one tip of NWI#8 (BFplp).

plant fibres, amorphous cellulose fragments, charcoal and small starch grains. Five of the seven artefacts functioned as incisors as attested to by the wear on the tips, often with scraping and cutting as adjunct functions. One artefact (NWI#17) appeared to have been used to burnish wood, with significant wear and residues on the obtuse angle ridge. Four artefacts exhibited evidence for hafting.

Only one artefact (NWI#11) was found to be associated with bone-working. Bone collagen and a putative blood residue in the form of a greasy proteinaceous film were the most significant residues, indicating either secondary or tertiary bone-working (Figure 6). Use-wear suggested a cutting action.

Feathers and ochre

NWI#3 was the only artefact with feather residues and, although three downy barbules were observed, more specific identification was not possible. The residues were mostly located on and below the obtuse angle ridge and in association with resin, and any inference regarding their use is purely speculative. They may have been the result of a specific processing task or alternatively they may relate directly to the hafting method, possibly as temper or decoration. The artefact was also used for general plant processing.

Nineteen artefacts exhibited ochre residues, with both red and yellow ochre occurring on six of these, and red alone on the remainder (Figure 7). In many instances (n=11) where quantities were slight and patchy, ochre appears to be the result of incidental transfer from the surrounding soil or possibly from the hands of the user or the excavator. On artefact NWI#24, ochre was scraped onto edges in sufficient quantity to suggest use of the artefact in pigment preparation or in a ritual or ceremonial context. On the remaining seven artefacts, ochre was possibly associated with decorative use, especially on or around the haft, although four of these artefacts were used for wood-working, which may also account for the presence of ochre if the wood was painted with ochre. Ochre also has a practical purpose as an additive to hafting resin and may be used as a filler to reduce the brittle character of the resin (Wadley *et al.* 2004b:670).

Six artefacts with ochre residues exhibited clear evidence for hafting, and a further three may have been hafted although the evidence is ambiguous. A variety of coloured ochre fragments (including red, yellow and white) were found throughout the excavated deposits at Native Well, and also on the ground adjacent to the shelter and in caches in association with stone tools and pieces of worked wood (Morwood 1979:165). A retouched stone flake with one heavily abraded 'corner' and splatters of adhering white and yellow ochre attesting to its use as an engraving implement was discovered cached in an area surrounded by engravings (Morwood 1979:171). Morwood (1979:180) deduced that artistic endeavours, including engraving and decoration of the walls with stencils and painting, have been a significant feature of the site since its initial use by Aboriginal people. The use of ochre for other aesthetic purposes in connection with stone artefacts was borne out by the current study.

Resin and hafting

Hafting was inferred on the basis of appearance, relative quantity and pattern of distribution of resin on 13 of the 26 tools examined. Only one artefact (NWI#19) had no visible resin but the remainder exhibited resin in some form, with resin or a resinous plant exudate usually the result of wood-working (n=3) or general or starchy plant processing (n=7). Resin/plant sap on the latter was frequently smeared and/or charred and was located on the tip or cutting edge in association with use-wear features such as striations and edge damage. On several artefacts, resin was in limited quantity and/or without any particular pattern of distribution preventing any confident inference regarding hafting.

Hafting resin on most artefacts was brownish-black but on two artefacts (NWI#2, NWI#12) it appeared red-brown and on one (NWI#5), yellow, indicating different sources. Resin was frequently charred and occasionally thick and 'crazed' (Figure 8). The hafting method appeared to vary with the morphology of the tool, with resin distribution on the symmetric artefacts (n=8) generally on the backed edge, obtuse angle ridge and one or both lateral margins (Figure 9). The symmetric artefacts may have been hafted in series. For the asymmetric artefacts (n=2), hafting resin was located principally on the backed edge at the proximal end, and for these the inferred function was incising which would require individual hafting. As expected, there was no general pattern for hafting of the backed pieces (n=3).

Figure 8. Smeared black/charred and 'mud-cracked' resin near the chord on NWI#8 (BFxp).

Figure 9. Resin distribution along the left lateral margin on the ventral face of NWI#15 (Wild).

Summary

Of the 26 Native Well I artefacts analysed, all exhibited some evidence of use although four were unable to be assigned a specific task/function, and a ceremonial or decorative purpose was tentatively proposed for one on the basis of residues. Eleven artefacts were used for general plant processing with cutting and/or scraping as the inferred functions. Five artefacts were allocated the inferred task association of processing cooked starchy plants, and seven had been used for wood-working, principally as incisors. Only two artefacts gave any indication of an association with animals: one artefact had been used for bone-working, with use-wear indicating a cutting function; and one had feather residues, although these were unable to be unequivocally associated with a particular task or function. Several artefacts were probably multipurpose although the inferred task associations overlap since they were all associated with plant working in some form. For example, the separation of starchy plant processing and general plant processing may be too prescriptive. Half of the artefacts had hafting evidence, and there is a possibility that some of the geometric artefacts were hafted in series as proposed historically. Ochre residues occurred on over 70% of the backed artefacts, which is to be expected for an art site.

Native Well II

Of the 14 artefacts analysed from Native Well II, 13 exhibited clear evidence for use although one (NWII#4) was unable to be assigned a specific task (Table 4). Ochre was present on all artefacts, although in various quantities and locations and the presence of which requires clarification. Hafting evidence is notable and is discussed in conjunction with resins. Figure 10 provides an overview of the results.

Plant processing

Three artefacts were allocated the inferred task association of general plant working on the basis of the presence and distribution of plant sap or exudate, amorphous and fibrous cellulose and small starch grains and, occasionally, resin. Associated use-wear features included small bending and occasional step flake scars along a slight to moderately rounded edge with occasional lineation in the residues. General plant processing includes functions such as cutting, shredding, and scraping bark from woody stems, and use-wear evidence on the tools was often slight. Only one of the artefacts (NWII#1) had evidence for hafting.

Wood and bone-working

Seven artefacts were allocated the inferred task association of wood-working based on presence and distribution of resin (frequently charred and smeared), plant sap, plant fibres and tissue, small 1-2 µm starch grains and charcoal. Ochre frequently occurred in conjunction with the plant residues. Use-wear features were generally obvious and included moderate to pronounced edge-rounding on the chord (often including the tip), edge-scarring in the form of a semi-continuous distribution of bending and step flake scars, and various types of striations, the latter often occurring as lineation in resin or plant exudate on the chord (Figure 11). For five of the artefacts, one of the functions was incising as attested to by the wear on one or both tips, although this was always in conjunction with cutting and/or scraping activities. Only three of the seven wood-working artefacts had sufficient evidence to infer hafting, although there is obvious overlap in hafting and wood-working residues which makes the differentiation difficult. NWII#8 also had significant feather and ochre residues indicating a possible decorative or ceremonial association.

Tertiary bone-working as a task was inferred for two artefacts (NWII#3, NWII#7), with residues consisting of granular bone collagen, collagen fibrils and red ochre in conjunction with moderate edge-rounding and a series of bending flake scars with rounded margins on the chord and some striations. NWII#7 also exhibited bevelling on the obtuse angle ridge, while NWII#3 had feather residues associated with hafting resin. Both artefacts were hafted, and use-wear indicated that they were used as scrapers with NWII#3 also used for cutting, although both appear to have had limited use, possibly a single episode.

Feathers and ochre

Feather residues were present on only two artefacts (NWII#3, NWII#8), with two downy barbules visible on the latter. NWII#8 is a backed piece and the downy barbules were located near the truncated end with both red and yellow ochre residues in association. It was not possible to further identify the feather barbules to Order or Family as the morphology of the one of the barbules is common to a number of taxa and a second barbule was fragmented and exhibited few diagnostic features. The combination of ochre and feather residues suggests a decorative or ceremonial purpose, although the artefact had been used for wood-working. On NWII#3, feather residues occurred in association with resin and may relate to hafting or other activities as discussed previously.

Red and/or yellow ochres were present on all Native Well II artefacts although the quantity and distribution varied, as did an association with use-wear and/or other residues. Ochre occasionally appeared greasy and was associated with vivianite, lipids and/or collagen residues and had obviously been mixed with other substances. Roth (1904: 466) notes that ochre was often 'fixed' with a variety of materials including spittle, water, gum-cement (*Leptospermum*

Table 4. Task association and function for Native Well II artefacts

NWII#	Artefact Type	Animal Task Association	Animal Function	Plant Task Association	Plant Function	Other Task Association	Hafting Assessment
1	Backed piece			general plant processing / fleshy	cutting		hafted
2	Geometric microlith			general plant processing / woody stems	cutting and scraping; possible incising		insufficient evidence
3	Geometric microlith	tertiary bone-working	scraping and cutting				hafted
4	Geometric microlith			uncertain	uncertain / possible cutting		hafted
5	Bondi point			general plant processing or wood-working / resinous	scraping and cutting		probably hafted
6	Geometric microlith			wood-working / resinous	incising and scraping		hafted
7	Bondi point	tertiary bone-working	scraping and incising				hafted
8	Backed piece			wood-working / resinous	incising and/or cutting	decorative / ceremonial	insufficient evidence
9	Backed piece			wood-working / resinous	cutting and scraping		hafted
10	Geometric microlith			wood-working / resinous	incising and/or burnishing		insufficient evidence
11	Backed piece			wood-working / resinous	incising and/or cutting		insufficient evidence
12	Backed piece			wood-working / resinous	scraping and cutting or incising		insufficient evidence
13	Backed piece						hafted
14	Geometric microlith			general plant processing	cutting		insufficient evidence

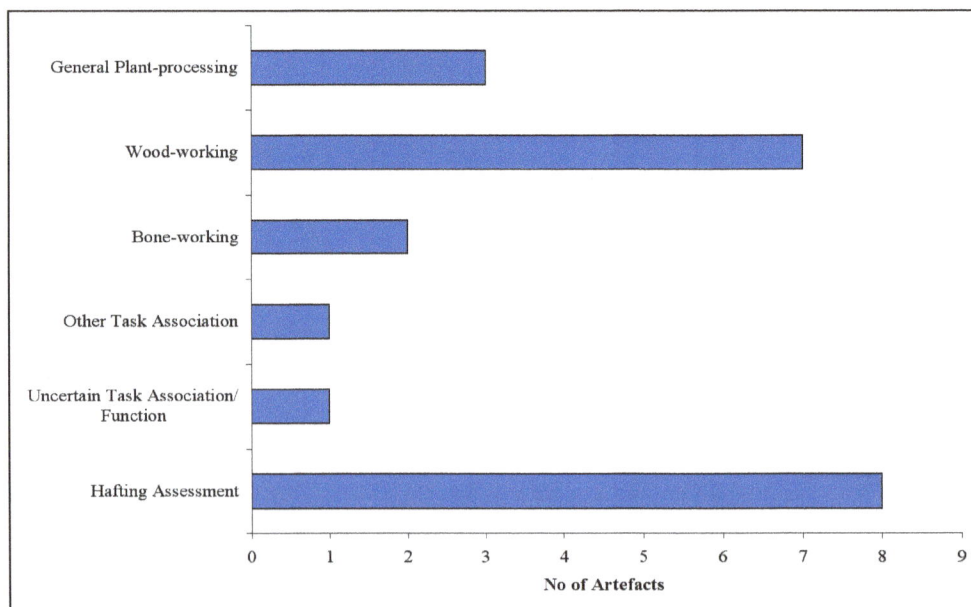

Figure 10. Inferred task associations and hafting assessment for Native Well II artefacts.

Figure 11. Bending flake scars on the rounded edge near the tip of NWII#8.

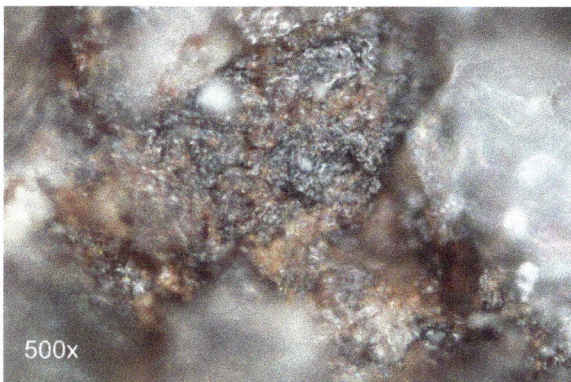

Figure 12. Red ochre and charred resin smeared on the obtuse angle ridge on NWII#12 (DF).

sp.), human or animal blood (especially on men's weapons and implements), honey (on women's digging sticks), candle-nut oil (*Aleurites moluccana*), and snake and iguana fat. In some instances ochre is associated with resin on hafted implements and may be either decorative or an additive to improve the hardening qualities of the resin (Robertson 2005:83; Wadley *et al.* 2004b) (Figure 12).

Several artefacts used for wood-working or tertiary bone-working had co-occurring smeared ochre residues indicating use on ochred materials or possibly concurrent application of ochre to the materials being worked. However, there is no doubt that some ochre residues on Native Well II artefacts also occurred as a result of incidental soil or hand transfer. The presence of ubiquitous ochre residues is not unexpected because Native Well II is a rock art gallery, and large quantities of ochre were recovered during the excavation of the site (Morwood 1979:213). However, because not all excavated pigments are represented on the gallery walls, Morwood (1979:213) argued that the pigments were applied not only to rock art but also to a diversity of artistic activities, and the results here tend to support his conclusion.

Resin and hafting

Based on the presence, form and pattern of distribution of resin, eight of the fourteen Native Well II artefacts provided distinct evidence of hafting. Hafting resin on these tools was consistently black or blackish-brown, often thick and crazed, and occasionally with lineation or striations visible in the residue. Resin was frequently associated with small (2-4 μm) starch grains and cellulose, both amorphous and fibrous, charcoal and ochre, and in one instance (NWII#3), feather residues. Three backed pieces (NWII#8, NWII#11, NWII#12) are missing the proximal end and although resin distribution appeared to relate to hafting as well as wood-working, an unequivocal inference could not be made. One artefact (NWII#2) had no visible resin, and the resin on another (NWII#14) was fragmented and distributed indiscriminately across the artefact.

Three of the hafted artefacts had been used for wood-working, two for bone-working and one for general plant processing. Artefact NWII#13 appeared to have been hafted but not used, while NWII#4 had been used but did not exhibit any identifiable use-related residues. The method of hafting was not totally clear, although the haft on the geometric microliths (NWII#3, NWII#4, NWII#6) encompassed the backed edge and obtuse angle ridge and parts of one or both lateral margins, and the haft on one backed piece (NWII#13) appeared to have been angled across part of the chord.

Summary

Of the fourteen artefacts analysed, seven had the inferred task association of wood-working, with one of these possibly also associated with a decorative or ceremonial activity. Three artefacts were used for general plant processing and two for tertiary bone-working, while one had not been used and another was used but its task association was unable to be determined. The wood-working tools have functioned as incisors and scrapers and occasionally also as knives, and those applied to bone-working and general plant processing were scraping and cutting tools. Eight artefacts exhibited evidence for hafting, although, as discussed previously, this did not indicate that the remaining artefacts had never been hafted. Of particular interest is the fact that blood residues were not observed on any of the artefacts. Bone collagen and fibrils with occasional vivianite and lipids were the only mammalian residues identified, although feather barbules occurred on two artefacts. Ochre residues are ubiquitous, which is a reasonable finding considering that the site includes a rock art gallery, and a large number and variety of ochre and pipe-clay pieces were recovered during excavation.

CONCLUSION

Although they comprise only a small percentage of the excavated stone artefacts at Native Well I and II, backed artefacts provide an interesting insight into some of the activities undertaken by the Aboriginal occupants of these two rockshelters during the mid-to-late Holocene. An integrated residue and use-wear analysis revealed that backed artefacts were used for a range of craft and subsistence activities. They functioned as knives, scrapers and/or incisors for wood-working and bone-working, and also as knives and scrapers for plant processing, including the processing of cooked starchy plants. Some artefacts with ochre and/or feather residues may have been used for ceremonial purposes, which, given that both sites have extensive galleries of rock art, is not unexpected. The presence and location of resin indicates more than half the artefacts were hafted at some stage of their use cycle, and some may have been hafted in series.

This research has made an important contribution to our knowledge of site activities at Native Well I and II. Prior to this work, interpretation of these activities generally depended on the analysis of faunal remains, the presence of ochre and shell fragments, inferences for stone tool use based on morphology, and ethnographic analogy. At these sites, although ochre was ubiquitous throughout the levels where backed artefacts were found, bone was poorly preserved and there is no ethnographic analogy for the use of backed artefacts in the Australian context. This study therefore begins the process of filling in the missing details of mid-Holocene Aboriginal life in the Queensland Central Highlands.

ACKNOWLEDGMENTS

My special thanks to Tom Loy, who introduced me to residue analysis, and was my teacher, mentor and friend throughout his time at the University of Queensland, and is sorely missed. My thanks also to both Dr Jay Hall (Reader in Archaeology, School of Social Science, University of Queensland), who initiated the project, and Dr Val Attenbrow (Principal Research Scientist, Anthropology, Australian Museum) for providing continuing support, advice and focus. I am grateful to my colleagues in the Archaeological Sciences Laboratory in the School of Social Science, especially Alison Crowther, Luke Kirkwood, Sue Nugent, and Michael Haslam. Their assistance and that of others in creating the comparative reference database used in my research, as well as their feedback on various aspects of this study, has been invaluable. Several aspects of my research required help from specialists and I particularly wish to thank Dr Peter Hiscock (Reader in Archaeology, School of Archaeology and Anthropology, Australian National University) and Dr Richard Fullagar (Honorary Research Associate, Department of Archaeology, University of Sydney). Permission to study the artefacts from Professor Mike Morwood's excavations in the Central Highlands, western Queensland was granted by the Queensland Museum, and I thank Dr Richard Robins for his support in accessing the collection.

REFERENCES

Akerman, K., R. Fullagar and A. Van Gijn 2002. Weapons and *wunan*: production, function and exchange of Kimberley points. *Australian Aboriginal Studies* 1:13-42.

Balme, J., G. Garbin and R.A. Gould 2001. Residue analysis and palaeodiet in arid Australia. *Australian Archaeology* 53:1-6.

Briuer, F.L. 1976. New clues to stone tool function: Plant and animal residues. *American Antiquity* 41(4):478-483.

Burroni, D., R.E. Donahue, A.M. Pollard and M. Mussi 2002. The surface alteration features of flint artefacts as a record of environmental processes. *Journal of Archaeological Science* 29:1277-1287.

Cattaneo, C., K. Gelsthorpe, P. Phillips and R. Sokol 1993. Blood residues on stone tools: indoor and outdoor experiments. *World Archaeology* 25(1):29-43.

Cooper, J. 2003. Tulas on the Surface: A (re)evaluation of surface collected lithics from Camooweal, northwest Queensland via microscopic residue and use-wear analysis. Unpublished BA (Honours) thesis. St Lucia: School of Social Science, The University of Queensland.

David, M. 1993. Postcards from the edge: An analysis of tasks undertaken during the late Pleistocene at Devils Lair, south-western Australia. Unpublished MA thesis. Armidale: University of New England.

Franceschi, V. and H. Horner 1980. Calcium oxalate crystals in plants. *The Botanical Review* 46:361-427.

Francis, V. 2000. What's the Point? An Investigation of the Bone Artefacts from Platypus Rockshelter, Southeast Queensland. Unpublished BA (Honours) thesis. St Lucia: Department of Anthropology, Sociology and Archaeology, The University of Queensland.

Francis, V. 2002. Twenty interesting points: An analysis of bone artefacts from Platypus Rockshelter. *Queensland Archaeological Research* 13:63-70.

Fullagar, R. 1986. Use-wear and residues on stone tools: functional analysis and its application to two southeastern Australian archaeological assemblages. Unpublished Ph.D. thesis, La Trobe University.

Fullagar, R. 1994. Objectives for use-wear and residue studies: views from an Australian microscope. *Helinium* XXXIV(2):210-224.

Fullagar, R. 1998. *A Closer Look. Recent Australian Studies of Stone Tools*, Sydney University Archaeological Methods Series. Sydney: Archaeological Computing Laboratory, University of Sydney.

Fullagar, R. and R. Jones 2004. Use-wear and residue analysis of stone artefacts from Enclosed Chamber, Rocky Cape, Tasmania. *Archaeology in Oceania* 39:79-93.

Gunning, B. and M. Steer 1975. *Ultrastructure and the Biology of Plant Cells*. London: Edward Arnold Ltd.

Haslam, M. 1999. What a Dump: Use-wear and Residue Analysis of Lithic Artefacts from Copan, Honduras. Unpublished BA (Honours) thesis. St Lucia: Department of Anthropology and Sociology, The University of Queensland.

Haslam, M. 2004. The decomposition of starch grains in soils: implications for archaeological residue analyses, *Journal of Archaeological Science* 31: 1715-1734.

Hiscock, P. 1993. Bondaian Technology in the Hunter Valley, New South Wales. *Archaeology in Oceania* 28(2):65-76.

Hiscock, P. 2002. Pattern and context in the Holocene proliferation of backed artefacts in Australia. In R.G. Elston and S.L. Kuhn (eds) *Thinking Small: Global Perspectives on Microlithization*, Archaeological Papers of the American Anthropological Association (AP3A). vol. 12, pp. 163-177.

Hiscock, P. and V. Attenbrow 1996. Backed into a corner. *Australian Archaeology* 42:64-65.

Horner, H. and B. Wagner 1995. Calcium oxalate formation in higher plants. In S.R. Khan (ed.) *Calcium Oxalate in Biological Systems*, pp. 53-71. Boca Raton, Florida: CRC Press.

Hurcombe, L.M. 1992. *Use Wear Analysis and Obsidian: Theory, Experiments and Results.* Sheffield Archaeological Monographs. Sheffield: J.R. Collis Publications.

Kamminga, J. 1982. *Over the Edge: Functional analysis of Australian stone tools.* Occasional Papers in Anthropology 12. St Lucia: Anthropology Museum, University of Queensland.

Kooyman, B. 2000. *Understanding Stone Tools and Archaeological Sites.* Calgary, Alberta: University of Calgary Press.

Langenheim, J. 2003. *Plant Resins: Chemistry, evolution, ecology, and ethnobotany.* Portland, OR: Timber Press.

Lombard, M. 2005. Evidence of hunting and hafting during the Middle Stone Age at Sibidu Cave, KwaZulu-Natal, South Africa: a multianalytical approach. *Journal of Human Evolution* 48(3):279-300.

Lombard, M. and L. Wadley 2005. The morphological identification of micro-residues on stone tools using light microscopy: progress and difficulties based on blind tests. *Journal of Archaeological Science* 34(1):155-165.

Loy, T.H. 1983. Prehistoric Blood Residues: Detection on Tool Surfaces and Identification of Species of Origin. *Science* 220:1269-1271.

Loy, T.H. 1985. Preliminary residue analysis: AMNH specimen 20.4/509. In D.H. Thomas (ed.) *The Archaeology of Hidden Cave*, Anthropological Papers of the American Museum of Natural History. vol. 61, (1), pp. 224-225. Washington: American Museum of Natural History.

Loy, T.H. 1990. Prehistoric Organic Residues: Recent advances in identification, dating, and their antiquity (paper presented at Archaeometry '90, Basel).

Loy, T.H. 1993. Prehistoric organic residue analysis: the future meets the past. In W. Ambrose, A. Andrews, R. Jones, A. Thorne, M. Spriggs and D. Yen (eds) *A Community of Culture*, pp. 56-72. Canberra: Department of Prehistory, Research School of Pacific Studies, Australian National University.

Loy, T.H. 1994. Residue Analysis of Artifacts and Burned Rock from the Mustang Branch and Barton Sites (41HY209 and 41 HY202). In R. Ricklis and M. Collins (eds) *Archaic and Late Prehistoric Human Ecology in the Middle Onion Creek Valley, Hays County, Texas*, Studies in Archaeology. vol. 1, pp. 607-27. Austin, Texas: Texas Archaeological Research Laboratory.

Loy, T.H. and B.G. Hardy 1992. Blood residue analysis of 90,000 year old stone tools from Tabun Cave, Israel. *Antiquity* 66:24-35.

Loy, T.H. and D.E. Nelson 1986. Potential applications of the organic residues on ancient tools. In J. Olin and J. Blackman (eds) *Proceedings of the 24th International Archaeometry Symposium*, pp. 179-185. Washington, D.C.: Smithsonian Institution Press.

Loy, T.H. and A.R. Wood 1989. Blood residue analysis at Cayönü Tepesi, Turkey. *Journal of Field Archaeology* 16:451-460.

Loy, T.H. and E.J. Dixon 1998. Blood residues on fluted points from Eastern Beringia. *American Antiquity* 63(1):21-46.

McCarthy, F.D. 1976. *Australian Aboriginal Stone Implements*. Second Edition (Revised): The Australian Museum Trust.

McCarthy, F.D., E. Bramell and H.V.V. Noone 1946. The Stone Implements of Australia. *Memoirs of the Australian Museum* IX (November).

McNiven, I. 1993. Tula adzes and bifacial points on the east coast of Australia. *Australian Archaeology* 36:22-31.

Morwood, M.J. 1979. Art and Stone. Unpublished PhD thesis, Australian National University.

Morwood, M.J. 1981. Archaeology of the Central Queensland Highlands: the stone component. *Archaeology in Oceania* 16:1-52.

Odell, G.H. 2004. *Lithic Analysis*Manuals in Archaeological Method, Theory and Techniques. New York: Kluwer Academic/Plenum Publishers.

Raven, P.H., R.F. Evert and S.E. Eichhorn 1999. *Biology of Plants*. Sixth edition. New York: W.H.Freeman and Co.

Robertson, G. 2002. Birds of a feather stick: microscopic feather residues on stone artefacts from Deep Creek Shelter, New South Wales. In S. Ulm, C. Westcott, J. Reid, A. Ross, I. Lilley, J. Prangnell and L. Kirkwood (eds) *Barriers, Borders, Boundaries: Proceedings of the 2001 Australian Archaeological Association Annual Conference*, pp. 175-182. Tempus 7. St Lucia: Anthropology Museum, The University of Queensland.

Robertson, G. 2005. Backed Artefact Use in Eastern Australia: A Residue and Use-Wear Analysis. Unpublished Ph.D. thesis, University of Queensland.

Robertson, G. 2006. Diatoms and sponge spicules as indicators of contamination on utilised backed artefacts from Turtle Rock, Central Highlands, western Queensland. In S. Ulm and I. Lilley (eds) *An Archaeological Life: Papers in honour of Jay Hall*, Aboriginal and Torres Strait Islander Studies Unit Research Report Series. vol. 7, pp. 125-140. Brisbane: Aboriginal and Torres Strait Islander Studies Unit, The University of Queensland, Brisbane.

Rots, V. 2003. Towards an understanding of hafting: the macro- and microscopic evidence. *Antiquity* 77(298):805-815.

Rots, V. and B.S. Williamson 2004. Microwear and residue analyses in perspective: the contribution of ethnoarchaeological evidence. *Journal of Archaeological Science* 31:1287-1299.

Schiffer, M.B. 1972. Archaeological context and systemic context. *American Antiquity* 37(2):156-165.

Sobolik, K. 1996. Lithic organic residue analysis: an example from the Southwestern Archaic. *Journal of Field Archaeology* 23(4):461-469.

Tomlinson, N. 2001. Residue analysis of segments, backed, and obliquely backed blades from the Howieson Poort layers of the Middle Stone Age site of Rose Cottage Cave, South Africa. Unpublished MA thesis. University of Witwatersrand.

Vaughan, P.C. 1985. *Use-wear analysis of flaked stone tools*. Tucson, Arizona: University of Arizona Press.

Wadley, L. and M. Lombard 2007. Small things in perspective: the contribution of our blind tests to micro-residue studies on archaeological stone tools. *Journal of Archaeological Science* 34(6):1001-1010.

Wadley, L., M. Lombard and B. Williamson 2004a. The first residue analysis blind tests: results and lessons learnt. *Journal of Archaeological Science* 31:1491-1501.

Wadley, L., B. Williamson and M. Lombard 2004b. Ochre in hafting in Middle Stone Age southern Africa: a practical role. *Antiquity* 78(301):661-675.

Wallis, L. and S. O'Connor 1998. Residues on a sample of stone points from the west Kimberley. In R. Fullagar (ed.) *A Closer Look. Recent Australian Studies of Stone Tools*, pp. 150-178. Sydney University Archaeological Methods Series 6. Sydney: Archaeological Computing Laboratory, University of Sydney.

Williamson, B.S. 2000. Prehistoric Stone Tool Residue Analysis from Rose Cottage Cave and other Southern African Sites. Unpublished PhD thesis. University of Witwatersrand.

19

Deadly weapons: backed microliths from Narrabeen, New South Wales

Richard Fullagar[1], Josephine McDonald[2], Judith Field[3] and Denise Donlon[4]

1. Scarp Archaeology
25 Balfour Road, Austinmer NSW 2515 Australia
Email: richard.fullagar@scarp.com.au

2. Jo McDonald Cultural Heritage Management and
Research School of Humanities, Australian National University
77 Justin St., Lilyfield, NSW 2040 Australia

3. Australian Key Centre for Microscopy and Microanalysis F09 and
School of Philosophical and Historical Inquiry
The University of Sydney, NSW 2006 Australia

4. Department of Anatomy and Histology
The University of Sydney, NSW 2006 Australia

ABSTRACT

A recently excavated skeleton dated to 3677 cal BP provides an extraordinary opportunity to determine the function of its associated backed artefacts. Seventeen stone artefacts were recovered during salvage excavation of an adult male Aboriginal skeleton from a sand dune in Narrabeen, a coastal suburb of Sydney. The skeletal and artefact evidence indicate death by spearing. Three artefacts were refitted, and, of the 14 near complete artifacts, 12 have been clearly backed. One backed artefact was found lodged between the L2 and L3 vertebrae with unhealed wounds, indicating spear penetration near the left hip. Other backed artefacts were found adjacent to or lodged in vertebrae suggesting two spears had penetrated from the back. Breakage and use-wear on most artefacts indicate use as barbs or 'lacerators'. In this study, we describe the use-wear and suggest possible hafting arrangements of these backed microliths, which probably functioned as piercing, cutting and lacerating elements of spears and knives.

KEYWORDS

use-wear, residues, backed artefacts, microliths, spears, knives

INTRODUCTION

A recently excavated skeleton dated to 3677 cal BP provides an extraordinary opportunity to determine the function of its associated backed artefacts (Figure 1). Seventeen stone artefacts (Figures 2 – 4; Table 1) were recovered during salvage excavation of an adult male Aboriginal skeleton, exposed during cable installations in a sand dune, 1.5 m below the present ground level in Narrabeen, a coastal suburb of Sydney (McDonald *et al.* 2007). The skeletal and artefact evidence indicates death by spearing.

A backed artefact (OON1; Figure 2) found during excavation was lodged between the second and third lumbar vertebrae in the region of the intervertebral disc, with major unhealed

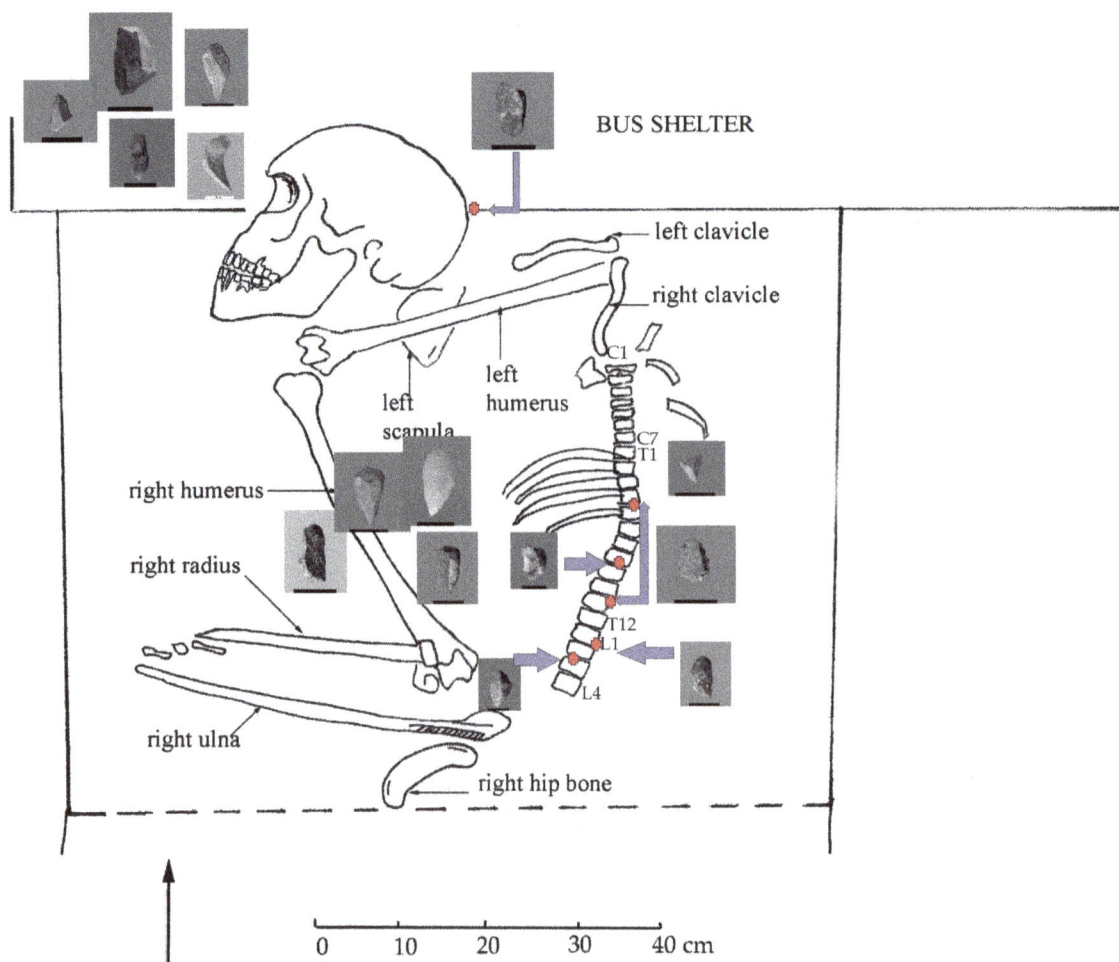

Figure 1. Schematic diagram of the upper torso and the location of the backed artefacts associated with the skeleton. The head has moved approximately 40 cm away from the vertebral column; however the mandible is still articulated. Dots on the spinal column indicate the location of the lodged backed artefacts (arrowed) and the placement of the other images indicates schematically the locations in which the artefacts were found (Reproduced from Antiquity [McDonald et al. 2007:879, Figure 1] with permission).

damage to the body of L2 and minor but unhealed damage to the body of L3 (McDonald *et al.* 2007). The artefact has bone residue (similar in colour and structure to the human vertebra) embedded in cracks at the crushed tip. If it were a spear barb, tip or lacerator – the latter a term employed by Kim Akerman (pers. comm.) to describe fragments of stone designed to release from the haft like shrapnel to aggravate haemorrhaging and other internal injuries (rather than to hold a spear in the wound as a barb might function) – this artefact would have entered the body on the left hand side, just above the blade of the left hipbone, assuming the body was in normal anatomical position and a horizontal entry wound. This spear probably passed through the large and small intestines and came close to the left renal artery and vein, and possibly the aorta. Backed artefact OON14 (Figure 3) was found in the position of the (missing) spinous process of the 11th thoracic vertebra. A tiny fragment (OON15, Figure 4) that refits to OON14 was later found in the vertebral canal of another thoracic vertebra (T4). Another backed artefact, OON16 (Figures 1 and 3), was found near L1 although this does not appear to have damaged the bone. Given their positions, these two artefacts (OON14 and 16) are likely to have been part of a spear that entered the back of the individual from the rear.

The artefacts were grouped in four areas around the skeleton (Figure 1): six stone artefacts (OON1, 11, 14+15, 16 and 17) found near the vertebral column; four (OON5,

Table 1. Summary of use-wear and residues on stone artefacts from the Narrabeen site (Reproduced from *Antiquity* [McDonald *et al.* 2007, Table 1 at http://www.antiquity.ac.uk/projgall/mcdonald] with permission)

OON No.	Type¹	Location	Refit with no.	Stone material	Length (mm)	Width (mm)	Thick. (mm)	TCSA²	Retouch	Damage	Usewear	Residues³	Hafting	Use⁴	Function⁵
1	Backed flake	Backed blade, vertebral column		Pink-red silcrete	17.4	10	8	40	Bi-direct. backing	Broken tips.	Rounding on backed edge; none on chord, use scar at proximal tip; cf. barb	Grey residue on backed edge	Probably hafted. Grey residue is possibly resin	3	[A] Impact, probably projectile barb
2	Backed flake	Around skull		Grey quartzite	21	11	5	27.5	Bi-direct. backing	No breakage	Rounding, polish, striae	Dark smears cellulose, starch on backed edge	Probably hafted. Dark smears are possibly resin.	3	[C] Piercing and slicing skin. Not from a projectile.
3⁶	Backed broken flake (tip)	Around skull	4	Pink silcrete	10	6.3	4.2	13.2	Bi-direct. backing	Crushed tip	Crushing at tip		Complex fracture, probably from hafting configuration	3	[B] Likely damage from projectile tip
4	Backed flake		3	Pink silcrete	16	10	11	55	Bi-direct. backing	Steps from break	Scarring on chord	Plant tissue, charcoal, carbonate		3	
5	Backed flake	West side vert column		Red silcrete	17.4	7.7	5.8	22	Uni-direct. backing initiated on ventral		Impact scar on tip		Probable	3	[B] Likely damage from projectile tip
6	Backed fragment	Underneath skull	8	Red silcrete	13.3	8.4	5.3	22.3	Bi-direct. backing but rare	Break is probably along 'old' fracture caused by backing.	Probable impact scar.	Impacted yellow tissue same colour as bone fragments.	Probable	2	[B] Likely damage from projectile tip
8	Fragment (tip)	Underneath skull	6	Red silcrete	6	4.6	2.4	na						2	

No.	Type	Context		Material				TCSA	Backing		Rounding and step scar on tip	Dark residues on backed edge	Dark residue is possibly hafting resin	Use	Function
7	Backed flake	Excavated around skull		Grey quartzite	18.8	8.6	4.6	19.8	Bi-direct. backing near tip		Rounding and step scar on tip			3	Uncertain, tip used
9	Backed flake			Quartzite	14.2	6.5	3.6	11.7	Bi-direct. backing		Possible impact scar			1	Uncertain, tip used
10	Bipolar piece	Around skull		Quartz	15.7	7	4.4	15.4	Backing not clear	Bipolar crushing and scars	Uncertain	Carbonised plaques – probably not from use		1	Uncertain, possible use of tip
11	Backed flake	Vertebral column		Quartzite	15.1	10.4	4	20.8	Bi-direct. backing, not very steep	Tip broken	Scars associated with broken tip		Probable	3	[A] Hard impact, possibly from projectile
12	Backed flake			Red silcrete	15	7.5	5.6	21	Uni-direct. backing, initiated on ventral	Tip broken	Rounding and longitudinal striations near tip			3	[C] Uncertain, probably not hard impact, awl or projectile?
13	Backed fragment	Dry sieved		Quartzite	19.7	10	4.2	21	Bi-direct. backing	Tip crushed or broken?	Scarring on chord, tip crushed	Black residue on backed edge	Uncertain	2	Uncertain, possible use of tip, awl or projectile?
14	Backed fragment	Dry sieved	15	Red silcrete	11	7.6	3.8	14.4	Bi-direct. backing	Tips broken	Rounding, bending scars along chord	Sediment, unidentified particles	Probably hafted	3	[A] Hard impact possibly from projectile; also considerable damage along chord
15	Backed fragment (tip)	Inside vertebral canal. Dry sieved	14	Red silcrete	4.5	3.6	2	na	Bi-direct. backing	Missing fragments.	Impact scar cf. barb			3	
16	Backed flake	Between L1 and L2		Red silcrete	18.8	8.8	3.8	na	Bi-direct. backing	Tip broken	Impact scar cf. barb, scarring		Probably hafted	3	[A] Hard impact possibly from projectile
17	Fragment			Grey quartzite	7.5	5.6	4.1	na	Possible backing, uni-direct.	Broken	Impact damage unclear			1	Uncertain, possible use of tip

1. Type refers to presence of backing and technological classes
2. Tip Cross Sectional Area after Shea (2006)
3. Observations are from incident light only. Residues have not yet been removed for study under transmitted light microscopy.
4. Use refers to confidence of interpretations, as follows: 0 (no traces of use), 1 (possible), 2 (probable) and 3 (definite).
5. Function refers to the likely cause of dominant usewear. [A] indicates protruding oblique impact; [B] indicates head-on impact; [C] indicates dominant function not from hard impact.
6. Grey shading indicates conjoin set.

Figure 2. Both sides of artefacts OON1 to 8 (in sequence from upper left to right, and down the page) (scale bar = 1 cm).

9, 12, and 13) were found near the right arm (humerus); five (00N2, 3+4, 7 and 10) were found near the front of the skull, and two (00N6+8) were found just behind and underneath the skull.

A total of 17 stone artefacts including three conjoin sets (Figures 4-6; Table 1) were found, resulting in 14 near complete artefacts. Of these, twelve have clear backing retouch and two others (OON10 and 17) have indistinct edge crushing, which may also be the result of deliberate backing retouch.

STONE PROJECTILES, POINTS, LACERATORS AND BARBS

Archaeological evidence for, and diagnostic indicators of, projectile tips have been important in tracking hunting technology and modern human evolution (Shea 2006). Diagnostic use-wear traces have been reported on experimental stone tipped arrows and spears (e.g. Boot 2005; Dockall 1997; Fischer *et al.* 1984; Lombard 2005; Odell 2004: 178-9; Odell and Cowan 1986). Dockall (1997) reviews the range of impact breaks, macrowear, and microwear that have been considered

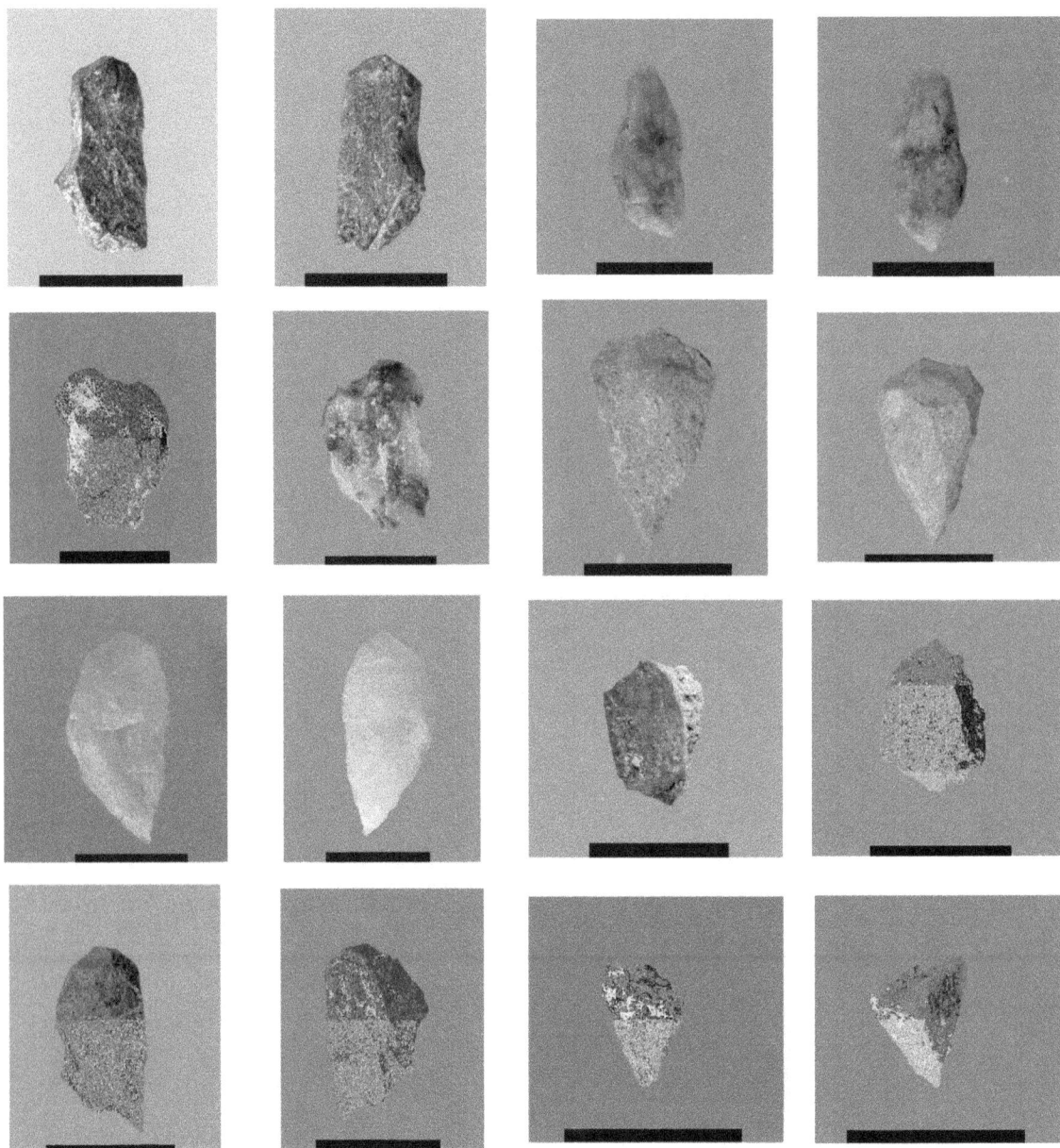

Figure 3. Both sides of artefacts OON9 to 17 (in sequence from upper left to right, and down the page). Note that the tiny fragment OON15 (which refits tip the tip of OON14) is excluded (scale bar = 1 cm).

diagnostic either alone or in combination with other traces of use such as linear polish, striae, edge rounding, longitudinal macroscars, lateral macroscars, distal breaks, distal crushing and spin-off factures. Using these categories, we provide a summary of the traces found on the Narrabeen artefacts (Table 2). Longitudinal macroscars and lateral macroscars were not found on these small backed artefacts. Step and feather terminated bending scars occurred along the backed margins of several artefacts, indicating head-on and oblique impact, depending on the force producing the fractures (Figure 5). Neither microscopic linear streaks of polish or 'MLIT' (Fischer *et al.* 1984; van Gijn 1990:45-46) nor edge rounding were distinctly visible on any artefacts with diagnostic impact damage, but this was perhaps because of the grainy stone material. However, rounding and weakly developed polish was observed on the tip of OON2 (Figure 6), the chord of OON14 (Figure 7) and near the tip of OON12 (Figure 8). Linear striations were also very rare and only visible at high magnification in the form of possible scratches on quartz crystals. Rounding was visible on the fragile tip of OON2 as well as OON7, 12 and 14.

Distal crushing and breaks were both common, the latter occurring mostly in the form of scars with bending initiations and step or feather terminations along an arris or main edge of the

Figure 4. Detail of conjoin OON14 and 15, showing narrow feather terminating bending scar down the backed edge.

backed margin (Table 2). Spin off fractures that appear to be initiated from the bending scars snapped from the tips were common (e.g. Figure 9), and are thought by Fischer *et al.* (1984) and Lombard (2005) to be a diagnostic impact fracture on points hafted as arrows or spears. It is uncertain whether the small robust backed artefacts (as in the Narrabeen assemblage) will break in quite the same way, although it seems likely. Further experiments are needed, particularly to model variables such as the effect of hafting arrangements and impact forces. Proximal damage was also rare and less marked than distal damage near the tips.

Few distinctive residue structures or films were observed directly on the artefacts or in extractions after aqueous sonication (Table 1). Embedded in cracks on some artifacts were fragments similar in colour and structure to bone; and on the backed edges of some artefacts there are dark, opaque smears (thought to be resin). Cellulose fibres and starch were noted on OON2. Presumably the open sandy environment was inimical to survival of blood and other tissues.

Table 2. Wear traces found on the Narrabeen artefacts. STBS back: step terminated bending scar on the backed surface; FTBS back: feather terminated bending scar on the backed surface. Artefact numbers with '*' indicate that backing retouch is not distinct. No longitudinal macroscars or lateral macroscars were observed (cf. Dockall 1997)

No.	Find Location	Linear polish	Striae	Edge rounding	Distal breaks	Distal crushing	Spin-off factures	Proximal damage
1	Spine	-	?	-	STBS back	-	-	
2	Skull front	x	?	x	-	-	-	
3-4	Skull front	-	-	-	STBS back	x	x	
5	Right humerus	-	-	-	STBS back	x	-	
6-8	Skull back	-	-	-	STBS lateral	x	x	
7	Skull front	-	-	x	Step? lateral	x	-	x
9	Right humerus	-	-	-	Steps back	x	x	-
10*	Skull front	-	-	-	FTBS	x	-	x
11	Spine	?	-	-	STBS back	-	x	
12	Right humerus	?	x	x	snap	x	-	-
13	Right humerus	-	-	-	-	?	-	-
14-15	Spine	?	-	x	FTBS back	-	xx	x
16	Spine	-	-	-	STBS back	-	x	-
17*	Spine	-	-	-	?	?	-	?

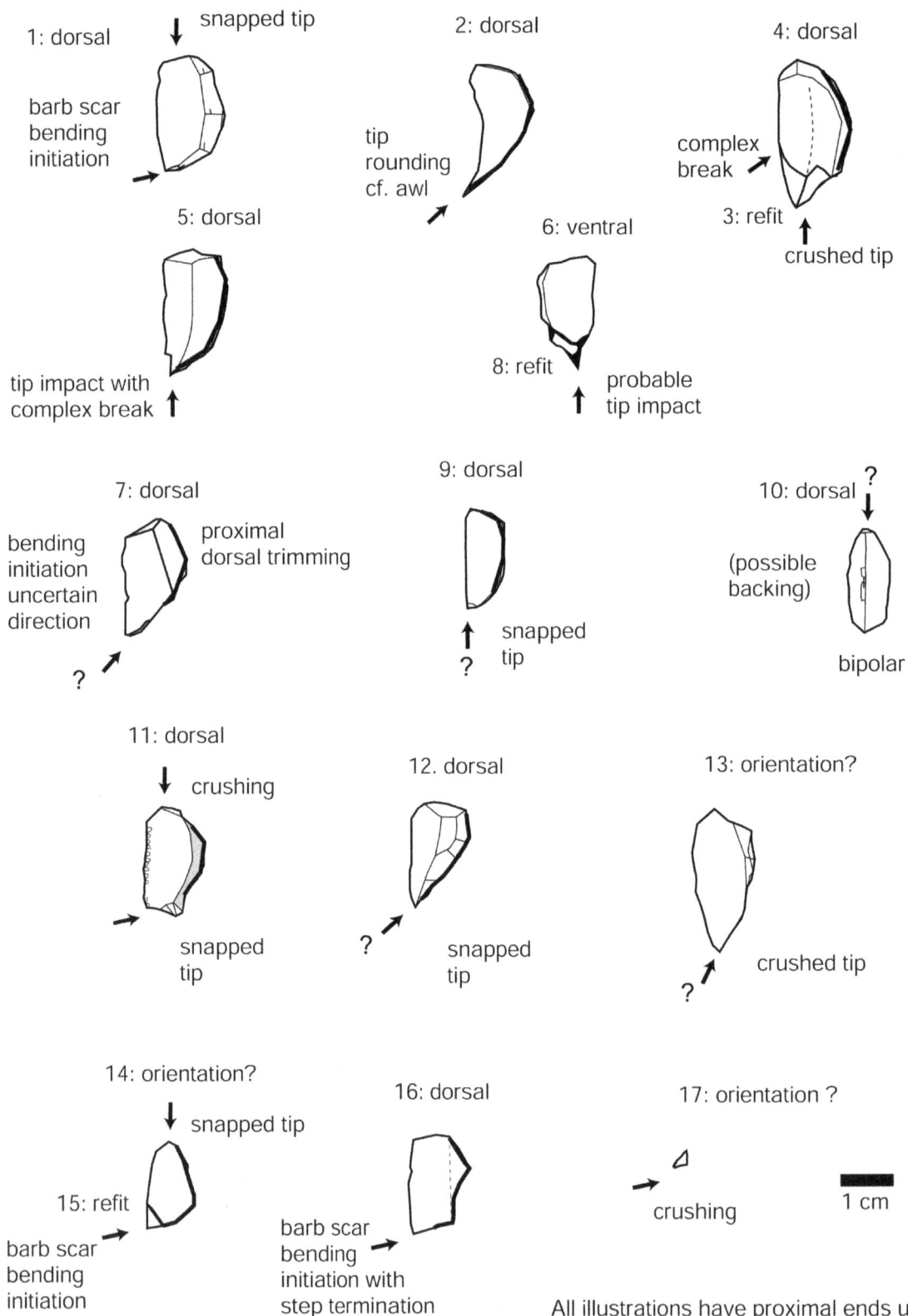

Figure 5: Narrabeen artefacts showing suggested impact direction and breaks.

Figure 6: Microwear on the fragile tip of OON2, with marked rounding and polish indicating function as a skin working implement (awl).

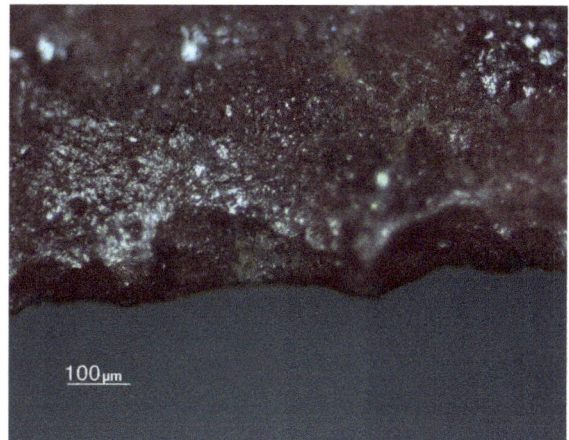

Figure 7: Edge scarring and rounding on the chord of OON14.

Figure 8: Rounding and faint striations near the tip of OON12.

Figure 9. Tip break of OON3+4, showing a long narrow fracture with a step termination that initiates a spin off fracture (with step termination).

DISCUSSION

Possible hafting arrangements (see Figure 10), given the fractures and wear traces, must account for hard impact on small asymmetric stone artefacts which have hafting traces in the form of dark smears similar to resin, and rounding along the backed edge. Such weapons, armed with lithic barbs, lacerators or tips, might have been thrown (e.g. spears), stabbed (e.g. spears or knives) or swung (e.g. clubs). The wear traces are all consistent with use as hafted elements of spears, knives or even 'barbed' clubs (i.e. clubs studded with backed artifacts). Ethnographic, experimental and contextual evidence indicate that spears and knives are likely. For example, OON1 is most likely to be from a spear simply because of the penetration requirements from left hip to spine. Conjoined artefacts OON14 and 15 together with OON16 could be from one or more spears or stabbing knives.

The absence of distinctive wear along the backed edge and the proximal end of each flake suggests that the backed artefacts were not firmly slotted into wooden or bone handles, which might be expected for reliable use as a knife or club, although 'taap' saw- knives restricted to southwest Western Australia were probably used for general butchering. However, these were 'resin hafted' and not slotted into their wooden handles (Kamminga 1982:32). On the other hand, ethnographic data (e.g. Akerman et al. 2002) show that stone lacerators and tips may be deliberately set in resin

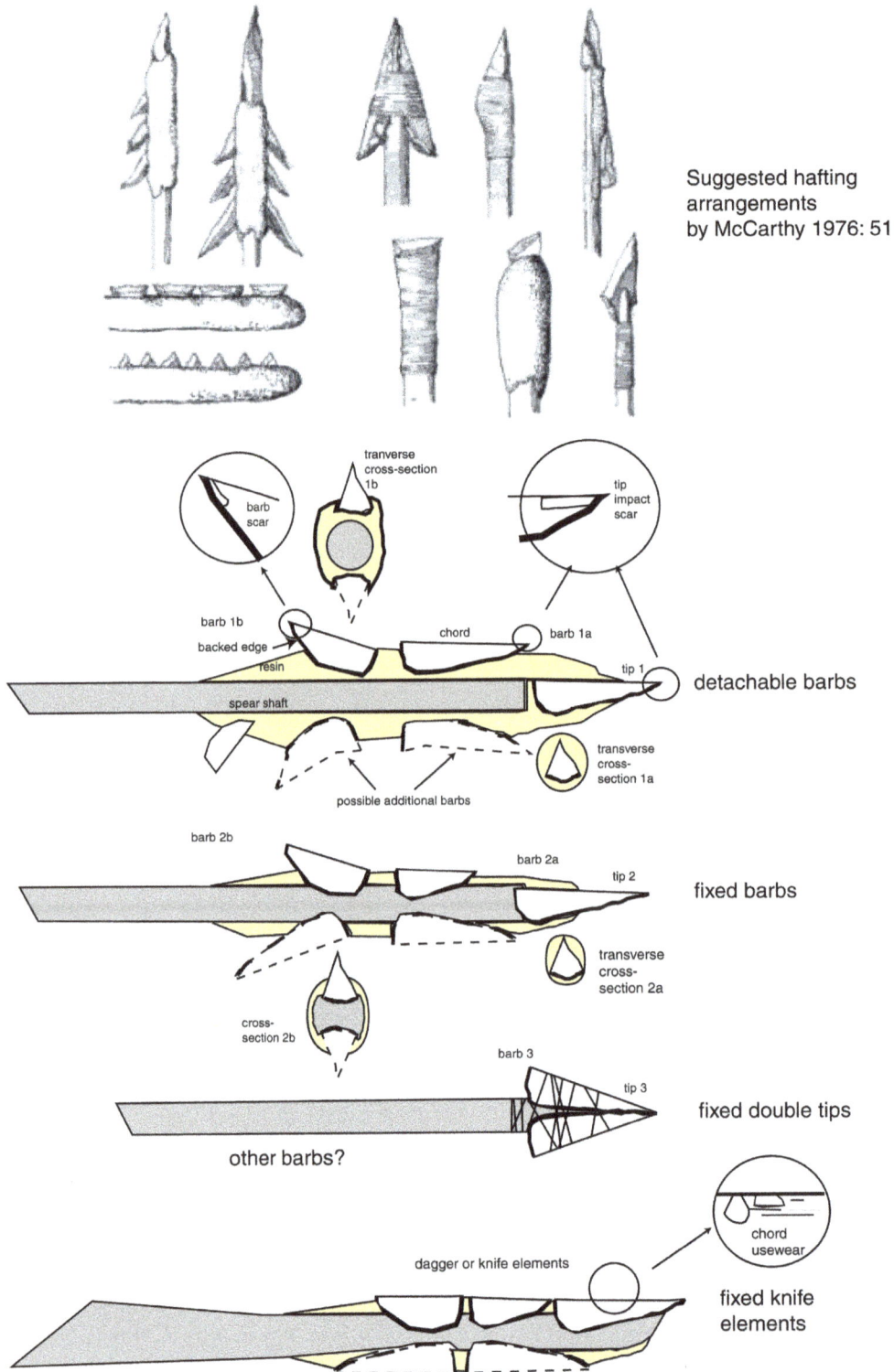

Suggested hafting
arrangements
by McCarthy 1976: 51

Figure 10. Possible hafting arrangements of backed artefacts. McCarthy's suggested hafting arrangements reproduced with permission from The Australian Museum.

away from direct contact with the spear shaft, so that the chipped stones could easily detach on impact (Akerman 1978). Kim Akerman (pers. comm. April 15th 2007) suggested that the effect of multiple detachable lacerators would be similar to the effect of shrapnel wounds. Several Narrabeen artefacts have edge rounding and other use-wear on the chord indicating use as knives, and there is one awl. Small stones including awls, edge elements of knives and other tools could have been re-cycled as lacerators, in much the same way as the stone chips of the ethnographically known death spear (e.g. Dortch 1984:53; see also an illustration of a death spear [Collector : Unknown (A4932)] on the South Australian Museum Website, 2007). It is also possible that detachable fore shafts (with *firmly* attached backed artefacts) might have been removed from spears and used for a variety of tasks including butchery, despite their primary function as projectile heads (see also the description by Davidson (1934:61) of reed shafts with (detachable?) hardwood heads armed with stone flakes).

CONCLUSION

The skeletal injuries, penetration depth, distribution of fragments and use-wear indicate a minimum of three weapons, and probably more, were used in the slaying of the Narrabeen man. Substantial proximal (tip) damage and spin off fractures thought to be diagnostic of projectile impact are found on artefacts in all find locations (skull – front and back, spine and right humerus), suggesting a minimum of three spears. Six artefacts may be barbs, lacerators or tips with spin-off fractures (depending on the possible orientation and hafting arrangements). These six indicate the maximum number of possible high impact contacts (e.g. with bone). If we assume that only the initial impact of each spear with the victim is likely to result in such damage (to lacerators, barbs or tips), then up to six spears each armed with one tip and two or three barbs or lacerators seems a likely configuration. Of course, there are many assumptions involved in such reconstructions, and we have outlined the logic of some possibilities. Trying to test and evaluate each possibility is fraught with difficulty, and Tom Loy (to whom this volume is dedicated, and who kept revising elements of Ötzi's alpine mummy mystery) would be familiar with such unfinished stories! We are planning further experimental work to evaluate likely hafting arrangements.

The Narrabeen artefacts provide the first Australian archaeological evidence of backed artefacts used for fighting, payback killing or other human violence, as distinct from hunting game, as commonly inferred (Kamminga 1980; McBryde 1985, 1986; McCarthy 1976). If the Narrabeen artefacts were recycled tools and hafted in similar fashion to the stone lacerators of the death spear, this evidence may also be consistent with the wide range of functions identified recently by Robertson (2005), Fullagar *et al.* (1994) and McDonald *et al.* (1994). The timing of this mid-Holocene occurrence of payback or other killing correlates with a widespread proliferation of backed artefacts in the archaeological record, particularly in south-eastern Australia (Hiscock & Attenbrow 2005a, 2005b; Jo McDonald Cultural Heritage Management 2005). While the slaying of the Narrabeen man may be related to climate change, increased stress, shifts in subsistence and settlement and an increased social proscription, it seems less and less likely that backed artefacts as a class have a dominant primary function. They appear to have been used for many purposes in different times and places. We have demonstrated here that one of these functions was as detachable lacerating elements of death spears.

ACKNOWLEDGEMENTS

We thank the Editors and Symposium organisers for the opportunity to present these results. We thank Allan Madden of the Metropolitan Local Aboriginal Land Council for his continued interest in this research and permission to publish details pertaining to the Narrabeen man's remains. We also thank several anonymous referees, as well as Annie Ross, Veerle Rots, Val Attenbrow, and also Kim Akerman for his specific comments on barbs, lacerators and hafting arrangements. The authors acknowledge both the facilities and scientific and technical assistance from the staff in the Australian Key Centre for Microscopy and Microanalysis Research Facility (AMMRF) and the Australian Key Centre for Microscopy and Microanalysis at the University of Sydney.

REFERENCES

Akerman, K. 1978. Notes on the Kimberley stone-tipped spear, focusing on the hafting mechanism. *Mankind* 11(4): 486.

Akerman, K., R. Fullagar and A. van Gijn 2002. Weapons and *wunan*: production, function and exchange of spear points from the Kimberley, northwestern Australia. *Australian Aboriginal Studies* 1:13-42.

Boot, P. 2005. Transverse snapping on stone artefacts. In I. McFarlane, M.J. Roberts and R. Paton (eds.) *Many Exchanges: Archaeology, History, Community and the Work of Isobel McBryde*, pp. 343-366. Canberra: Aboriginal History Monograph..

Davidson, D.S. 1934. Australian spear-traits and their derivations. *Journal of the Polynesian Society* 43(1):41-72.

Dockall 1997. Wear traces and projectile impact: a review of the experimental and archaeological evidence. *Journal of Field Archaeology* 24(3): 321-331.

Dortch, C. 1984. *Devil's Lair, A Study in Prehistory*. Perth: Western Australian Museum.

Fischer, A., P.V. Hensen and P. Rasmussen 1984. Macro and micro wear traces on lithic projectile points. *Journal of Danish Archaeology* 3:19-46.

Fullagar, R., J. Furby and L. Brass 1994. Stone artefacts from Bulga, Hunter valley. Unpublished report for M. Koettig. [Fullagar, R., J. Furby and L. Brass 1994. Use-wear and residue analysis of stone tools from Bulga. In Bulga Lease Authorisation 219 salvage excavations. A report to Saxonvale Coal Pty Ltd, by M. Koettig, Vol. 5, pp. 26-105].

Hiscock, P. and Attenbrow V. 2005a. Reduction continuums and tool use. In C. Clarkson & L. Lamb 2005. (eds) *Lithics Down Under: Recent Australian Approaches to Lithic Reduction, Use and Classification*, pp. 43-55. BAR International Monograph Series S1408. Oxford: Archaeopress..

Hiscock, P. and V. Attenbrow 2005b. *Australia's Eastern Regional Sequence revisited: technology and change at Capertee 3*. BAR International Monograph Series 1397. Oxford: Archaeopress.

Jo McDonald Cultural Heritage Management 2005. *Salvage Excavation of Six Sites along Caddies, Second Ponds, Smalls and Cattai Creeks in the Rouse Hill Development Area, NSW*. Australian Archaeological Consultancy Monograph Series, Volume 1 [Available online: http://www.aacai.com.au/monograph/index.html].

Kamminga, J. 1980 A functional investigation of Australian microliths. *The Artefact* 5:1-18.

Kamminga, J. 1982. *Over the Edge. Functional Analysis of Australian Stone Tools*. Occasional Papers In Anthropology 12. St Lucia: Anthropology Museum, University of Queensland.

Lombard, M. 2005. A method for identifying stone age hunting tools. *South African Archaeological Bulletin* 60(182):115-120.

McBryde, I. 1985. Backed blade industries from the Graman rock shelters, New South Wales: some evidence on function, in V. N. Misra and P. Bellwood (ed.) Recent advances in Indo-Pacific Prehistory, pp. 231-249. New Delhi: Oxford and IBH Publishing.

McBryde, I. 1986. The broken artefact and functional studies. In G. Ward (ed.) *Archaeology at ANZAAS, Canberra*, pp.203-209Canberra: Canberra Archaeological Society.

McCarthy, F. 1976. *Australian Aboriginal Stone Implements, Including Bone, Shell and Tooth Implements*. Sydney: The Australian Museum Trust.

McDonald, J., Rich, E., and Barton, H. 1994 The Rouse Hill Infrastructure Project (Stage 1) on the Cumberland Plain, western Sydney. In M.E. Sullivan, S. Brockwell and A. Webb (eds) *Archaeology in the North: Proceedings of the 1993 Australian Archaeological Association Conference*, pp. 259-293. Darwin: North Australian Research Unit, Australian National University.

McDonald, J. D. Donlon, J. Field, R. Fullagar, J. Brenner Coltrain, P. Mitchell and M. Rawson 2007. The first archaeological evidence for death by spearing in Australia. *Antiquity* 81:877-885.

Odell, G. 2004. *Lithic Analysis*. New York: Kluwer Academic/Blackwell.

Odell, G. H. and F. Cowan 1986. Experimentation with spears and arrows using animal targets. *Journal of Field Archaeology* 13:195-212.

Robertson, G. R. 2005. Backed artefact use in Eastern Australia: a residue and use-wear analysis. Unpublished PhD thesis. St Lucia: School of Social Science, The University of Queensland.

Shea J. 2006. The origins of lithic projectile point technology: evidence from Africa, the Levant, and Europe. *Journal of Archaeological Science* 33 (6):823-846.

South Australian Museum Website [URL: http://www.samuseum.sa.gov.au/ngurunderi/ng6htm.htm; Accessed May 2007].

van Gijn, A. L. 1990 The wear and tear of flint: principles of functional analysis applied to Dutch Neolithic assemblages. *Analecta Praehistorica Leidensia* 22.

EDITOR BIOGRAPHIES

Michael Haslam is a Postdoctoral Fellow in the Leverhulme Centre for Human Evolutionary Studies at the University of Cambridge. He is interested in the fine-scale reconstruction of past technical and social activities through microscopic residue and use-wear analysis, and the integration of these findings into broader archaeological and evolutionary perspectives on human behaviour. He obtained his PhD from The University of Queensland. Recent and current collaborative projects include analysis of the stone tools associated with *Homo floresiensis* in Indonesia (with colleagues at The University of Queensland), study of the Gibraltar Neanderthal stone artefacts, and examination of the effects of the 74kyr BP Toba super-eruption on hominins in India.

Gail Robertson is a Research Associate in the School of Archaeology and Anthropology, Australian National University. She obtained a PhD from The University of Queensland where she is currently an Honorary Research Advisor in the School of Social Science. Her research specialties include microscopic analysis of archaeological residues and use-wear and subsequent interpretation of stone tool use. A recent research project involved analysis of stone artefacts found in association with *Homo floresiensis* in Indonesia. She is currently collaborating in an Australian Research Council funded research project "Evolution of Technology and Tool Use in 10,000 years of Aboriginal history" with colleagues at the ANU and Australian Museum.

Alison Crowther is a Marie Curie Research Fellow in Archaeobotany at the Department of Archaeology, University of Sheffield. She is completing her doctoral degree in the School of Social Science, The University of Queensland, on the analysis of starch residues on Lapita pottery from the western Pacific. Her research integrates microbotanical evidence for food preparation and consumption activities into models for prehistoric subsistence and the development of agriculture, involving projects in the Pacific Islands, Europe and the Near East.

Suzanne Nugent is a PhD candidate in the School of Social Science at The University of Queensland, studying Aboriginal Australian wooden spears to determine if they retain traces of use. She is the author of "Applying use-wear and residue analyses to digging sticks" (*Memoirs of the Queensland Museum, Culture and Heritage Series,* Vol 4, 2006). Suzanne is the current editor of the World Archaeological Congress e-Newsletter.

Luke Kirkwood is a cultural heritage consultant with Environmental Resources Management Pty Ltd. Currently completing his doctorate in comparative functional evolutionary genetics at the Institute for Molecular Bioscience at The University of Queensland, Luke's research has focussed on analysing the genetic component of phenotypic differences between humans and the great apes with the aim of understanding hominid evolution. His active involvement within the Australian archaeological community has also been recognised with a Life Membership Award by the Australian Archaeological Association.